Particle Methods for Multi-Scale and Multi-Physics

Particle Methods for Multi-Scale and Multi-Physics

M B Liu

Peking University, China

G R Liu

University of Cincinnati, USA

World Scientific

NEW JERSEY • LONDON • SINGAPORE • BEIJING • SHANGHAI • HONG KONG • TAIPEI • CHENNAI • TOKYO

Published by

Imperial College Press
57 Shelton Street
Covent Garden
London WC2H 9HE

Distributed by

World Scientific Publishing Co. Pte. Ltd.
5 Toh Tuck Link, Singapore 596224
USA office: 27 Warren Street, Suite 401-402, Hackensack, NJ 07601
UK office: 57 Shelton Street, Covent Garden, London WC2H 9HE

Library of Congress Cataloging-in-Publication Data
Liu, Mou-Bin, author.
Particle methods for multi-scale and multi-physics / by Mou-Bin Liu, (Peking University, China), Gui-Rong Liu, (University of Cincinnati, USA).
pages cm
Includes bibliographical references and index.
ISBN 978-9814571692 (hardback : alk. paper)
1. Particles--Mathematical models. 2. Scaling laws (Statistical physics). 3. Dynamics of a particle--Mathematics. 4. Collisions (Physics)--Statistical methods. I. Liu, G. R. (Gui-Rong), author. II. Title.
TA418.78.L58 2016
530.15'828--dc23

2015031987

British Library Cataloguing-in-Publication Data
A catalogue record for this book is available from the British Library.

In-house Editor: Amanda Yun

To Huiqi,

Siyao, Jianing,

and my family

for their love, support and encouragement.

\- M. B. Liu

To Zuona,

Yun, Kun, Run,

and my extended family

for their love, time and support.

\- G. R. Liu

Preface

Background

With the development of computer hardware and software, computer modeling has been playing an increasingly important role in providing tests and examinations for theories, offering insights to complex physics, and assisting in the interpretation and even the discovery of new phenomena. For a typical problem in engineering or the sciences that involves ordinary differential equations (ODEs) or partial differential equations (PDEs) governing the concerned physics, grid or mesh based numerical methods — such as the finite difference methods (FDM), finite volume methods (FVM) and the finite element methods (FEM) — are usually used to discretize the computational domain into discrete small sub-domains via a process termed as discretization or meshing. In these grid-based numerical methods, individual grid points (or nodes) are connected together in a pre-defined manner by a topological map, which is termed as a mesh (or grid). The meshing results in elements in FEM, cells in FVM, and grids in FDM. A mesh or grid system consisting of nodes, and cells or elements must be defined to provide the relationship between the nodes before the approximation process for the differential or partial differential equations. Based on a properly pre-defined mesh, the governing equations can be converted to a set of algebraic equations with nodal unknowns for the field variables. So far, the grid-based numerical models have achieved remarkably, and they are currently the dominant methods in numerical simulations for solving practical problems in engineering and sciences.

However, grid-based numerical methods suffer from some difficulties in some aspects, which limit their applications in many complex problems. Firstly, the entire formulation is based on the mesh, and generating a high quality mesh is a time-consuming and costly process. Secondly, the use of mesh can lead to difficulties in dealing with free surface, deformable boundary, moving interface (for FDM), and extremely large deformation (for FEM). Thirdly, grid-based methods may not be valid when the spatial scale gradually reduces, as the grid-based methods are usually based on material models with continuum assumption.

This is especially true for meso-, micro- or even nano-scale problems or problems with multiple scale physics. Last but not the least, grid-based methods are usually not suitable for problems with discontinuous physics, even in macro-scale. Typical examples include granular flows in environmental, geophysical, chemical and bio-engineering; landslide and mudflow in environmental disasters; and transport and storage of granular materials (corns, chemicals, debris, etc.).

Recently, particle-based methods (or 'particle methods' for abbreviation) have been attracting more and more researchers, as these methods possess different features from grid-based methods either in physical description or in computational modeling. As such, there are basically two aspects for a particle-based method. The first one originated from *physical descriptions*, in which particles are used to represent the state of a system. For example, depending on the scale of the model, a particle may vary in size, from a single atom or molecule in the molecular dynamics (MD) method in atomistic scale, to a small cluster of atoms or molecules in the dissipative particle dynamics (DPD) method in meso-scale, to an infinitesimal macroscopic region in the smoothed particle hydrodynamics (SPH) method in macro-scale. Each particle can be associated with a set of field variables such as mass, momentum, energy, position, charge, vorticity, etc. Also the particles are usually of Lagrangian nature, which makes the particles follow the motion of the simulated medium. This is appealing in dealing with free surfaces, moving interfaces or deformable boundaries. In this aspect, particles function as *material points*. The other aspect of a particle method involves *computational modeling*, in which particles are used as *interpolation* or *approximation points* for solving an ODE or PDE. This concept results in the meshfree or meshless methods, which modify the internal structure of the grid-based FDM and FEM with a set of arbitrarily distributed nodes (or particles). As there is no mesh or grid providing the connectivity of these nodes or particles, the meshfree methods are expected to be more adaptive, versatile and robust, and thus can be more attractive in modeling problems with large deformations or discontinuous physics such as cracks. For more details on meshfree methods, readers are encouraged to refer to some related monographs listed in Chapter 1. The two aspects of a specific particle method can be integrated together and hence a particle can act both as a material point and as an approximation point. This makes the particle method more attractive.

This book

As any material fundamentally consists of particles, it is natural and attractive to use a particle method with integrated features from both physical description and computational modeling to numerically simulate the behavior of either simple or complex systems. There are a number of such particle methods such as MD, DPD, SPH, the moving particle semi-implicit (MPS) method, the material point method (MPM), the discrete (or distinct) Element Method (DEM), and many others.

In this book, for the very first time, three typical particle methods, MD, DPD and SPH, will be addressed together in detail. All these three particle methods are meshfree, particle methods of pure Lagrangian nature. A particle in MD, DPD and SPH acts both as a material point and as an approximation point, though the particle can be a single atom or molecule in MD in the atomistic scale, a small cluster of atoms or molecules in DPD in meso-scale, and a very small region in SPH in macro-scale.

The book is written for senior university students, graduate students, researchers and professionals, both in computational engineering and the sciences. The presented methodologies, techniques and intriguing applications will be useful to students from mechanical, civil, chemical and bio-engineering, and to researchers and professionals in computational physics, and computational fluid and solid dynamics.

The authors and their research teams started the work on particle methods since 1997, from SPH, when they were searching for an alternative numerical approach for simulating the explosion of high explosives, underwater explosions, etc. As we found that it was very common for conventional grid-based methods to encounter unexpected terminations during the computation due to mesh distortion related problems, the authors attempted the feasibility of applying SPH to modeling problems with intensive loadings and large deformations. The theoretical background, numerical techniques and code implementation issues of SPH were also investigated with many different applications. This led to the first monograph on SPH[1] — a popular publication that attracts many fellow researchers. One noticeable point is that the work presented in the monograph is basically based on the conventional SPH, which is known to have poor accuracy especially for irregularly distributed particles. Later on, after around 10 years of development, the SPH method has been intensively investigated, especially on the kernel and particle consistency. This has led to many modified SPH methods with better accuracy. Also the SPH method, either modified or conventional, has been extended to many new and diversified applications. *The essence of the conventional SPH, as well as the latest developments in methodologies and applications, will be addressed in this book.*

In 2004, when studying multiphase flows in pores and fractured porous media, the DPD method was used, as it is a coarse-grained molecular dynamics method, and is suitable for modeling meso-scale static and dynamic fluid behaviors. An interaction potential with short-range repulsion and long-distance attraction and a more efficient boundary treatment algorithm were integrated into

[1] Liu, G. R. and Liu, M. B. (2003) *Smoothed particle hydrodynamics: A meshfree particle method*. World Scientific.

DPD for modeling multiphase flows in pores and fractured porous media. The DPD method has been extended to other applications such as the movement and suspension of macromolecules, movement and deformation of cells due to external loads, and some others. *This book contains a short description of classic molecular dynamics, and a comprehensive overview on the DPD method with basic concepts, latest developments and diversified applications.*

Though researchers have been using MD for investigating different problems at the atomic scale for quite a long period of time, numerical simulations using other particle methods such as SPH and DPD are relatively new, and are still under development. There are problems awaiting further improvements in SPH, DPD and other particle methods. These problems in turn offer ample opportunities for researchers to develop more advanced particle methods — the next generation of numerical methods. The authors hope that the methodologies and application examples in this book can serve the purpose of providing a smoother start for readers to efficiently learn, test, practice and further develop particle-based methods.

In our first monograph on SPH (Liu and Liu, 2003), we provided a 3D SPH source code, which has been appealing due to its readability (easily understood), applicability (usable by varying applications) and extendibility (easy to modify. During the last decades, a lot of open-source codes based on particle methods have been developed. For example, for MD, many open-source codes are easily available both online and in different monographs, while DPD source codes can be obtained from modifying the MD source codes. Also with the fast development and applications of SPH, a number of open-source codes of SPH have also emerged. One of them is SPHYSICS, which is a good open-source code mainly for modeling free surface flows[2]. Another good example is LAMMPS (acronym for Large-scale Atomic/Molecular Massively Parallel Simulator) by Sandia National Laboratories. LAMMPS was originally intended for classical molecular dynamics simulation, but has since been extended as a parallel particle simulator at the atomic, meso or continuum scale. DPD and SPH solvers are also available in LAMMPS[3]. These open-source codes are generally associated with comprehensive content (e.g., different time integration techniques, optimized particle-interaction searching algorithms and diversified application modules) for selection. More importantly, the open-source codes are widely available, are generally well-structured for extendibility, and well-parallelized for high performance computing; for such reasons, the authors did not provide any source codes in this book.

---

[2] https://wiki.manchester.ac.uk/sphysics/index.php/Main_Page

[3] http://lammps.sandia.gov/

Outline of the book

This book provides comprehensive overviews on DPD and SPH in a systematic way. It is organized in a total of seven chapters, described in brief as follows.

Chapter 1 introduces some background knowledge on numerical simulation. The features and limitations of grid-based numerical methods are discussed. The basic ideas of meshfree and particle methods are described.

Chapter 2 provides a short description of the molecular dynamics method, and the knowledge in this chapter can serve as a pre-requisite for introducing the DPD method in Chapter 3 for meso-scale applications in Chapter 4.

Chapter 3 introduces the dissipative particle dynamics methodology. The basic concepts of DPD are introduced, including governing equations, time integration algorithms, determination of DPD coefficients, and the computational procedure of DPD simulation. Some numerical aspects of DPD are addressed, including the assessment of dynamic properties, solid boundary treatment, conservative interaction potential and spring-bead chain models for simulating macromolecules. In particular, a generic algorithm for treating complex solid boundaries, and a novel approach for constructing conservative interaction potential with short-rang repulsion and long-distance attraction are addressed in detail. Moreover, the DPD method in modeling complex physics and reproducing the continuum hydrodynamic behavior are demonstrated with a number of benchmark numerical examples.

Chapter 4 provides an overview on DPD in diversified applications with special focuses on micro drop dynamics (including DPD modeling of the formation of drop with co-existing liquid-vapor, large-amplitude oscillation of a liquid drop and flow transition in controlled drug delivery), multiphase flows in pore-scale fracture network and porous media, movement and suspension of macromolecules in micro channels and movement, and the deformation of a single cell due to external loads.

Chapter 5 introduces the smoothed particle hydrodynamics methodology. Firstly, the basic ideas of the numerical approximations of the SPH are discussed. These include the kernel and particle approximations of a field function and its derivatives in conventional SPH; techniques to deriving SPH formulations for partial differential equations such as the Navier–Stokes (N–S) equations. Secondly, the basic properties of a typical smoothing function are discussed and the constructing conditions of smoothing functions are generalized. Thirdly and most importantly, the consistency concept of SPH is introduced with consistency conditions on kernel and particle approximations. Some particle consistency restoring approaches are reviewed, and a restored particle consistency through reconstructing the smoothing function is described. Lastly, a finite particle method, which can be regarded as a generalized version of SPH, is introduced and compared with the conventional SPH and some other modified SPH.

Chapter 6 provides an overview on SPH in diversified applications followed by special interests on SPH modeling incompressible fluid flows in hydrodynamics and ocean engineering. This is different from the authors' previous SPH monograph, in which the applications are basically focused on modeling problems with intensive loadings and large material deformations. In this chapter, a detailed comparison of the weakly compressible SPH (WCSPH) model and the incompressible SPH (ISPH) model for modeling incompressible flows is provided. Some typical applications in SPH modeling — free surface flows (dam break, surge front and etc.), free surface flows with rigid (liquid sloshing, water entry and exit and etc.) and elastic solid objects (head-on collision of two rubber rings, dam break with an elastic gate and water impact onto a forefront elastic plate) — are provided in detail.

Chapter 7 introduces three popular macro-scale particle based methods including the particle-in-cell (PIC) method, material point method (MPM) and moving-particle semi-implicit (MPS) method. The similarities and differences of PIC, MPM and MPS are comparatively discussed.

Mou-Bin Liu
Gui-Rong Liu

August, 2015

Acknowledgments

The authors' work in the area of computer modeling using particle methods has been profoundly influenced by the works of Profs. M. P. Allen, D. C. Rapaport, and their research teams on MD; Profs. D. Frenkel, B. Smit, P. J. Hoogerbrugge, P. Espanol, G. E. Karniadakis, and their research teams on DPD; Profs. Ted Belytschko, W. K. Liu, E. Onate, S. Koshizuka, Z. Chen, and their research teams on meshfree and particle methods; Prof. J. J. Monaghan, Drs. L. D. Libersky, L. Lucy, W. Benz, J. K. Chen, P. W. Cleary, X. Y. Hu, A. Colagrossi and many other fellow researchers on SPH. Without their significant contributions, writing this book would not be possible.

In preparing this book, many colleagues and students have supported and contributed to the writing of this book. The authors would like to express their sincere thanks to all of them. Special thanks are given to Profs. K. Y. Lam, X. Han, J. Z. Chang, Z. Zong, Y. T. Gu, Z. R. Li, H. T. Liu, Drs. P. Meakin, H. Huang, Y. Chen, X. F. Yang, J. R. Shao, Mr. Z. Chen, L. W. Zhou and many others. They have helped either in preparing some examples, or in reading the drafts of the book, or have been involved in carrying out a number of projects related to the SPH and DPD methods.

Some works presented in this book have been supported by the National Natural Science Foundation of China (NSFC) (Grant No. 10942004, 11172306, 11302237 and 50976108), NSAF (Grant No. U1530110), Peking University, (PKU), China, Chinese Academy of Sciences (CAS), China State Key Laboratory for Turbulence and Complex Systems, Agency for Science, Technology and Research (A*STAR), Singapore, and the National University of Singapore. The authors would like to thank these funding organizations for their partial financial sponsorship for research projects undertaken by the authors and their teams related to the topic of this book.

About the Authors

Mou-Bin Liu

College of Engineering &
Institute of Ocean Research &
State Key Laboratory for Turbulence and Complex Systems, Peking University
Beijing, 100871, China

M. B. Liu received his B.E. and M.E. degrees from Xi'an JiaoTong University (XJTU), China, in 1993 and 1996, respectively, and received his Ph.D. from the National University of Singapore (NUS) in 2003. He is currently a professor at the College of Engineering and the Institute of Ocean Research at Peking University, China. He has authored more than 100 reviewed technical papers and one monograph, with more than 2000 SCI citations. He has received a number of prestigious awards from universities and scientific organizations worldwide, including the **100 Talent Program Award** from the Chinese Academy of Sciences (CAS) in 2009, the **Young Investigator Award** from the Asia Pacific Association of Computational Mechanics (APACM) in 2007 and the **Lee Kuan Yew Fellow Award** from Nanyang Technological University (NTU), Singapore, in 2005. He is interested in the computer modeling of complex flows with different coupling effects, and has developed some original numerical methods, both grid-based and meshfree particle-based, for modeling fluid dynamics and fluid-solid interactions, including high energy explosive detonation and explosion, underwater explosion, high strain hydrodynamics with material strength, wave breaking and interaction with structures (e.g., sloshing and slamming), and water entry and exit.

Gui-Rong Liu

Aerospace Engineering and Engineering Mechanics
Ohio Eminent Scholar (State Endowed Chair)
University of Cincinnati, Cincinnati,
OH 45221-0070, USA

G. R. Liu received his Ph.D. from Tohoku University, Japan, in 1991. He was a Postdoctoral Fellow at Northwestern University, USA, from 1991–1993. He is currently a Professor and Ohio Eminent Scholar (State Endowed Chair) at the Department of Aerospace Engineering and Engineering Mechanics in the University of Cincinnati, USA. He served as the School Faculty Chair of the School of Aerospace Systems, University of Cincinnati; a Deputy Head of the Department of Mechanical Engineering, and the Director of the Centre for Advanced Computations in Engineering Science (ACES), National University of Singapore; and the President of the Association for Computational Mechanics. He was the President of the Asia-Pacific Association for Computational Methods, and is now an Executive Council member of the International Association for Computational Mechanics. He authored a large number of international journal papers and books including two bestsellers: *Mesh Free Method: moving beyond the finite element method* and *Smoothed Particle Hydrodynamics: a Meshfree Particle Method*. He recently authored a book on *Smoothed Finite Element Methods* as well as *The Smoothed Point Interpolation Methods — G Space Theory and Weakened Weak Forms* — both of which are now used for various types of problems. He is the Editor-in-Chief of the *International Journal of Computational Methods* (IJCM), Associate Editor of the international technical journal *Inverse Problems in Science and Engineering* (IPSE), and serves as an editorial member of five other journals, including the *International Journal of Numerical Methods in Engineering* (IJNME). He is the recipient of numerous awards, including the Singapore Defence Technology Prize, the NUS Outstanding University Researcher Award, the APACM Computational Mechanics Awards, the ICACM Computational Mechanics Awards, the JSME Computational Mechanics Awards, and the ASME Ted Belytschko Applied Mechanics Award. He is listed as one of the top 1% of *The World's Most Influential Scientific Minds [in] 2014* (Highly Cited Researchers in Engineering) by Thomson Reuters.

Contents

Chapter 1

Introduction

1.1 Computer modeling

1.1.1 Computer modeling and its general solution procedure

Computer modeling (or numerical simulation using computers) has increasingly become a very important approach for solving and analyzing complex practical problems in engineering and sciences. A general procedure of computer modeling includes translating important phenomena of a physical problem into a discrete form of mathematical description, recasting the problem in discrete numerical equations, solving the equations on a computer, and then revealing the phenomena virtually according to the requirements of the analysts.

Computer modeling follows a similar procedure to serve a practical purpose. There are in principle some necessary steps in the procedure, as shown in **Figure 1.1**. From the physical phenomena observed, *mathematical models* are established with some possible simplifications and assumptions. These mathematical models are generally expressed in the form of *governing equations* defined in the problem domain with proper *boundary conditions* (BC) and/or *initial conditions* (IC). The governing equations may be a set of *ordinary differential equations* (ODE), *partial differential equations* (PDE), *integral equations* or equations in any other possible forms of physical laws. Boundary and/or initial conditions are necessary for determining the field variables in space and/or time.

To numerically solve the governing equations, the involved geometry of the problem domain needs to be divided into discrete finite number of parts, for which numerical approximations can be easily made. A computational frame is then formed known traditionally as a set of mesh, which consists of cells, grids or nodes. The grids or nodes are the locations where the field variables are evaluated, and their relations are defined by some kind of *nodal connectivity* defined by the mesh. Accuracy of the numerical approximation is closely related to the mesh density and pattern.

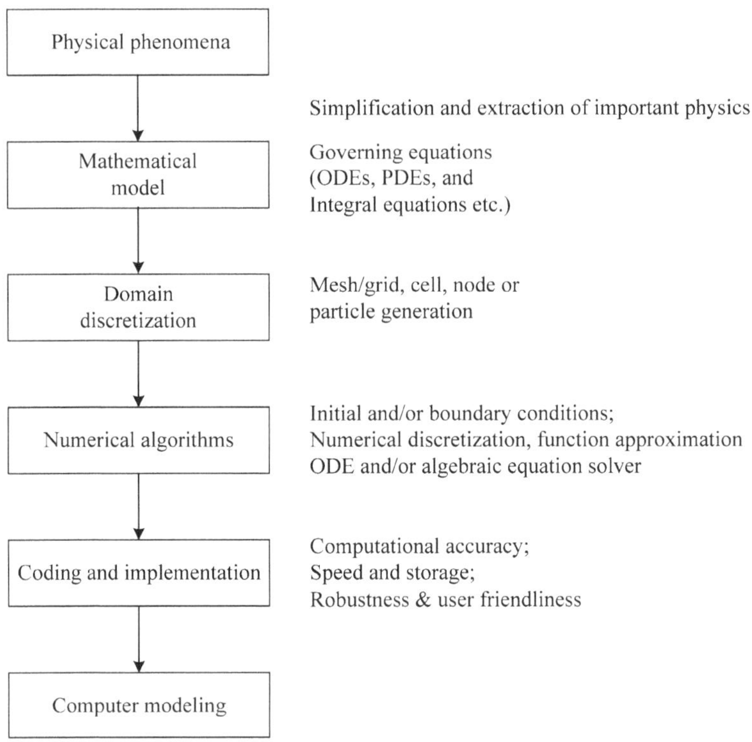

Figure 1.1 Procedure of conducting a computer modeling.

Numerical discretization provides means to change the spatial (integral or derivative) operators in the governing equations to discrete representations at the grids or nodes. Such a numerical discretization is based on one of the theories of *function approximations* (Liu, 2002). After the numerical discretization, the original physical equations are changed into a set of *algebraic equations* or ordinary differential equations, which can be solved using the existing numerical routines. In the process of establishing the algebraic or ODE equations, the so-called strong or weak forms (Liu and Gu, 2003), or weakened weak form (Liu, 2009) formulation can be used These forms of formulation can also be combined together to take the full advantages of both weak and strong form formulations.

Implementation of a numerical simulation involves translating the domain decomposition and numerical algorithms into a computer code in some programming language(s). In coding a computer program, the *accuracy*, and *efficiency* (speed and storage) are two very important considerations. Other considerations include *robustness* of the code (consistency check, error trap), *user-friendliness* of the code (easy to read, use and even to modify), and etc. Before performing a practical numerical simulation, the code should be tested against theoretical solutions, or the exact results from other established methods for benchmark problems, or the experimental data from actual engineering problems. In other words, a computer modeling needs *verification and validation* (V&V), as will be further discussed in Section 1.1.3.

For numerical simulations of problems in fluid mechanics, the governing equations can be established from the conservation laws, which state that field variables such as the *mass*, *momentum* and *energy* must be conserved during the evolution process of the flow. These three fundamental principles of conservation, together with additional information concerning the specification of the nature of the material/medium, conditions at the boundary, and conditions at the initial stage determine the behavior of the fluid system.

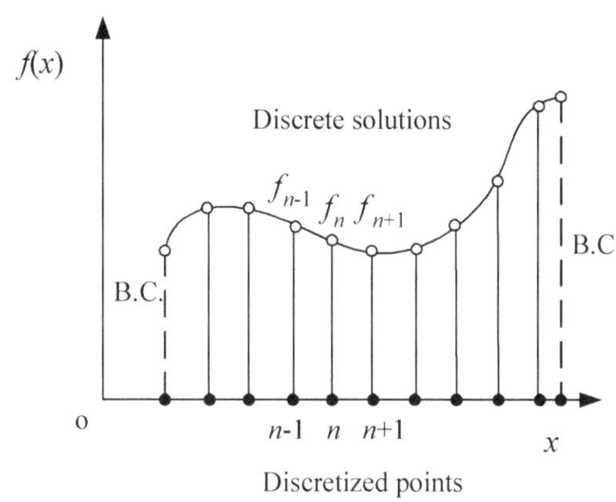

Figure 1.2 Domain and numerical discretization for computer modeling of a field function $f(x)$ defined in one-dimensional space.

Except for a few circumstances of very simple settings, it is very difficult to obtain analytical solution of these integral equations or partial differential equations. *Computational fluid dynamics* (CFD) deals with the techniques of spatially approximating the integral or the differential operators in the integral or differential equations into a set of simple algebraic summations (or ODEs with respect to time only), which can be solved to obtain numerical values for field functions (such as density, pressure, velocity, etc.) at discrete points in space and/or time **Figure 1.2**). A typical computer modeling of a CFD problem deals with

1. governing equations,
2. proper boundary conditions and/or initial conditions,
3. domain discretization technique,
4. numerical discretization technique,
5. numerical technique to solve the resultant algebraic equations or ordinary differential equations.

1.1.2 Computer modeling, theory and experiment

Rather than adopting the traditional theoretical practice of constructing layers of assumptions and approximations, computer modeling attacks the original problems in detail with minimum assumptions, with the help of the increasing computer power. It provides an alternative tool of scientific investigation, instead of carrying out expensive, time-consuming or even dangerous experiments in laboratories or on site. The numerical tools are often more useful than the traditional experimental methods in terms of providing insightful and complete information that cannot be directly measured or observed, or difficult to acquire via other means. Computer modeling plays a valuable role in providing verifications for theories, offers insights to the experimental results and assists in the interpretation or even the discovery of new phenomena. It acts also as a bridge between the experimental models and the theoretical predictions.

Figure 1.3 shows the connection between the computer modeling, theory and experiment. With the rapid development of computer hardware and software,

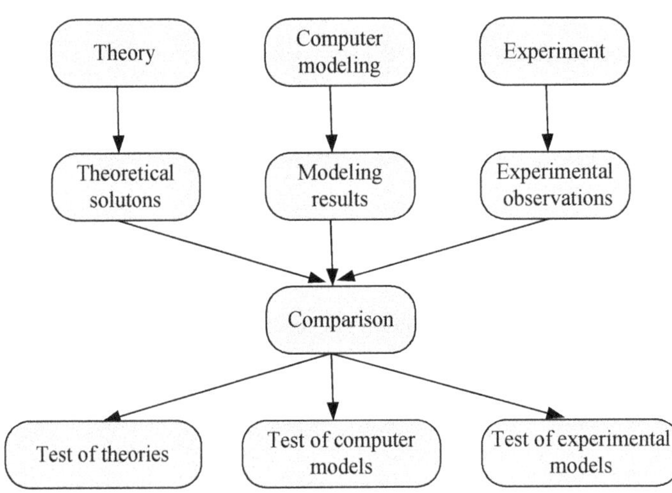

Figure 1.3 Connection between computer modeling, theory and experiment.

computer modeling is increasingly playing a more and more important role in conducting scientific investigations. However, this does not mean we do not need experimental and theoretical works any more. It must be clearly pointed out that experimental phenomena and theoretical analyses are usually the fundaments of computer modeling and the modeling results also need to be verified and validated.

1.1.3 Verification and validation

Computer modeling today can server both as a research and a design tool for many important engineering and scientific projects. One typical example is the computational fluid dynamics, which is a branch of fluid mechanics that uses numerical methods to solve and analyze fluid mechanics problems. With the advent of high performance computers together with advanced numerical algorithms, open source codes and commercial CFD software are easily accessible. As such, CFD now plays a more and more important role in understanding fluid flows. The accuracy of CFD codes need to be demonstrated so that the CFD codes may be used with confidence for practical applications and the results can be considered credible for decision making in design.

Early in 1979, the Society of Computer Simulation (SCS) first defined the term "verification" and "validation" (Schlesinger, 1979), and provided two related terms, i.e., *computerized model* and *conceptual model*. In 1998, the American Institute of Aeronautics and Astronautics (AIAA) provided a guide for the verification and validation of computational fluid dynamics simulations (Reston, 1998). The guide clearly defined the key terms, discussed fundamental concepts, and specified general procedures for conducting verification and validation of CFD simulations. In 2002, Oberkampf and Trucano presented an extensive review of the literature in V&V from members of the operations research, statistics, and CFD communities and discussed methods and procedures for assessing V&V in CFD (Oberkampf and Trucano, 2002).

According to SCS's definition, *model verification* substantiates that a computerized model represents a conceptual model within specified limits of accuracy, and *model validation* substantiates that a computerized model within its domain of applicability possesses a satisfactory range of accuracy consistent with the intended application of the model. SCS also defined a term "*model qualification*", which is the determination of adequacy of the conceptual model to provide an acceptable level of agreement for the domain of intended application (reality).

Figure 1.4 shows the connection between the reality, conceptual model and computerized model. It is seen that from "conceptual model" to "computerized model", computer programming is required, and the process needs "model verification" to ensure that the computer code accurately mimics the original conceptual model. From "computerized model" to "reality", computer modeling

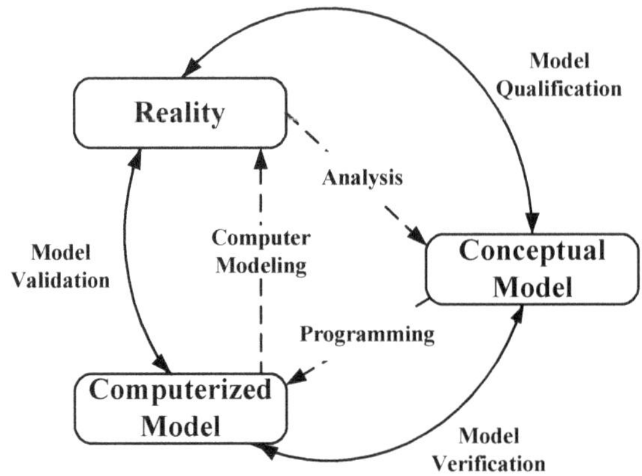

Figure 1.4 Phases of computer modeling and the role of V&V.

is conducted, and this process needs "model validation" to ensure that the computerized model possesses a satisfactory range of accuracy consistent with reality. From "reality" to "conceptual model", some analysis works are necessary and this process needs "model qualification" to ensure that the conceptual model is consistent with reality.

1.2 Governing equations

Obtaining the basic equations of fluid motion is the process from reality (physics) to conceptual model as shown in **Figure 1.4**, while the conceptual model in CFD includes governing equations for the conservation of mass, momentum and energy, and some auxiliary equations such as turbulence model, chemical reaction model and cavitation model. In classic molecular dynamics, the governing equation (equation of motion) is based on Newton's second law, while the force can be obtained from the inter-atomic potential which is in general a function of the position vector of all the atoms (Rapaport, 2004). In continuum scale fluid mechanics, the process of obtaining the basic equations of fluid motion is similar. We first need to choose the appropriate fundamental physical principles from the law of physics such as *mass, momentum and energy conservation*, then apply the physical principles to a suitable fluid model, and finally extract the mathematical equations which represent the physical principles.

1.2.1 Eulerian and Lagrangian descriptions

In obtaining the basic equations of fluid motion, the selection of fluid model is important, as when a fluid is in motion, the state can be different at different locations of the fluid. Mathematically, there are two approaches for describing the governing equations for fluids, the *Eulerian description* and *Lagrangian description*. The Eulerian description is a spatial description, whereas the Lagrangian description is a material description that employs the total time derivative as the combination of local derivative and convective derivative.

Consider a closed volume with finite dimensions in a fluid flow as shown in **Figure 1.5**. This volume defines a *control volume* V bounded by a closed *control surface* S. In the Eulerian description, this control volume is fixed in space while the fluid moving through it. In the Lagrangian description, this control volume moves together with the fluid flow such that the same material of fluid is always staying inside the control volume. Therefore, though the fluid flow may result in expansion, compression, and deformation of the Lagrangian control volume, the mass of the fluids contained in the Lagrangian control volume remains unchanged. The Lagrangian control volume is reasonably large with finite dimensions in the flow system and the governing conservation laws can be directly applied to the fluids inside the control volume. Applying the conservation laws to the fluids to Lagrangian finite control volume can result in a set of governing equations in integral form (Anderson, 2002; Chung, 2002).

Another approach to obtain governing equations is to use the concept of *infinitesimal fluid cell*. The infinitesimal fluid cell (illustrated in **Figure 1.6**) can be regarded as a very small clump of fluids associated with a very small control volume δV and a very small control surface δS surrounding δV. At

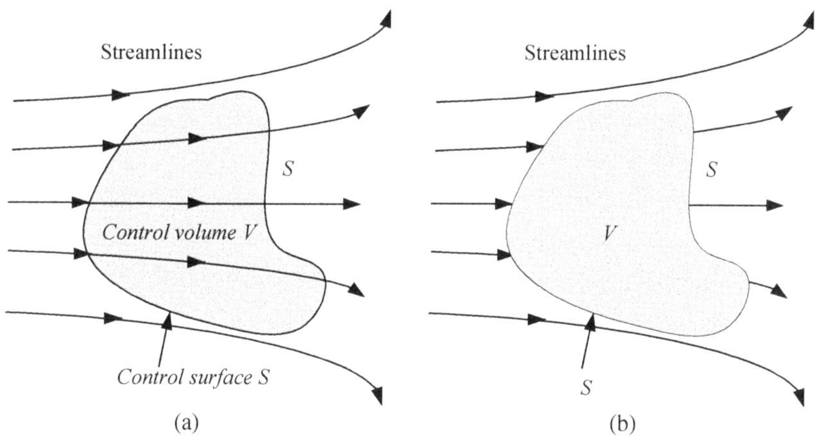

Figure 1.5 A finite control volume in Eulerian (a) and Lagrangian (b) descriptions.

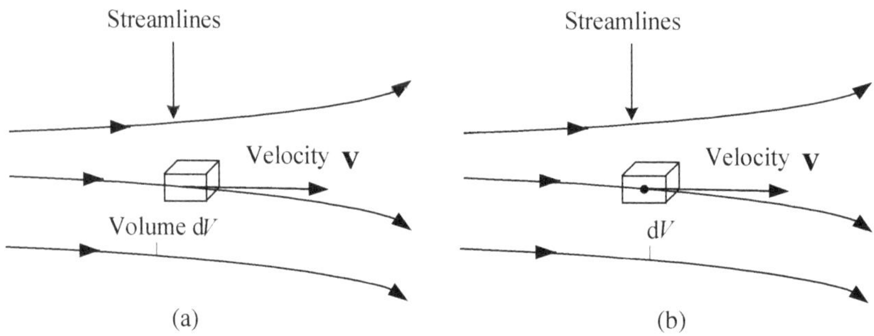

Figure 1.6 An infinitesimal fluid cell in Eulerian (a) and Lagrangian (b) descriptions.

the limit, δV and δS can be the differential volume dV and the differential surface dS. This infinitesimal fluid cell, on one hand, is large enough so that the assumptions of continuum mechanics are valid, on the other hand, is small enough so that a field property inside it can be regarded as the same throughout the entire cell. Similarly, an infinitesimal fluid cell in Eulerian description is fixed in space with fluid moving through it, and an infinitesimal fluid cell in Lagrangian description moves with the same material of the fluid staying inside it. Within the Lagrangian description, the infinitesimal fluid cell can move along a streamline with a vector velocity $\mathbf{v} = (v_x, v_y, v_z)$ equal to the flow velocity at that point. Applying the conservation laws to the Lagrangian infinitesimal fluid cell, governing equations in the form of partial different equation can be established (Anderson, 2002; Chung, 2002).

1.2.2 Control volume, surface and velocity divergence

For a Lagrangian control volume, the movement of the fluids inside the control volume V leads to the change of the control surface S. The change of the control surface again results in a volume change of the control volume. As illustrated in **Figure 1.6**(b), the volume change of the control volume due to the movement of dS over a time increment Δt is

$$dV = \mathbf{v}\Delta t \cdot \mathbf{n} dS \,, \tag{1.1}$$

where $\mathbf{n}$ is the unit normal vector perpendicular to the surface dS.

The total volume change of the entire Lagrangian control volume is therefore the integral over the control surface S

$$\Delta V = \int_S \mathbf{v}\Delta t \cdot \mathbf{n} dS \,. \tag{1.2}$$

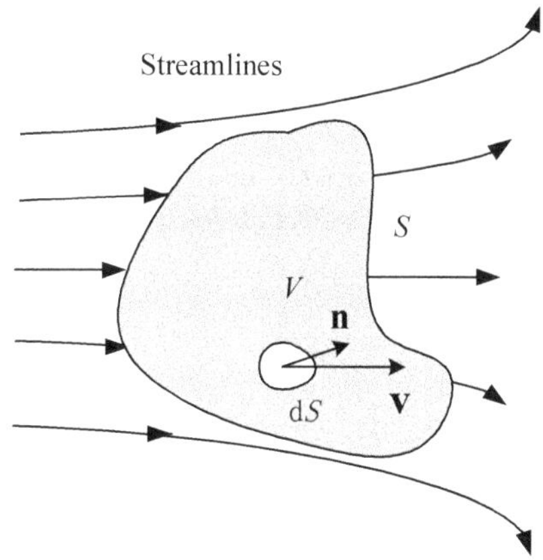

Figure 1.7 Volume change of the Lagrangian infinitesimal fluid cell.

Dividing both sides in equation (1.2) by Δt and applying the divergence theorem yield

$$\frac{\Delta V}{\Delta t} = \int_V \nabla \cdot \mathbf{v} dV , \tag{1.3}$$

where ∇ is the gradient operator. If the Lagrangian control volume is downgraded (shrunk) to an infinitesimal fluid cell with volume of δV, so that the field state and property are equal throughout δV, the following equation can be obtained as

$$\frac{\Delta(\delta V)}{\Delta t} = (\nabla \cdot \mathbf{v}) \int_V d(\delta V) = (\nabla \cdot \mathbf{v}) \delta V . \tag{1.4}$$

Therefore, the time rate of volume change for the infinitesimal fluid cell is

$$\frac{D(\delta V)}{Dt} = (\nabla \cdot \mathbf{v}) \delta V . \tag{1.5}$$

From equation (1.5), the velocity divergence becomes

$$\nabla \cdot \mathbf{v} = \frac{1}{\delta V} \frac{D(\delta V)}{Dt} . \tag{1.6}$$

It shows that the velocity divergence can be physically interpreted as the time rate of volume change per unit volume.

1.2.3 Navier-Stokes equations in Lagrangian frame

Continuity equation

The *continuity equation* is based on the conservation of mass. For a Lagrangian infinitesimal fluid cell with volume of δV, the mass contained in the control volume is

$$\delta m = \rho \delta V , \tag{1.7}$$

where m and ρ are mass and density, respectively.

Since the mass is conserved in the Lagrangian fluid cell, the time rate of mass change is zero. Therefore, we have

$$\frac{D(\delta m)}{Dt} = \frac{D(\rho \delta V)}{Dt} = \delta V \frac{D\rho}{Dt} + \rho \frac{D(\delta V)}{Dt} = 0 . \tag{1.8}$$

Equation (1.8) can be rewritten as

$$\frac{D\rho}{Dt} + \rho \frac{1}{\delta V} \frac{D(\delta V)}{Dt} = 0 . \tag{1.9}$$

Considering equation (1.6), and replacing the second term in equation (1.9) with the velocity divergence, the continuity equation or the mass conservation equation in Lagrangian form is obtained as

$$\frac{D\rho}{Dt} = -\rho \nabla \cdot \mathbf{v} . \tag{1.10}$$

Momentum equation

The *momentum equation* is based on the conservation of momentum, which in the continuum mechanics, is represented by Newton's second law which states that the net force on a Lagrangian fluid cell equals to its mass multiplying the acceleration of that fluid cell.

As illustrated in **Figure 1.8**, the position vector is $\mathbf{x} = (x, y, z)$, and the accelerations of the infinitesimal fluid cell in the three directions are $\frac{Dv_x}{Dt}$, $\frac{Dv_y}{Dt}$ and $\frac{Dv_z}{Dt}$, respectively. The net force on the fluid cell consists of body forces and surface forces. The body force may be the gravitational force, magnetic forces and other possible forces acting on the body of the entire fluid cell. The

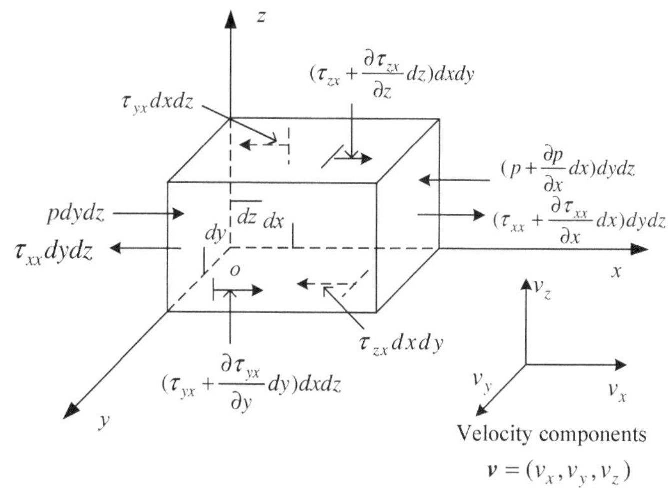

Figure 1.8 Forces in the x direction on a Lagrangian infinitesimal fluid cell.

surface force includes

1) the pressure, which is imposed by the outside fluids surrounding the concerned fluid cell,
2) the shear and normal stress, which result in shear deformation and volume change, respectively.

In the x direction, all the forces acting on the Lagrangian infinite fluid cell are

$$\begin{aligned}
&-[(p+\frac{\partial p}{\partial x}dx)-p]dydz+\\
&[(\tau_{xx}+\frac{\partial\tau_{xx}}{\partial x}dx)-\tau_{xx}]dydz+\\
&[(\tau_{yx}+\frac{\partial\tau_{yx}}{\partial y}dy)-\tau_{yx}]dxdz+\\
&[(\tau_{zx}+\frac{\partial\tau_{zx}}{\partial z}dz)-\tau_{zx}]dxdy\\
&=-\frac{\partial p}{\partial x}dxdydz+\frac{\partial\tau_{xx}}{\partial x}dxdydz+\frac{\partial\tau_{yx}}{\partial y}dxdydz+\frac{\partial\tau_{zx}}{\partial z}dxdydz,
\end{aligned}\tag{1.11}$$

where p is pressure, τ_{ij} is the stress in the j direction exerted on a plane perpendicular to the i axis. If the body force per unit mass in the x direction is

F_x, Newton's second law can be written as

$$
\begin{aligned}
m\frac{dv_x}{dt} &= \rho dxdydz\frac{dv_x}{dt} \\
&= -\frac{\partial p}{\partial x}dxdydz \\
&\quad +\frac{\partial \tau_{xx}}{\partial x}dxdydz+\frac{\partial \tau_{yx}}{\partial y}dxdydz+\frac{\partial \tau_{zx}}{\partial z}dxdydz \\
&\quad +F_x(\rho dxdydz).
\end{aligned}
\tag{1.12}
$$

Therefore the momentum equation in the x direction is

$$\rho\frac{Dv_x}{Dt} = -\frac{\partial p}{\partial x}+\frac{\partial \tau_{xx}}{\partial x}+\frac{\partial \tau_{yx}}{\partial y}+\frac{\partial \tau_{zx}}{\partial z}+\rho F_x. \tag{1.13}$$

Similarly, the momentum equations in the y and z directions are

$$\rho\frac{Dv_y}{Dt} = -\frac{\partial p}{\partial y}+\frac{\partial \tau_{xy}}{\partial x}+\frac{\partial \tau_{yy}}{\partial y}+\frac{\partial \tau_{zy}}{\partial z}+\rho F_y. \tag{1.14}$$

$$\rho\frac{Dv_z}{Dt} = -\frac{\partial p}{\partial z}+\frac{\partial \tau_{xz}}{\partial x}+\frac{\partial \tau_{yz}}{\partial y}+\frac{\partial \tau_{zz}}{\partial z}+\rho F_z. \tag{1.15}$$

For Newtonian fluids, the stress should be proportional to the strain rate denoted by ε through the dynamic viscosity μ

$$\tau_{ij} = \mu\varepsilon_{ij}, \tag{1.16}$$

where

$$\varepsilon_{ij} = \frac{\partial v_j}{\partial x_i}+\frac{\partial v_i}{\partial x_j}-\frac{2}{3}(\nabla\cdot\mathbf{v})\delta_{ij}. \tag{1.17}$$

where δ_{ij} is the Dirac delta function.

Energy equation

The *energy equation* is based on the conservation of energy, which is a representation of the first law of thermodynamics. The energy equation states that the time rate of energy change inside an infinitesimal fluid cell should equal to the summation of the net heat flux into that fluid cell, and the time rate of work done by the body and surface forces acting on that fluid cell. If neglecting the heat flux, and the body force, the time rate of change of the internal energy e of the infinitesimal fluid cell consists of following two parts.

1) the work done by the isotropic pressure multiplying the volumetric strain
2) the energy dissipation due to the viscous shear forces

Therefore, the energy equation can be written as follows.

$$\begin{aligned}\rho\frac{De}{Dt} &= -p\left(\frac{\partial v_x}{\partial x}+\frac{\partial v_y}{\partial y}+\frac{\partial v_z}{\partial z}\right)\\ &+\tau_{xx}\frac{\partial v_x}{\partial x}+\tau_{yx}\frac{\partial v_x}{\partial y}+\tau_{zx}\frac{\partial v_x}{\partial z}\\ &+\tau_{xy}\frac{\partial v_y}{\partial x}+\tau_{yy}\frac{\partial v_y}{\partial y}+\tau_{zy}\frac{\partial v_y}{\partial z}\\ &+\tau_{xz}\frac{\partial v_z}{\partial x}+\tau_{yz}\frac{\partial v_z}{\partial y}+\tau_{zz}\frac{\partial v_z}{\partial z}\end{aligned}. \tag{1.18}$$

In summary, the governing equations for dynamic fluid flows can be written as a set of partial differential equations in Lagrangian description. The set of partial differential equations is the well-known *Navier-Stokes (N-S) equations*, which state the conservation of mass, momentum and energy. If the Greek superscripts α and β are used to denote the coordinate directions, the summation in the equations is taken over repeated indices, and the total time derivatives are taken in the moving Lagrangian frame, the Navier-Stokes equations consist of the following set of equations.

1) The continuity equation

$$\frac{D\rho}{Dt} = -\rho\frac{\partial v^\beta}{\partial x^\beta}. \tag{1.19}$$

2) The momentum equation (in the case of free external force)

$$\frac{Dv^\alpha}{Dt} = \frac{1}{\rho}\frac{\partial \sigma^{\alpha\beta}}{\partial x^\beta}. \tag{1.20}$$

3) The energy equation

$$\frac{De}{Dt} = \frac{\sigma^{\alpha\beta}}{\rho} \frac{\partial v^{\alpha}}{\partial x^{\beta}} . \tag{1.21}$$

In the above equations σ is the total stress tensor. It is made up of two parts, one part of isotropic pressure p and the other part of viscous stress τ.

$$\sigma^{\alpha\beta} = -p\delta^{\alpha\beta} + \tau^{\alpha\beta} . \tag{1.22}$$

For Newtonian fluids, the viscous shear stress should be proportional to the shear strain rate denoted by ε through the dynamic viscosity μ.

$$\tau^{\alpha\beta} = \mu\varepsilon^{\alpha\beta} , \tag{1.23}$$

where

$$\varepsilon^{\alpha\beta} = \frac{\partial v^{\beta}}{\partial x^{\alpha}} + \frac{\partial v^{\alpha}}{\partial x^{\beta}} - \frac{2}{3}(\nabla \cdot \mathbf{v})\delta^{\alpha\beta} . \tag{1.24}$$

If separating the isotropic pressure and the viscous stress, the energy equation can be rewritten as

$$\frac{De}{Dt} = -\frac{p}{\rho} \frac{\partial v^{\beta}}{\partial x^{\beta}} + \frac{\mu}{2\rho} \varepsilon^{\alpha\beta} \varepsilon^{\alpha\beta} . \tag{1.25}$$

1.3 Grid-based methods

As discussed in Section 1.2, there are two fundamental frames for describing the physical governing equations: the Eulerian description and the Lagrangian description. The Eulerian description is a spatial description, and is typically represented by the *finite difference method* (FDM) (Hirsch, 1988; Anderson, 1995; Wilkins, 1999; Anderson, 2002). The Lagrangian description is a material description, and is typically represented by the *finite element method* (FEM) (Zienkiewicz and Taylor, 2000; Liu and Quek, 2003). For example, in fluid mechanics, if the viscosity and the heat conduction as well as the external forces are neglected (see Section 1.2.3), the conservation equations in PDE form for these two descriptions are very much different, as listed in **Table 1.1**.

Table 1.1 Conservation equations in PDE form in the Lagrangian and Eulerian descriptions.

Conservation	Lagrangian description	Eulerian description
Mass	$\frac{D\rho}{Dt}=-\rho\frac{\partial v^\beta}{\partial x^\beta}$	$\frac{\partial\rho}{\partial t}+v^\beta\frac{\partial\rho}{\partial x^\beta}=-\rho\frac{\partial v^\beta}{\partial x^\beta}$
Momentum	$\frac{Dv^\beta}{Dt}=-\frac{1}{\rho}\frac{\partial p}{\partial x^\beta}$	$\frac{\partial v^\beta}{\partial t}+v^\alpha\frac{\partial v^\beta}{\partial x^\alpha}=-\frac{1}{\rho}\frac{\partial p}{\partial x^\beta}$
Energy	$\frac{De}{Dt}=-\frac{p}{\rho}\frac{\partial v^\beta}{\partial x^\beta}$	$\frac{\partial e}{\partial t}+v^\beta\frac{\partial e}{\partial x^\beta}=-\frac{p}{\rho}\frac{\partial v^\beta}{\partial x^\beta}$

In **Table 1.1**, ρ, e, $\mathbf{v}$ and $\mathbf{x}$ are density, internal energy, velocity and position vector respectively. The Greek superscripts α and β are used to denote the coordinate directions, while the summation in the equations is taken over repeated indices. It is seen that the differences between the two sets of equations are inherited in the definition of the *total time derivative* as the combination of the *local derivative* and the *convective derivative*, i.e.,

$$\frac{D}{Dt}=\frac{\partial}{\partial t}+v^\alpha\frac{\partial}{\partial x^\alpha} \tag{1.26}$$

where D/Dt is the total time derivative (or *substantial derivative*, *material derivative*, or *global derivative*) that is physically the time rate of change following a moving fluid elements; $\partial/\partial t$ is the local derivative that is physically the time rate of change at a fixed point; $v^\alpha\,\partial/\partial x^\alpha$ is the convective derivative that is physically the change due to the movement of the fluid element from one location to another in the flow field where the flow properties are spatially different. Therefore, the total time derivative describes that the flow property of the fluid element is changing, as a fluid element sweeps passing a point in the flow. This is because 1) at that point, the flow field property itself may be fluctuating with time (the local derivative); 2) the fluid element is on its way to another location in the flow field where the flow property may be different (the convective derivative).

The Eulerian and Lagrangian descriptions correspond to two disparate kinds of grid of domain discretization: the Eulerian grid and the Lagrangian grid. Both of them are widely used in computer modeling with preferences on types of problems, and hence are briefed in the followings.

1.3.1 Lagrangian grid

In Lagrangian grid-based methods such as the well-known and widely used FEM (Zienkiewicz and Taylor, 2000; Liu and Quek, 2003), the *Lagrangian grid* is fixed to or attached on the material in the entire computation process, and therefore it moves with the material as illustrated in **Figure 1.9**.

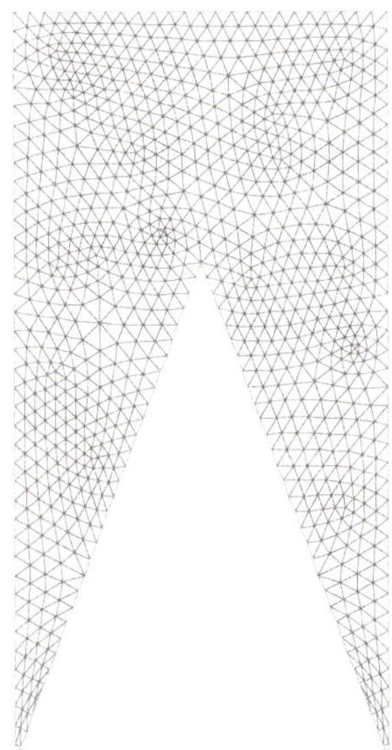

Figure 1.9 Lagrangian mesh/cells/grids for the computer modeling of the detonation and explosion process of a shaped charge. The triangular cells and the entire mesh of cell move with the material.

Since each grid node follows the path of the material at the grid point, the relative movement of the connecting nodes may result in expansion, compression and deformation of a mesh cell (or element). Mass, momentum and energy are transported with the movement of the mesh cells. Because the mass within each cell remains fixed, no mass flux crosses the mesh cell boundaries. When the material deforms, the mesh deforms accordingly.

The Lagrangian grid-based methods have several advantages.

1. Since no convective term exists in the related partial differential equations, the code is conceptually simpler and should be faster as no computational effort is necessary for dealing with the convective terms.
2. Since the grid is fixed on the moving material, the entire time history of all the field variables at a material point can be easily tracked and obtained.

3. In the Lagrangian computation, some grid nodes can be placed along boundaries and material interfaces. The boundary conditions at free surfaces, moving boundaries, and material interfaces are automatically imposed, tracked and determined simply by the movement of these grid nodes.
4. Irregular or complicated geometries can be conveniently treated by using an irregular mesh.
5. Since the grid is required only within the problem domain, no additional grids beyond the problem domain is required, and hence the Lagrangian grid-based methods are computationally efficient.

Due to these advantages, Lagrangian methods are very popular and successful in solving computational solid mechanics (CSM) problems, where the deformation is not as large as that in the fluid flows.

However, Lagrangian grid-based methods are practically very difficult to apply for cases with extremely distorted mesh, because their formulation is always based on mesh. When mesh is heavily distorted, accuracy of the formulation and hence the solution will be severely affected, especially when mapping is involved (Liu, 2010a). In addition, the time step, which is controlled by the smallest nodal spacing, can become too small to be efficient for the time marching, and may even lead to the breakdown of the computation.

A possible option to enhance the Lagrangian computation is to *rezone* the mesh or re-mesh the problem domain. The mesh rezoning involves overlaying of a new, undistorted mesh on the old, distorted mesh, so that the following-up computation can be performed on the new undistorted mesh. The physical properties in the new mesh cells are approximated from the old mesh cells through calculating the mass, momentum and energy transport in an Eulerian description. Adaptive rezoning techniques are quite popular for simulations of impact, penetration, explosion, fragmentation, turbulence flows, and fluid-structure interaction problems. The rezoning procedure in Lagrangian computations can be tedious and very time-consuming. Moreover, with each rezoning, some material diffusion occurs and material histories may be lost. In addition, the Lagrangian codes under frequent re-mesh turn to resemble an Eulerian code in an overall sense. Therefore, even though there are some very good advantages in Lagrangian grid-based methods, the disadvantages can result in numerical difficulties when simulating events of extremely large deformation (Anderson Jr, 1987; Benson, 1992; Mair, 1999).

A Lagrangian numerical method, whose solution does not depend on a mesh and hence is not affected by the heavy movement of the nodes, is indeed desirable.

1.3.2 Eulerian grid

Contrary to the Lagrangian grid, the *Eulerian grid* is fixed on the space, in which the simulated object is located and moves across the fixed mesh cells in the grid (illustrated in **Figure 1.10**). Therefore, all grid nodes and mesh cells remain spatially fixed in space and do not change with time while the materials are flowing across the mesh. The flux of mass, momentum and energy across mesh cell boundaries are simulated to compute the distribution of mass, velocity, energy, etc. in the problem domain. The shape and volume of the mesh cell remain unchanged in the entire process of the computation.

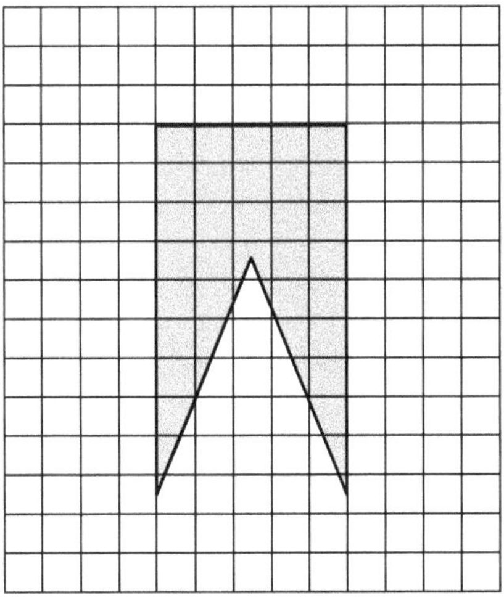

Figure 1.10 Eulerian mesh/cells/grids for the computer modeling of the detonation and explosion process of a shaped charge. The mesh/grid is fixed in space and does not move or deform with time. The material moves/flows across the fixed mesh cells.

Since the Eulerian grid is fixed in space and with time, large deformations in the object do not cause any deformations in the mesh itself and therefore do not cause the same kind of numerical problems as in the Lagrangian grid-based methods. *Eulerian methods* are therefore dominant in the area of computational fluid dynamics, where the flow of the material dominates. In principle, all hydrodynamic problems can be numerically solved using a multi-material Eulerian method that calculates the mass, momentum and energy flux across the fixed Eulerian mesh cell boundaries. Early simulations of problems with large deformation such as explosion and high velocity impacts were usually performed using some kind of Eulerian methods (Anderson Jr, 1987; Benson, 1992; Mair, 1999). However, there are some disadvantages associated with Eulerian grid-based methods.

1. It is very difficult to analyze the time history of field variables at a fix point on the material, because the movement of the material cannot be tracked using a fixed mesh. One can only have the time history of field variables at fixed-in-space Eulerian grid.
2. It is not easy to treat the irregular or complicated geometries of material/media in the Eulerian grid-based methods. A complicated mesh generation procedure to convert the irregular geometry of problem domain into a regular computational domain is usually necessary. Sometimes, expensive numerical mapping is required.
3. The Eulerian methods track the mass, momentum and energy flux across the mesh cell boundaries, while the position of free surfaces, deformable boundaries, and moving material interfaces are difficult to be determined accurately.
4. Since the Eulerian methods require a grid over a computational domain, which should be large enough to cover the entire area to which the material can possibly flow. It sometimes requires the modeler to use a very coarse grid for computational efficiency at the expenses of the resolution of domain discretization and the accuracy of the solution.

The features of both the Lagrangian and Eulerian methods are summarized in **Table 1.2**.

Table 1.2 Comparisons of Lagrangian and Eulerian methods.

	Lagrangian methods	**Eulerian methods**
Grid	Attached on the moving material	Fixed in the space
Track	Movement of any point on materials	Mass, momentum, and energy flux across grid nodes and mesh cell boundary
Time history	Easy to obtain time-history data at a point attached on materials	Difficult to obtain time-history data at a point attached on materials
Moving boundary and interface	Easy to track	Difficult to track
Irregular geometry	Easy to model	Difficult to model with good accuracy
Large deformation	Difficult to handle	Easy to handle

1.3.3 Combined Lagrangian and Eulerian grids

The different but complementary features of the Lagrangian and Eulerian descriptions suggest that it would be computationally beneficial to combine these two descriptions so as to strengthen their advantages and to avoid their disadvantages. This idea has led to the development of two complicated approaches that apply both the Lagrangian and Eulerian descriptions: the *Coupled Eulerian Lagrangian* (CEL) (Mair, 1999) and the *Arbitrary Lagrange Eulerian* (ALE) (Liu et al., 1986; Benson, 1992; Hirt et al., 1997). The CEL approach employs both the Eulerian and Lagrangian methods in separate (or with some overlap) regions of the problem domain. One of the most common practices is to discretize solids in a Lagrangian frame, and fluids (or materials behaving like fluids) in a Eulerian frame. The Lagrangian region and Eulerian region continuously interact with each other through a coupling module in which computational information is exchanged either by mapping or by special interface treatments between these two sets of grid.

The ALE is closely related to the rezoning techniques for Lagrangian mesh, and aims to move the mesh independently of the materials so that the mesh distortion can be minimized. In an ALE, Lagrangian motion is computed at every time step in the beginning, followed by a possible rezoning stage in which the mesh is either not rezoned (pure Lagrangian description), or rezoned to the original shape (Eulerian description), or rezoned to some more advantageous shape (somewhat between the Lagrangian and Eulerian description).

These two approaches of combining Eulerian and Lagrangian descriptions receive much research interest and have achieved a lot in obtaining more stable solutions. Many commercial hydrocodes such as MSC/Dytran (MSC/Dytran, 1997), DYNA2D and DYNA3D (Hallquist, 1988, 1998), and AUTODYN (Century Dynamics Incorporated, 1997) have incorporated CEL or/and ALE for coupled analyses of dynamic phenomena with fluid solid interaction behavior. Unfortunately, even with the CEL and ALE formulations a highly distorted mesh can still introduce severe errors in numerical simulations (Benson, 1992; Hirt et al., 1997).

1.3.4 Limitations of the grid-based methods

Conventional grid-based numerical methods such as FDM and FEM have been widely applied to various areas of CFD and CSM, and currently are the dominant methods in numerical simulations of domain discretization and numerical discretization. Despite the great success, grid-based numerical methods suffer from some inherent difficulties in many aspects, which limit their applications to many problems.

In grid-based numerical methods, mesh generation for the problem domain is a prerequisite for the numerical simulations. For the Eulerian grid methods like FDM, constructing a regular grid for irregular or complex geometry has never been an easy task, and usually requires additional complex mathematical transformation that can be even more expensive than solving the problem itself. Determining the precise locations of the inhomogeneities, free surfaces, deformable boundaries and moving interfaces within the frame of the fixed Eulerian grid is also a formidable task. The Eulerian methods are also not well suited to problems that need monitoring the material properties in fixed volumes, e.g. particulate flows. For the Lagrangian grid methods like FEM, mesh generation is necessary for the objects being simulated, and usually occupies a significant portion of the computational effort. Treatment of large deformation is an important issue in a Lagrangian grid-based method. It usually requires special techniques like rezoning. Mesh rezoning, however, is tedious and time consuming, and may introduce additional inaccuracy into the solution.

The difficulties and limitations of the grid-based methods are especially evident when simulating hydrodynamic phenomena such as *explosion* and *high velocity impact* (HVI). In the whole process of an explosion, there exist special features such as large deformations, large inhomogeneities, moving material interfaces, deformable boundaries, and free surfaces. These special features pose great challenges to numerical simulations using the grid-based methods. High velocity impact problems involve shock waves propagating through the colliding or impacting bodies that behave like fluids. Analytically, the equations of motion and a high-pressure equation of state are the key descriptors of material behavior. In HVI phenomena, there exist large deformations, moving material interfaces, deformable boundaries, and free surfaces, which are, again, very difficult for grid-based numerical methods. As can be seen from many existing literatures, simulation of hydrodynamic phenomena such as explosion and HVI by methods without using a mesh is a very promising alternative.

The grid-based numerical methods are also not suitable for situations where the main concern of the object is a set of discrete physical particles rather than a continuum, e.g., the interaction of stars in astrophysics, movement of millions of atoms in an equilibrium or non-equilibrium state, dynamic behavior of protein molecules, and etc. Simulation of such discrete systems using the continuum grid-based methods may not always be a good choice.

1.4 Meshfree methods

1.4.1 Types of methods

A recent strong interest is focused on the development of the next generation computational methods — *meshfree methods*, which are expected to be superior to the conventional grid-based FDM and FEM for many applications. The key idea of the meshfree methods is to provide accurate and stable numerical solutions for integral equations or PDEs with a proper set of boundary conditions, using a set of arbitrarily distributed nodes (or particles) without any mesh that defines the connectivity of these nodes or particles. Details on many existing meshfree methods can be found in recent monographs (Liu, 2010a; Liu and Zhang, 2013). One important goal of the initial research is to modify the internal structure of the grid-based FDM and FEM to become more adaptive, versatile and robust. Much effort is concentrated on problems to which the conventional FDM and FEM are difficult to apply, such as problems with free surface, deformable boundary, moving interface (for FDM), large deformation (for FEM), complex mesh generation, mesh adaptivity, and multi-scale resolution (for both FDM and FEM). Recently, a number of meshfree methods have been proposed for analyzing solids and structures as well as fluid flows. These meshfree methods share some common features, but are different in the means of function approximation and the implementation process.

Smoothed particle hydrodynamics (SPH) (Gingold and Monaghan, 1977; Lucy, 1977), as a meshfree and particle method, was originally invented for modeling astrophysical phenomena, and later widely extended for applications to problems of continuum solid and fluid mechanics. The SPH method and its different variants are the major type of particle methods, and have been incorporated into many commercial codes. The first monograph dedicated on SPH was published in 2003 (Liu and Liu, 2003).

Liszka and Orkisz proposed a generalized finite difference method that can deal with arbitrary irregular grids (Liszka and Orkisz, 1980). Nayroles et al. are the first to use moving least square approximations in a Galerkin method to formulate the so-called diffuse element method (DEM) (Nayroles et al., 1992; Onate et al., 1996). Based on diffuse element method, Belytschko et al. advanced remarkably the element free Galerkin (EFG) method (Belytschko et al., 1994). The EFG is currently one of the most popular meshfree methods, and applied to many solid mechanics problems with the help of a background mesh for integration (Belytschko et al., 1996). Atluri and Zhu have originated the Meshless Local Petrov-Galerkin (MLPG) method that requires only local background cells for the integration (Atluri and Zhu, 1998). Because the MLPG does not need a global background mesh for integration, it has been applied to the analysis of beam and plate structures (Gu and Liu, 2001c; Atluri and Shen, 2002; Long and Atluri, 2002), fluid flows (Lin and Atluri, 2001). Detailed descriptions of the MLPG and its applications can be found in the monograph by Atluri and Shen (Atluri and Shen, 2002).

W. K. Liu and his co-workers, through revisiting the consistency and reproducing conditions in SPH, proposed a reproducing kernel particle method (RKPM) which improves the accuracy of the SPH approximation especially around the boundary (Liu and Chen, 1995; Liu et al., 1995; Liu et al., 1996b). There are good literatures available on RKPM and its applications (Liu et al., 1996a; Li and Liu, 2002).

G. R. Liu and his colleagues in a series of papers developed the point interpolation method (PIM) and some variants (Gu and Liu, 2001b; Liu and Gu, 2001b, a; Gu and Liu, 2002; Liu, 2002). Their struggle has been on the singularity issue in the polynomial PIMs, and different ways to solve the problem have been attempted. The use of radial basis function (or together with the polynomials) has well resolved the problem for both the local Petrov-Galerkin weak-form (Gu and Liu, 2001c) and the global Galerkin weak-form (Wang and Liu, 2002). Later, a meshfree weak-strong (MWS) form formulation based on a combined weak and strong forms (Liu and Gu, 2003; Liu et al., 2004) has been proposed. The MWS method uses both MLS and the radial PIM shape functions, and needs only a local background mesh for nodes that is near the natural boundaries of the problem domain.

The most recent developments on meshfree methods maybe the smoothed point interpolation method (S-PIM) (Liu and Zhang, 2013). The formulation of S-PIM uses the so-called weakened weak-form (W2) based on the G space theory (Liu, 2010b) established using the generalized gradient smoothing techniques that allows the use of discontinuous functions (Liu, 2008). The S-PIM has a family of different models, according to how the smoothing domains are formed, including the NS-PIM that uses the node-based smoothed domains (SD), and ES-PIM using edge-based SD, FS-PIM using face-based SD, CS-PIM using cell-based SD, and S-PIM using mixed cell and nodal based SD. These S-PIMs have been found to be superior in a number of ways. For example, they all work particularly well with T-meshes (triangular for 2D and tetrahedral for 3D). The NS-PIM can produce upper bound solutions (Liu and Zhang, 2008b) and ES-PIM offers ultra-accurate solutions (Liu and Zhang, 2008a). These advances in W2 formulations open a wide window for developing the next generation of computational methods. A simpler version of S-PIM is known as the smoothed finite element method (S-FEM) that uses the FEM shape functions, instead of the PIM shape functions. The S-FEM is much simpler than meshfree S-PIMs, but possesses some of the excellent properties of the S-PIM. Detailed formulations and examination of S-FEM can be found in the S-FEM monograph (Liu and Nguyen, 2010).

Other notable representatives of meshfree methods include the HP-cloud method (Duarte and Oden, 1996), and free mesh method (FMM) (Yagawa and Yamada, 1996, 1998). Some typical meshfree methods either in strong or weak form are listed in **Table 1.3**.

1.4.2 Applications

Comprehensive investigations on meshfree methods are closely related to the applications to complex computational solid and fluid mechanics problems. Since the computational frame in the meshfree methods is a set of arbitrarily distributed nodes rather than a system of pre-defined mesh/grid, the meshfree

Table 1.3 Some typical meshfree methods in chronological order.

Methods	References	Methods of approximation
Smoothed particle hydrodynamics (SPH)	(Gingold and Monaghan, 1977; Lucy, 1977)	Integral representation
Finite point method (FPM)	(Liszka and Orkisz, 1980; Onate et al., 1996)	Finite difference representation
Diffuse element method (DEM)	(Nayroles et al., 1992)	Moving least square (MLS) approximation Galerkin method
Element free Galerkin (EFG)	(Belytschko et al., 1994; Belytschko et al., 1996)	MLS approximation Galerkin method
Reproduced kernel particle method (RKPM)	(Liu et al., 1995; Liu et al., 1996b)	Integral representation Galerkin method
HP-cloud method	(Duarte and Oden, 1996)	MLS approximation, Partition of unity
Free mesh method	(Yagawa and Yamada, 1996, 1998)	Galerkin method
Meshless local Petrov-Galerkin (MLPG) method	(Atluri and Zhu, 1998; Gu and Liu, 2001c; Atluri and Shen, 2002; Long and Atluri, 2002)	MLS approximation Petrov-Galerkin method
Point interpolation method (PIM)	(Gu and Liu, 2001b; Liu and Gu, 2001b, a; Gu and Liu, 2002; Liu, 2002)	Point interpolation, (Radial and Polynomial basis), Galerkin method, Petrov-Galerkin method
Meshfree weak-strong form (MWS)	(Liu and Gu, 2003; Liu et al., 2004)	MLS, PIM, radial PIM (RPIM), Collocation plus Petrov-Galerkin
Smoothed Point interpolation method (S-PIM)	(Liu and Zhang, 2008a, b; Liu, 2010b; Liu and Zhang, 2013)	Point interpolation, (Radial and Polynomial basis), Weakened weak form (W2), G space theory

methods are attractive in dealing with problems that are difficult for traditional grid-based methods. The interesting applications of meshfree methods include large deformation analyses in solids (Chen et al., 1996; Chen et al., 1997, 1998; Jun et al., 1998; Li et al., 2000a, b; Li and Liu, 2002; Liu, 2002), vibration analyses especially for plates and shells (Gu and Liu, 2001c; Liu and Chen, 2001; Liu and Gu, 2001a), structure buckling problems (Liu and Chen, 2002), piezoelectric structure simulations (Liu et al., 2002a, b), non-linear foundation consolidation problems (Wang et al., 2001), incompressible flows (Lin and Atluri, 2001; Liu et al., 2004), singular boundary-value problems (Liu et al., 2002c), and impact and explosion simulations (Johnson and Beissel, 1996; Randles and Libersky, 1996; Liu et al., 2003a; Liu et al., 2003b; Liu et al., 2006; Zhou et al., 2007).

Meshfree methods have also been developed for boundary integral equations to develop boundary meshfree methods, in which only the boundary of the problem domain needs to be represented with nodes. The formulation developed by Mukherjee and Mukherjee was based on the formulation of EFG using the MLS approximation (Mukherjee and Mukherjee, 1997a, b). Boundary point interpolation methods (BPIM) were developed by Gu and Liu, using polynomial PIM and radial PIM interpolations (Gu and Liu, 2002), which give a set of much smaller discretized system equations due to the delta function property of the PIM shape functions.

In practical applications, a meshfree method can be coupled with other meshfree methods or a conventional numerical method to take the full advantages of each method. Examples include SPH coupling with FEM (Attaway et al., 1993a; Attaway et al., 1993b; Attaway et al., 1994; Johnson, 1994; Zhang et al., 2011; Hu et al., 2014), EFG coupling with boundary element method (BEM) (Liu and Gu, 2000a; Gu and Liu, 2001a), MPLG coupling with BEM or FEM (Liu and Gu, 2000c), and SPH coupling with BEM for modeling transient fluid-structure interaction and applications in underwater impacts (Zhang et al., 2013). A meshfree method can also be coupled with another meshfree method for particular applications (Liu and Gu, 2000b). An adaptive stress analysis package based on the meshfree technology, MFree2D©, has been developed (Liu, 2002; Liu and Tu, 2002).

There are basically three types of meshfree methods: methods based on strong form formulations, methods based on weak form formulations, and particle methods. The *strong form method* such as the *collocation method*, has attractive advantages of being simple to implement, computationally efficient and "truly" meshfree, because no integration is required in establishing the discrete system equations. However, they are often unstable and less accurate, especially when irregularly distributed nodes are used for problems governed by partial differential equations with Neumann (derivative) boundary conditions, such as solid mechanics problems with stress (natural) boundary conditions. On the other hand, *weak form method* such as the EFG, MLPG and PIM has the advantages of excellent stability, accuracy. The Neumann boundary conditions can be naturally satisfied due to the use of the weak form that involves smoothing (integral) operators. However, the weak form method is said not to be "truly" meshfree, as a background mesh (local or global) is required for the integration of the weak forms.

Recently, a novel *meshfree weak-strong* (MWS) form of method is proposed by Liu and his colleagues based on a combined formulation of both the strong form and the local weak form (Liu and Gu, 2003; Liu et al., 2004). In the MWS method, the strong form formulation is used for all the internal nodes and the nodes on the essential boundaries. The local weak form (Petrov-Galerkin weak form) is used only for nodes near the natural boundaries. Hence, there is no need for numerical integrations for all the internal nodes and the nodes on the essential boundaries. The numerical integration is performed locally only for the nodes on the natural boundaries and thus only local background cells for the nodes near the natural boundaries are required. The locally supported radial point interpolation and the moving least squares approximation have been used to construct the meshfree shape functions for the MWS. The final system matrices are sparse and banded for computational efficiency. The MWS method is, so far, the meshfree method that uses least meshes in the entire computation and produce stable solutions even for solid mechanics problems using irregularly distributed nodes. It is one more step close to realize the dream of the "truly" meshfree method that is capable of producing stable and accurate solutions for solid mechanics problems using irregularly distributed nodes.

The S-PIM and S-FEM have been developed since 2005 and applied to solve various problems, such as the elastic solid problems, fracture mechanics, hyper-elastic problems, viscoelastoplasticity, plate analysis, piezoelectric structures, heat transfer problems, acoustics problems, etc. The details are given, respectively, in the S-PIM monograph (Liu and Zhang, 2013) and the S-FEM monograph (Liu and Nguyen, 2010).

For more details on meshfree methods, the readers may refer to some excellent review papers (Belytschko et al., 1996; Liu et al., 1996a; Li and Liu, 2002) and some published monographs (Atluri and Shen, 2002; Liu, 2002; Liu and Liu, 2003; Li and Liu, 2004; Liu and Gu, 2005).

1.4.3 Particle methods — a special class of meshfree methods

Particle-based methods (or *particle methods* for abbreviation) in general refers to a special class of meshfree methods that employ a set of finite number of discrete particles to represent the state of a system and to record the movement of the system. Each particle can either be directly associated with one discrete physical object, or be generated to represent a part of the continuum problem domain. The particles can range from very small (nano or micro) scale, to meso scale, to macro scale, and even to astronomical scale. For CFD problems, each particle possesses a set of field variables such as mass, momentum, energy, position etc, and other variables (e.g., charge, vorticity, etc.) related to the specific problem. The evolution of the physical system is

Table 1.4 Some typical particle methods.

Methods	References
Molecular dynamics (MD)	(Alder and Wainright, 1957; Rahman, 1964; Stillinger and Rahman, 1974; Liu et al., 2005)
Monte Carlo (MC)	(Metropolis and Ulam, 1949; Binder, 1988, 1992)
Direct simulation Monte Carlo (DSMC)	(Bird, 1994; Pan et al., 1999; Pan et al., 2000)
Dissipative particle dynamics (DPD)	(Hoogerbrugge and Koelman, 1992; Espanol and Warren, 1995; Duong-Hong et al., 2008; Yan et al., 2012; Liu et al., 2014)
Lattice gas cellular automaton (CA)	(Pomeau and Frisch, 1986; McNamara and Zanetti, 1988)
Lattice Bolztmann equation (LBE)	(Ladd, 1994; He and Luo, 1997b, a; Chen and Doolen, 1998; Luo, 1998)
Particle-in-Cell (PIC)	(Evans and Harlow, 1957; Harlow, 1957)
Marker-and-Cell (MAC)	(Harlow and Welch, 1965)
Fluid-in-Cell (FLIC)	(Gentry et al., 1966)
Moving particle semi-implicit (MPS)	(Koshizuka et al., 1995; Koshizuka and Oka, 1996; Heo et al., 2001)
Discrete element method (DEM)	(Cundall, 1987)
Vortex methods	(Chorin, 1973; Leonard, 1980)
Smoothed particle hydrodynamics (SPH)	(Gingold and Monaghan, 1977; Lucy, 1977; Liu and Liu, 2003)

determined by the conservation of mass, momentum and energy. Some typical particle methods or particle-like methods are listed in **Table 1.4**.

Based on the *length scale*, the meshfree particle methods can be roughly divided into three classes: atomistic/microscopic scale meshfree particle methods, mesoscopic meshfree particle methods, and macroscopic meshfree particle methods. A typical atomistic MPM is the *molecular dynamics* (MD) method, either *ab inito* or classic that uses force potential functions (Kresse and Hafner, 1993; Kresse and Hafner, 1994; Frenkel and Smit, 2002; Xu and Liu,

2003; Rapaport, 2004; Xu et al., 2004; Yao et al., 2004; Liu et al., 2005; Yao et al., 2005; Wu et al., 2006).

Mesoscopic MPMs include *dissipative particle dynamics* (DPD) (Hoogerbrugge and Koelman, 1992; Espanol and Warren, 1995; Liu et al., 2014), lattice gas Cellular Automata (CA) (Pomeau and Frisch, 1986; McNamara and Zanetti, 1988). Macroscopic MPMs includes SPH (Gingold and Monaghan, 1977; Lucy, 1977; Liu and Liu, 2003), *particle-in-cell* (PIC) (Evans and Harlow, 1957; Harlow, 1957), *fluid-in-cell* (FLIC) (Gentry et al., 1966), *vorticity and stream function method* (Fromm and Harlow, 1963), *marker-and-cell* (MAC) (Harlow and Welch, 1965), *material point method* (MPM) (Sulsky et al., 1994; Sulsky et al., 1995), and *moving particle semi-implicit method* (Koshizuka et al., 1995; Koshizuka and Oka, 1996; Koshizuka et al., 1998; Heo et al., 2001).

Many particle methods were initially developed for systems with discrete particles. Examples include SPH simulation of the interaction of stars in astrophysics, MD simulation of movement of millions of atoms in an equilibrium or non-equilibrium state, *discrete element method* (DEM) simulation of soils and sands, etc. Particle methods have also been modified, extended, and applied to system of continuum media. In such cases, an additional operation is required to generate a set of particles to represent the continuum media. Each particle represents a part of the problem domain, with some attributes such as mass, position, momentum and energy concentrated on the mass or geometric center of the sub-domain. Examples include SPH, PIC, MAC, FLIC, MPS, vortex methods and many others.

According to the mathematical models used, particle methods can be deterministic or probabilistic. The deterministic particle methods deal directly with the governing system equations of physical law. In the deterministic particle methods, once the initial and boundary conditions are given, the particle evolution in the later time stages can be, in theory, precisely determined based on the physical laws that govern the problem. Many particle methods are inherent with probabilistic nature based on statistical principles. Main representatives of the probabilistic particle methods include MD, MD based MC, DPD, DSMC; lattice gas Cellular Automata, and lattice Bolztmann equation (LBE) and etc.

It is noted that some particle methods can have mixed features. One typical example is the SPH method. It was originally invented to simulate astrophysical problems, and is currently being applied to macroscopic continuum problems of computational solid and fluid mechanics, and to atomistic scale simulations (Español, 1997; Hoover, 1998; Nitsche and Zhang, 2002; Hoover and Hoover, 2003). The SPH method is now widely used for both discrete particle systems and continuum systems. It was actually initially developed as a probabilistic particle method, and was later modified and applied as a deterministic meshfree particle method.

There are some other points that need to be further clarified. Firstly, most particle methods are inherently Lagrangian methods, in which the particles represent the physical system moving in the Lagrangian frame according to the internal interactions and external force, and thus evolve the system in time. There are also examples in which the particles are fixed in the Eulerian space as interpolation points rather than moving objects (Laguna, 1995). Secondly, most particle methods use explicit methods for the time integration. Some exceptions, however, use implicit or semi-implicit procedures such as the MPS, in which the pressure term in the momentum equations are implicitly determined. Thirdly, most of the particle methods are basically meshfree methods, in which the particles form the computational frame for the field variable approximation. However, there are some particle methods that still use some kind of mesh (e.g. PIC, LBE etc.) for background interpolation, or for other purposes. As such, particle methods sometime are referred to as meshfree particle methods (MPMs). Lastly, in using particles to represent a continuum domain, some kind of mesh may be needed to generate the initial distribution of particles.

The advantages of the particle methods over conventional grid-based numerical methods can be summarized as follows:

1. In particle methods, the problem domain is discretized with particles without a fixed connectivity. Treatment of large deformation is relatively much easier;
2. Discretization of complex geometry for the particle methods is relatively simpler as only an initial discretization is required;
3. Refinement of the particles is expected much easier to perform than the mesh refinement;
4. It is easy to obtain the features of the entire physical system through tracing the motion of the particles. Therefore, identifying free surfaces, moving interfaces and deformable boundaries is no longer a tough task. Time history of field variables at any point on the material can also be naturally obtained.

1.5 Solution strategy of particle methods

For atomistic and meso-scale particle methods such as MD and DPD, the solution procedure include the initialization (the coordinates of initial particles, their velocities and the target system temperature), force computation (usually from the inter-particle potential), time integration (to update the positions and velocities of all particles according to a specific time integration algorithm), and data analysis (to evaluate desired physical quantities). Detailed descriptions of the computational procedures in MD and DPD can be found in Chapters 2 and 3 respectively.

Macro-scale particle methods (such as SPH, MPS and MPM) aim to perform numerical analyses for complex problems without the use of a mesh pre-defined using a connectivity of nodes. Similar to the simulations with the conventional grid-based numerical methods, a typical computer modeling using particle methods also involves

1) governing equations with proper boundary conditions (BC) and/or initial conditions (IC),
2) domain discretization technique for creating particles,
3) numerical discretization technique (weak form, strong form, particle methods),
4) and numerical technique to solve the resultant algebraic equations or ordinary differential equations (ODE).

It is seen that the differences between the grid-based methods and particle methods lie in:

- The problem domain is discretized with or represented by particles (spatial discretization with *particle representation*);
- Functions, derivatives and integrals in the governing equations are approximated using the particles rather than over a mesh (numerical discretization with *particle approximation*).

1.5.1 Particle representation

In particle methods, there is no need to prescribe the connectivity between the particles. All one needs is an initial distribution or generation of the particles that represent the problem domain, if the problem domain is not initially in discrete particle form. Different ways of generating particles for continuous domains can be employed. Since mesh generation algorithms (e.g. triangulation algorithm) are readily available for both the 2D and 3D space, one simple approach is to deploy particles in the mass or geometric centers of the mesh cells, as shown in **Figure 1.11** and **Figure 1.12**.

The triangular or tetrahedron meshes are preferred, because they can always be generated automatically for domains of complex geometry using existing mesh generation methods that are even commercially available. For domains of simple geometry, quadrilateral or hexahedron meshes can also be used for particle generation.

Note that when the particles are placed at the mass centers of the corresponding cells, the problem domain is approximated, and the original smoothing surface becomes a rough surface. Locating the particles at the nodes of the mesh can provide an apparently smoother surface. Place particles at both the centers and the nodes of the cells may be another alternative to provide a good representation for the continuum.

1.5.2 Particle approximation

Numerical discretization in particle methods involves approximating the values of functions, derivatives and integrals at a particle with particle approximations using the information at all the neighboring particles that have influence on the particle. The area of influence of a particle is determined by the so-called *influence domain* or *support domain*. The neighboring particles within the support domain of a particle provide all the necessary and sufficient information for the field variable approximations at the particle.

For example, the field variable (e.g., a component of the velocity) u for a particle located at $\mathbf{x} = (x, y, z)$ within the problem domain can be approximated using the information on the particles within the support domain of the particle at $\mathbf{x}$ (**Figure 1.13**):

$$u(\mathbf{x}) = \sum_{i=1}^{N} \phi_i(\mathbf{x}) u_i \tag{1.27}$$

where N is the number of particles within the support domain of the particle at $\mathbf{x}$; u_i is the field variable at particle i, ϕ_i is the shape function at particle i constructed using the information on all particles within the support domain of the particle at $\mathbf{x}$. The shape functions can then be used for establishing a set of discretized system equations using weak form, strong form, or both.

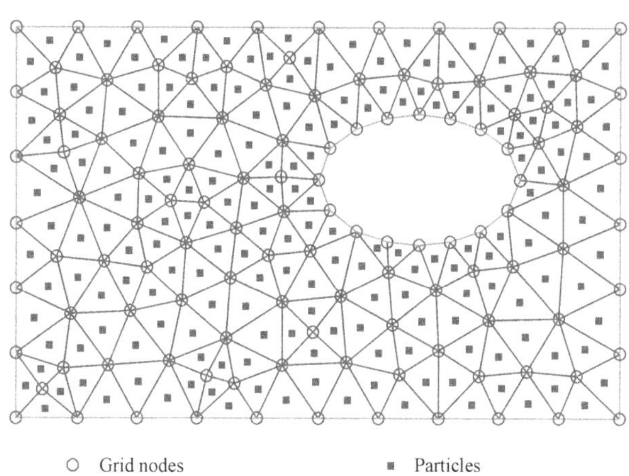

Figure 1.11 Initial particle generation for a continuum using a triangular mesh in two-dimensional space.

(a)

(b)

Figure 1.12 Initial particle generation for a continuum using a tetrahedron mesh in three-dimensional space. (a) tetrahedron mesh, (b) generated particles.

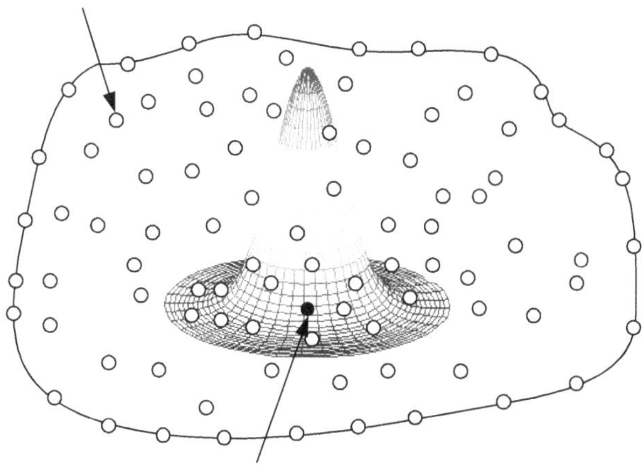

Figure 1.13 The support domain of a particle at x.

1.5.3 Solution procedure

The procedure in a MPM simulation is mainly similar to that in a grid-based numerical simulation, except for the particle representation and the particle approximation. A typical procedure of computer modeling using particle methods for dynamic fluid flow problems can be briefed as follows:

1. Represent the problem domain with particles so that the computational information is known at the discrete particles at an initial instant t with a proper treatment on the boundary conditions;
2. Discretize the derivatives or integrals in the governing equations with proper particle approximations;
3. From the given velocity and/or position, calculate the strain rate and/or strain, and then calculate the stress at each discrete particles at the instant t;
4. Calculate the acceleration at each discrete particles using the calculated stress;
5. Use the acceleration at the instant t to calculate the new velocities and the new positions at time instant $t+\Delta t$, where Δt is the incremental time step;
6. From new velocities and/or new positions, calculate the new strain rate and/or new strain at time instant $t+\Delta t$, and then calculate the new stress at time instant $t+\Delta t$. Repeat step 4, 5 and 6 to march forward in time until the final specified time instant.

References

Alder, B. J. and Wainright, T. E. (1957) Phase transition for a hard sphere system, *Journal of Chemical Physics* **27**: 1208–1209.

Anderson, J. D. (1995) Computational fluid dynamics: The basics with applications. McGraw-Hill, New York.

Anderson, J. D. (2002) Computational fluid dynamics: The basics with applications. McGraw-Hill, New York.

Anderson Jr., C. E. (1987) An overview of the theory of hydrocodes, *International Journal of Impact Engineering* **5**(1–4): 33–59.

Atluri, S. and Zhu, T. (1998) A new meshless local petrov–galerkin (MLPG) approach in computational mechanics, *Computational Mechanics* **22**(2): 117–127.

Atluri, S. N. and Shen, S. (2002) The meshless local petrov–galerkin (mlpg) method. Tech Science Press, Encino, CA.

Attaway, S. W., Heinstein, M. W., Mello, F. J. and Swegle, J. W. (1993a). Coupling of smooth particle hydrodynamics with pronto, *Annual Winter Meeting of the American Society of Mechanical Engineers*, New Orleans, LA.

Attaway, S. W., Heinstein, M. W. and Swegle, J. W. (1993b). Coupling of smooth particle hydrodynamics with the finite element method, *IMPACT-4: Structural Mechanics in Reactor Technology Conference*, Berlin, Germany.

Attaway, S. W., Heinsteinb, M. W. and Sweglec, J. W. (1994) Coupling of smooth particle hydrodynamics with the finite element method, *Nuclear Engineering and Design* **150**(2–3): 199–205.

Belytschko, T., Krongauz, Y., Organ, D., Fleming, M. and Krysl, P. (1996) Meshless methods: An overview and recent developments, *Computer Methods in Applied Mechanics and Engineering* **139**(1–4): 3–47.

Belytschko, T., Lu, Y. Y. and Gu, L. (1994) Element-free galerkin methods, *International Journal for Numerical Methods in Engineering* **37**(2): 229–256.

Benson, D. J. (1992) Computational methods in lagrangian and eulerian hydrocodes, *Computer Methods in Applied Mechanics and Engineering* **99**(2–3): 235–394.

Binder, K. (1988) The monte carlo method in condensed matter physics. Springer, Berlin.

Binder, K. (1992) The monte carlo method in statistical physics. Springer Berlin.

Bird, G. A. (1994) Molecular gas dynamics and the direct simulation of gas flow. Oxford University Press, Oxford.

Century Dynamics Incorporated (1997). Autodyn release notes version 3.1. AUTODYN™ Interactive Non-Linear Dynamic Analysis Software.

Chen, J. S., Pan, C. and Wu, C. T. (1997) Large deformation analysis of rubber based on a reproducing kernel particle method, *Computational Mechanics* **19**: 153–168.

Chen, J. S., Pan, C. and Wu, C. T. (1998) Application of reproducing kernel particle methods to large deformations and contact analysis of elastomers, *Rubber Chemistry and Technology* **7**: 191–213.

Chen, J. S., Pan, C., Wu, C. T. and Liu, W. K. (1996) Reproducing kernel particle methods for large deformation analysis of nonlinear structures, *Computer Methods in Applied Mechanics and Engineering* **139**: 195–227.

Chen, S. Y. and Doolen, G. D. (1998) Lattice boltzmann method for fluid flows, *Annual Review of Fluid Mechanics* **30**: 329–364.

Chorin, A. J. (1973) Discretization of a vertex sheet, with an example of roll-up, *Journal of Computational Physics* **13**: 423–429.

Chung, T. J. (2002) Computational fluid dynamics. Cambridge University Press, Cambridge.

Cundall, P. A., Ed. (1987). Distinct element models of rock and soil structure. Analytical and computational methods in engineering rock mechanics. London.

Duarte, C. A. and Oden, J. T. (1996) An HP adaptive method using clouds, *Computer Methods in Applied Mechanics and Engineering* **139**: 237–262.

Duong-Hong, D., Wang, J. S., Liu, G. R., Chen, Y. Z., Han, J. Y. and Hadjiconstantinou, N. G. (2008) Dissipative particle dynamics simulations of electroosmotic flow in nano-fluidic devices, *Microfluidics and Nanofluidics* **4**(3): 219–225.

Español, P. (1997) Fluid particle dynamics: A synthesis of dissipative particle dynamics and smoothed particle dynamics, *Europhysics Letters* **39**(6): 605.

Espanol, P. and Warren, P. (1995) Statistical mechanics of dissipative particle dynamics, *Europhysics Letters* **30**(4): 191–196.

Evans, M. W. and Harlow, F. H. (1957). The particle-in-cell method for hydrodynamic calculations, Los Alamos National Laboratory Report LA-2139.

Frenkel, D. and Smit, B. (2002) Understanding molecular simulation: From algorithms to applications. Academic Press.

Fromm, J. and Harlow, F. H. (1963) Numerical solution of the problem of vortex sheet development, *Physics of Fluids* **6**: 975.

Gentry, R. A., Martin, R. E. and Daly, B. J. (1966) An eulerian differencing method for unsteady compressible flow problems, *Journal of Computational Physics* **1**: 87–118.

Gingold, R. A. and Monaghan, J. J. (1977) Smoothed particle hydrodynamics-theory and application to non-spherical stars, *Monthly Notices of The Royal Astronomical Society* **181**: 375–389.

Gu, Y. T. and Liu, G. R. (2001a) A coupled element free galerkin/boundary element method for stress analysis of two-dimensional solids, *Computer Methods in Applied Mechanics and Engineering* **190**(34): 4405–4419.

Gu, Y. T. and Liu, G. R. (2001b) A local point interpolation method for static and dynamic analysis of thin beams, *Computer Methods in Applied Mechanics and Engineering* **190**(42): 5515–5528.

Gu, Y. T. and Liu, G. R. (2001c) A meshless local petrov–galerkin (MLPG) formulation for static and free vibration analyses of thin plates, *Computer Modeling in Engineering & Sciences* **2**(4): 463–476.

Gu, Y. T. and Liu, G. R. (2002) A boundary point interpolation method for stress analysis of solids, *Computational Mechanics* **28**(1): 47–54.

Hallquist, J. O. (1988). User's manual for dyna2d — an explicit two-dimensional hydrodynamic finite element code with interactive rezoning and graphic display. Lawrence Livermore National Laboratory report UCID — 18756, Rev. 3.

Hallquist, J. O. (1998). Ls-dyna theoretical manual. Livermore Software Technology Corporation.

Harlow, F. H. (1957) Hydrodynamic problems involving large fluid distortion, *Journal of the Association for Computing Machinery* **4**: 137.

Harlow, F. H. and Welch, J. E. (1965) Numerical calculation of time-dependent viscous incompressible flow of fluids with free surface, *Phyics of Fluids* **8**: 22182.

He, X. Y. and Luo, L. S. (1997a) A priori derivation of the lattice boltzmann equation, *Physical Review E* **55**(6): R6333.

He, X. Y. and Luo, L. S. (1997b) Theory of the lattice boltzmann method: From the boltzmann equation to the lattice boltzmann equation, *Physical Review E* **56**(6): 6811.

Heo, S., Koshizuka, S., Oka, Y. and Obata, H. (2001) Numerical analysis of boiling on high heat-flux and high subcolloing condition using mps-mafl, *International Journal of Heat and Mass Transfer* **45**: 2633–2642.

Hirsch, C. (1988) Numerical computation of internal & external flows: Fundamentals of numerical discretization. John Wiley & Sons, New York, NY, USA.

Hirt, C. W., Amsden, A. A. and Cook, J. L. (1997) An arbitrary Lagrangian-Eulerian computing method for all flow speeds, *Journal of Computational Physics* **135**(2): 203–216.

Hoogerbrugge, P. J. and Koelman, J. (1992) Simulating microscopic hydrodynamic phenomena with dissipative particle dynamics, *Europhysics Letters* **19**: 155.

Hoover, W. G. (1998) Isomorphism linking smooth particles and embedded atoms, *Physica A* **260**(3): 244–254.

Hoover, W. G. and Hoover, C. G. (2003) Links between microscopic and macroscopic fluid mechanics, *Molecular Physics* **101**(11): 1559–1573.

Hu, D., Long, T., Xiao, Y., Han, X. and Gu, Y. (2014) Fluid-structure interaction analysis by coupled FE-SPH model based on a novel searching algorithm, *Computer Methods in Applied Mechanics and Engineering* **276**: 266–286.

Johnson, G. R. (1994) Linking of lagrangian particle methods to standard finite element methods for high velocity impact computations, *Nuclear Engineering and Design* **150**(2–3): 265–274.

Johnson, G. R. and Beissel, S. R. (1996) Normalized smoothing functions for sph impact computations, *International Journal for Numerical Methods in Engineering* **39**(16): 2725–2741.

Jun, S., Liu, W. K. and T., B. (1998) Explicit reproducing kernel particle methods for large deformation problems, *International Journal for Numerical Methods in Engineering* **41**: 137–166.

Koshizuka, S., Nobe, A. and Oka, Y. (1998) Numerical analysis of breaking waves using the moving particle semi-implicit method, *International Journal for Numerical Methods in Fluids* **26**(7): 751–769.

Koshizuka, S. and Oka, Y. (1996) Moving-particle semi-implicit method for fragmentation of incompressible fluid, *Nuclear science and engineering* **123**(3): 421–434.

Koshizuka, S., Tamako, H. and Y., O. (1995) A particle method for incompressible viscous flow with fluid fragmentation, *Journal of Computational Fluid Dynamics* **4**: 29–46.

Kresse, G. and Hafner, J. (1993) Ab initio molecular dynamics for liquid metals, *Physical Review B* **47**(1): 558.

Kresse, G. and Hafner, J. (1994) Ab initio molecular-dynamics simulation of the liquid-metal–amorphous-semiconductor transition in germanium, *Physical Review B* **49**(20): 14251.

Ladd, A. J. (1994) Numerical simulations of particulate suspensions via a discretized boltzmann equation. Part 2. Numerical results, *Journal of Fluid Mechanics* **271**(1): 311–339.

Laguna, P. (1995) Smoothed particle interpolation, *Astrophysical Journal* **439**(2): 814–821.

Leonard, A. (1980) Vortex methods for flow simulation, *Computational Mechanics* **37**: 289–335.

Li, S. F., Hao, W. and Liu, W. K. (2000a) Meshfree simulation of shear banding in large deformation, *International Journal of Solids and Structures* **37**: 7185–7206.

Li, S. F., Hao, W. and Liu, W. K. (2000b) Numerical simulation of large deformation of thin shell structures using meshfree methods, *Computational Mechanics* **25**: 102–116.

Li, S. F. and Liu, W. K. (2002) Meshfree and particle methods and their applications, *Applied Mechanics Reviews* **55**(1): 1–34.

Li, S. F. and Liu, W. K. (2004) Meshfree particle methods. Springer, Berlin.

Lin, H. and Atluri, S. N. (2001) Analysis of incompressible navier-stokes flows by the meshless MPLG method, *Computer Modeling in Engineering & Sciences* **2**(2): 117–142.

Liszka, T. and Orkisz, J. (1980) The finite difference method at arbitrary irregular grids and its application in applied mechanics, *Computers & Structures* **11**: 83–95.

Liu, G., R. and Nguyen, T. T. (2010) Smoothed finite element methods. CRC Press, Florida.

Liu, G. R. (2002) Meshfree methods: Moving beyond the finite element method. CRC Press, Florida.
Liu, G. R. (2008) A generalized gradient smoothing technique and the smoothed bilinear form for galerkin formulation of a wide class of computational methods, *International Journal of Computational Methods* **5**(2): 199–236.
Liu, G. R. (2009) A g space theory and a weakened weak (w2) form for a unified formulation of compatible and incompatible methods: Part I theory, *International Journal for Numerical Methods in Engineering* **81**(9): 1093–1126.
Liu, G. R. (2010a) Meshfree methods: Moving beyond the finite element method. CRC Press, Florida.
Liu, G. R. (2010b) A weakened weak (w2) form for a unified formulation of compatible and incompatible methods, part I: Theory and part II: Applications to solid mechanics problems, *International Journal for Numerical Methods in Engineering* **81**: 1093–1126.
Liu, G. R. and Chen, X. L. (2001) A mesh-free method for static and free vibration analyses of thin plates of complicated shape, *Journal of Sound and Vibration* **241**(5): 839–855.
Liu, G. R. and Chen, X. L. (2002) Buckling of symmetrically laminated composite plates using the element-free galerkin method, *International Journal of Structural Stability and Dynamics* **2**(3): 281–294.
Liu, G. R., Cheng, Y., Mi, D. and Li, Z. R. (2005) A study on self-insertion of peptides into single-walled carbon nanotubes based on molecular dynamics simulation, *International Journal of Modern Physics C* **16**(8): 1239–1250.
Liu, G. R., Dai, K. Y., Lim, K. M. and Gu, Y. T. (2002a) A point interpolation mesh free method for static and frequency analysis of two-dimensional piezoelectric structures, *Computational Mechanics* **29**(6): 510–519.
Liu, G. R., Dai, K. Y., Lim, K. M. and Gu, Y. T. (2002b) A radial point interpolation method for simulation of two-dimensional piezoelectric structures, *Smart Materials and Structures* **12**(2): 171–180.
Liu, G. R. and Gu, Y. T. (2000a) Coupling of element free galerkin and hybrid boundary element methods using modified variational formulation, computational mechanics, *Computational Mechanics* **26**(2): 166–173.
Liu, G. R. and Gu, Y. T., Eds. (2000b). Coupling of element free galerkin method with boundary point interpolation method. Advances in Computational Engineering & Science 2000, Los Angles.
Liu, G. R. and Gu, Y. T. (2000c) Meshless local petrov–galerkin (mlpg) method in combination with finite element and boundary element approaches, *Computational Mechanics* **26**: 536–546.
Liu, G. R. and Gu, Y. T. (2001a) A local radial point interpolation method (LRPIM) for free vibration analyses of 2-D solids, *Journal of Sound and Vibration* **246**(1): 29–46.

Liu, G. R. and Gu, Y. T. (2001b) A point interpolation method for two-dimensional solids, *International Journal for Numerical Methods in Engineering* **50**(4): 937–951.

Liu, G. R. and Gu, Y. T. (2003) A meshfree method: Meshfree weak strong (mws) form method, for 2-D solids, *Computational Mechanics* **33**(1): 2–14.

Liu, G. R. and Gu, Y. T. (2005) An introduction to meshfree methods and their programming. Springer Dordrecht.

Liu, G. R. and Liu, M. B. (2003) Smoothed particle hydrodynamics: A meshfree particle method. World Scientific, Singapore.

Liu, G. R. and Quek, S. S. (2003) The finite element method: A practical course. Butterworth-Heinemann, Oxford.

Liu, G. R. and Tu, Z. H. (2002) An adaptive procedure based on background cells for meshless methods, *Computer Methods in Applied Mechanics and Engineering* **191**: 1923–1943.

Liu, G. R., Wu, Y. L. and Ding, H. (2004) Meshfree weak-strong (MWS) form method and its application to incompressible flow problems, *International Journal for Numerical Methods in Fluids* **46**(10): 1025–1047.

Liu, G. R. and Zhang, G. Y. (2008a) Edge-based smoothed point interpolation methods, *International Journal of Computational Methods* **5**(4): 621–646.

Liu, G. R. and Zhang, G. Y. (2008b) Upper bound solution to elasticity problems: A unique property of the linearly conforming point interpolation method (LC-PIM), *International Journal for Numerical Methods in Engineering* **74**: 1128–1161.

Liu, G. R. and Zhang, G. Y. (2013) The smoothed point interpolation methods — g space theory and weakened weak forms. World Scientific.

Liu, M. B., Liu, G. R. and Lam, K. Y. (2006) Adaptive smoothed particle hydrodynamics for high strain hydrodynamics with material strength, *Shock Waves* **15**(1): 21–29.

Liu, M. B., Liu, G. R., Lam, K. Y. and Zong, Z. (2003a) Computer simulation of shaped charge detonation using meshless particle method, *Fragblast* **7**(3): 181–202.

Liu, M. B., Liu, G. R., Zhou, L. V. and Chang, J. Z. (2014) Dissipative particle dynamics (DPD): An overview and recent developments, *Archives of Computational Methods in Engineering*.

Liu, M. B., Liu, G. R., Zong, Z. and Lam, K. Y. (2003b) Computer simulation of high explosive explosion using smoothed particle hydrodynamics methodology, *Computers & Fluids* **32**(3): 305–322.

Liu, W. K., Belytschko, T. and Chang, H. (1986) An arbitrary lagrangian-eulerian finite element method for path-dependent materials, *Computer Methods in Applied Mechanics and Engineering* **58**(2): 227–245.

Liu, W. K. and Chen, Y. (1995) Wavelet and multiple-scale reproducing kernel methods, *International Journal For Numerical Methods In Fluids* **21**: 901–931.

Liu, W. K., Chen, Y., Jun, S., Chen, J. S., Belytschko, T., Pan, C., Uras, R. A. and Chang, C. T. (1996a) Overview and applications of the reproducing kernel particle methods, *Archives of Computational Methods in Engineering* **3**(1): 3–80.

Liu, W. K., Chen, Y., Uras, R. A. and Chang, C. T. (1996b) Generalized multiple scale reproducing kernel particle methods, *Computer Methods in Applied Mechanics and Engineering* **139**(1): 91–157.

Liu, W. K., Jun, S., Li, S. F., Adee, J. and Belytschko, T. (1995) Reproducing kernel particle methods for structural dynamics, *International Journal for Numerical Methods in Engineering* **38**(10): 1655–1679.

Liu, X., Lee, C. K. and Fan, S. C. (2002c) On using enriched cover function in the partition-of-unity method for singular boundary-value problems, *Computational Mechanics* **29**: 212–225.

Long, S. Y. and Atluri, S. N. (2002) A meshless local petrov–galerkin (mlpg) method for solving the bending problem of a thin plate, *Computer Modeling in Engineering & Sciences* **3**(1): 53–64.

Lucy, L. B. (1977) A numerical approach to the testing of the fission hypothesis, *Astronomical Journal* **82**(12): 1013–1024.

Luo, L. S. (1998) Unified theory of lattice boltzmann models for nonideal gases, *Physical Review Letters* **81**(8): 1618.

Mair, H. U. (1999) Review: Hydrocodes for structural response to underwater explosions, *Shock and Vibration* **6**(2): 81–96.

McNamara, G. R. and Zanetti, G. (1988) Use of the boltzmann equation to simulate lattice-gas automata, *Physical Review Letters* **61**(20): 2332.

Metropolis, N. and Ulam, S. (1949) The monte carlo method, *Journal of American Statistical Association* **44**: 335–341.

MSC/Dytran (1997). User's manual, version 4. The MacNeal-Schwndler Corporation, USA.

Mukherjee, Y. X. and Mukherjee, S. (1997a) Boundary node method for potential problems, *International Journal for Numerical Methods in Engineerin* **40**: 797–815.

Mukherjee, Y. X. and Mukherjee, S. (1997b) On boundary conditions in the element-fee galerkin method, *Computational Mechanics* **19**: 264–270.

Nayroles, B., Touzot, G. and Villon, P. (1992) Generalizing the finite element method: Diffuse approximation and diffuse elements, *Computational Mechanics* **10**(5): 307–318.

Nitsche, L. C. and Zhang, W. D. (2002) Atomistic SPH and a link between diffusion and interfacial tension, *AIChE Journal* **48**(2): 201–211.

Oberkampf, W. L. and Trucano, T. G. (2002) Verification and validation in computational fluid dynamics, *Progress in Aerospace Sciences* **38**: 209–272.

Onate, E., Idelsohn, S., Zienkiewicz, O. C. and L., T. R. (1996) A finite point method in computational mechanics applications to convective transport and fluid flow,, *International Journal for Numerical methods in Engineering* **39**(22): 3839–3866.

Pan, L. S., Liu, G. R., Khoo, B. C. and Song, B. (2000) A modified direct simulation monte carlo method for low-speed microflows, *Journal of Micromechanics and Microengineering* **10**: 21–27.

Pan, L. S., Liu, G. R. and Lam, K. Y. (1999) Determination of slip coefficient for rarefied gas flows using direct simulation monte carlo, *Journal of Micromechanics and Microengineering* **9**(1): 89.

Pomeau, B. H. Y. and Frisch, U. (1986) Lattice-gas automata for the navier-stokes equation, *Physical Review Letters* **56**(14): 1505.

Rahman, A. (1964) Correlations in the motion of atoms in liquid argon, *Physics Review* **136**: 405–411.

Randles, P. W. and Libersky, L. D. (1996) Smoothed particle hydrodynamics: Some recent improvements and applications, *Computer Methods in Applied Mechanics and Engineering* **139**(1): 375–408.

Rapaport, D. C. (2004) The art of molecular dynamics simulation. Cambridge University Press, Cambridge.

Reston, V. A. (1998). Guide for verfication and validation of computational fluid dynamics simulations, AIAA Paper G-077.

Schlesinger, S. (1979) Terminology for model credibility, *Simulation* **32**(3): 103–104.

Stillinger, F. H. and Rahman, A. (1974) Improved simulation of liquid water by molecular dynamics, *Journal Of Chemical Physics* **60**: 1545–1557.

Sulsky, D., Chen, Z. and Schreyer, H. L. (1994) A particle method for history-dependent materials, *Computer Methods in Applied Mechanics and Engineering* **118**: 179–196.

Sulsky, D., J., Z. S. and Schreyer, H. L. (1995) Application of the particle-in-cell method to solid mechanics, *Computer Physics Communications* **87**: 236–252.

Wang, J. G. and Liu, G. R. (2002) On the optimal shape parameters of radial basis functions used for 2-d meshless methods, *Computer Methods In Applied Mechanics And Engineering* **191**(23–24): 2611–2630.

Wang, J. G., Liu, G. R. and Wu, Y. G. (2001) A point interpolation method for simulating dissipation process of consolidation, *Computer Methods in Applied Mechanics and Engineering* **190**: 5907–5922.

Wilkins, M. L. (1999) Computer simulation of dynamic phenomena. Springer-Verlag, Berlin Heidelberg.

Wu, H. A., Liu, G. R., Han, X. and Wang, X. X. (2006) An atomistic simulation method combining molecular dynamics with finite element technique, *Chaos Solitons & Fractals* **30**(4): 791–796.

Xu, Y. G. and Liu, G. R. (2003) Fitting interatomic potentials using molecular dynamics simulations and inter-generation projection genetic algorithm, *Journal of Micromechanics and Microengineering* **13**(2): 254–260.

Xu, Y. G., Liu, G. R., Behdinan, K. and Fawaz, Z. (2004) Stepwise-equilibrium and adaptive molecular dynamics simulation for fracture toughness of single crystals with defects, *Journal of Intelligent Material Systems and Structures* **15**(12): 933–939.

Yagawa, G. and Yamada, T. (1996) Free mesh method: A new meshless finite element method, *Computational Mechanics* **18**: 383–386.
Yagawa, G. and Yamada, T. (1998) Meshless method on massively parallel processor with application to fracture mechanics, *Key Engineering Materials* **1145-149**: 201–210.
Yan, K., Chen, Y. Z., Han, J. Y., Liu, G. R., Wang, J. S. and Hadjiconstantinou, N. G. (2012) Dissipative particle dynamics simulation of field-dependent DNA mobility in nanoslits, *Microfluidics and Nanofluidics* **12**(1–4): 157–163.
Yao, Z. H., Wang, J. S., Li, B. W. and Liu, G. R. (2005) Thermal conduction of carbon nanotubes using molecular dynamics, *Physical Review B* **71**(8): 085417.
Yao, Z. H., Wang, J. S., Liu, G. R. and Cheng, M. (2004) Improved neighbor list algorithm in molecular simulations using cell decomposition and data sorting method, *Computer Physics Communications* **161**(1–2): 27–35.
Zhang, A. M., Ming, F. R. and Wang, S. P. (2013) Coupled SPHS-BEM method for transient fluid-structure interaction and applications in underwater impacts, *Applied Ocean Research* **43**: 223–233.
Zhang, Z. C., Qiang, H. F. and Gao, W. R. (2011) Coupling of smoothed particle hydrodynamics and finite element method for impact dynamics simulation, *Engineering Structures* **33**(1): 255–264.
Zhou, C. E., Liu, G. R. and Lou, K. Y. (2007) Three-dimensional penetration simulation using smoothed particle hydrodynamics, *International Journal of Computational Methods* **4**(4): 671–691.
Zienkiewicz, O. C. and Taylor, R. L. (2000) The finite element method. Butterworth-Heinemann, Oxford.

Chapter 2

Molecular Dynamics

Molecular dynamics (MD) is a deterministic computational method originally developed for modeling the physical movements of atoms and molecules, and it is therefore a typical particle method at atomistic or micro scale with each particle representing an atom or a molecule. Particles (atoms or molecules) in an MD model interact with each other dictated by a known force *potential*. The trajectories of the particles are governed essentially by the Newton's equations of motion, which can be determined by numerically integrating these simultaneous equations with respect to time. Since its invention in late 1950s, MD techniques have been extensivcly developed, becoming an extremely powerful computational tool, which has been widely applied to various areas, especially in material sciences, biophysics and biochemistry.

There are now numerous references providing theoretical basics, numerical aspects, and diversified applications of MD. In this chapter, we only briefly introduce the basic procedures of MD, so as to provide necessary knowledge for the dissipative particle dynamics (DPD) method that is a coarse-grained MD to be presented in Chapters 3 and 4.

This chapter is outlined as follows.

- In Section 2.1, some background knowledge of MD is provided.
- Section 2.2 presents the detailed MD formulation including the equations of motion, interaction potential function, time integration, and the use of periodic boundary condition. The implementation procedure of classic MD and the simulation of Poiseuille flow with MD are provided.
- In Section 2.3, the concepts of multi-scale modeling with MD and other macro-scale numerical methods are briefed. Details include coupling MD with FEM, FDM and SPH.

- In Section 2.4, as an application example, MD simulation method is used to provide a microscopic view of peptide-CNT (carbon nanotube) interaction based on the atomic level. Intensive numerical simulation has been carried out for a large number of different peptides. Self-insertion of peptides into single-walled carbon nanotubes (SWCNTs) and binding of peptides to the outer surface of SWCNT are simulated. The energetics of interaction, as well as the conformational change of peptides is discussed.
- Section 2.5, concludes this chapter with some remarks.

2.1 Introduction

The molecular dynamics (MD) was first introduced in 1957 by Alder and Wainright to study the solid-fluid transition in a system composed of hard spheres interacting by instantaneous collisions (Alder and Wainright, 1957). Gibson et al. employed the continuous repulsive interaction potential in the MD simulation of radiation damage in a Cu target (Gibson et al., 1960). In 1964, Rahman carried out the first MD simulation using the *Lennard-Jones* potential to describe both attractive and repulsive interaction in a system of 864 Argon atoms (Rahman, 1964). The first MD simulation for a realistic system of liquid water was done in 1974 (Stillinger and Rahman, 1974).

As one of the most important and most widely used particle methods, molecular dynamics has rapidly found more and more applications in both engineering and science. In the study of liquids, molecular dynamics is used to investigate the transport properties such as viscosity and heat conductivity either using equilibrium or non-equilibrium techniques (Allen and Tildesley, 1987; Rapaport, 2004). It is also being employed to investigate fluid dynamics with the purpose of studying complex fluid behavior. In the study of solids, molecular dynamics is frequently used to investigate surface effects and material defects (Ciccootti et al., 1987; Meyer and Pontikis, 1991) from point defects (vacancies, interstitials etc.), linear defects (dislocations etc.) to planar defects (grain boundaries, stacking faults etc.). Investigations on the fracture/crack initiation and propagation by molecular dynamics are also providing some profound insights (Ruth and Lynden, 1994). Xu, Liu and their co-workers have used MD for the simulation of fracture toughness of single crystals (Xu et al., 2004). Yao et al. have used MD for the prediction of thermal conduction of carbon nanotubes (Yao et al., 2005). Molecular dynamics is currently finding a rapidly increasing number of applications in biology, chemistry and medicine since it facilitates the study of dynamics of large macromolecules including biological systems such as proteins, nucleic acids (DNA, RNA), and membranes (Allen and

Tildesley, 1993; Fraga et al., 1995). Drug design in pharmaceutical industry involves repeatedly testing the properties of a molecule on the computer without expensive real synthesis. Liu et al. studied the self-insertion procedure of peptides into single-walled carbon nanotubes (Liu et al., 2005).

In principle, for a molecular system with the presence of interacting nuclei and electrons, it is necessary to solve the Schrödinger equation and find a total wave function which tells the complete behavior of the system. The complexity in solving the Schrödinger equation is greatly simplified using the Born-Oppenheimer approximation, which treats the dynamics of the nuclei and electrons separately due to the very large difference of the masses of the nuclei and electrons. The Born–Oppenheimer approximation thus leads to the following two steps. The first step is to calculate the motion of electrons for fixed nuclei by solving the electronic Schrödinger equation for a specific set of nuclear variables. The second step is to calculate the motion of the nuclei by solving the nuclear Schrödinger equation with the energy obtained from the solution of the electronic Schrödinger equation as the interactomic potential for the interaction of the nuclei. There are various strategies to obtain the interatomic potential of a molecular system as a function of nuclear coordinates. The most fundamental approach is to calculate the potential energy from first principles by directly solving the electronic Schrödinger equation. Since this first principle or *ab initio* molecular dynamics is quite expensive, simplifications are usually necessary to approximate the most time consuming parts in the *ab initio* approach by choosing various empirical parameters. This is a semi-empirical approach, which uses analytical, semi-empirical potential. To further simplify the problem, one can assume an analytical potential function with properly tuned parameters, which can reproduce a set of experimental data or the exact solutions using more accurate approaches such as *ab initio* approach. This gives the so-called empirical potential functions in the classic molecular dynamics. For example, using inverse techniques (Liu and Han, 2003), Xu and Liu have constructed an inter-atomic potential using a genetic algorithm (Xu and Liu, 2003).

Another important issue in MD implementation is the search for atoms for computing the interactions, and many works have been conducted. For example, Yao et al. have suggested an improved neighbor list algorithm in molecular simulations using cell decomposition and data sorting method (Yao et al., 2004). Efficient searching algorithms are not only important for MD, but also for all other particle methods (Liu and Liu, 2003).

Molecular dynamics method is essentially a deterministic technique. After the initial positions and velocities of the atoms are given and the potential is known, the evolution of the system in time is in principle completely determined. Molecular dynamics simulation can also be used as a statistical mechanics method since it generates a set of configurations at the microscopic level that are distributed according to the statistical distribution functions. The connection of the microscopic information to macroscopic observables such as pressure, energy, heat capacities, etc., requires the use of the treatments in the statistical mechanics (Hoover, 1991). Molecular dynamics simulations provide the means to solve the equation of motion of the atoms and evaluate the corresponding mathematical formulas. Statistical mechanics provides the rigorous mathematical expressions that relate the macroscopic properties to the distribution and motion of the atoms and molecules of the system. According to statistical mechanics, a macroscopic property observable in experiments is the average of the corresponding ensemble. The ensemble is a collection of all possible phase states of the system with different microscopic states but an identical macroscopic state. The fundamental axiom of statistical mechanics, the *ergodic hypothesis*, which is based on the idea that if the system is allowed to evolve in time infinitely, the system will eventually pass through all possible states, describes that the ensemble average equals to the time average obtained by molecular dynamics simulation. Experimentally observable structural, dynamic and thermodynamic properties can therefore be calculated using molecular dynamics simulation if the run time is sufficiently long enough to generate enough representative configurations of the system.

2.2 Classic Molecular Dynamics

2.2.1 Equations of motion

In classic molecular dynamics, the equation of motion is based on Newton's second law. After giving the initial conditions (initial atomic positions and velocities), the force potential is used for deriving the forces between all the atoms within the *cutoff distance*. The derived forces will be used in the equation of motion, which will be then integrated forward to yield new atomic positions and velocities at the next time step. For a particular particle i, the Newton's equation of motion can be written in accelerations as

$$\mathbf{F}_i = m_i \mathbf{a}_i , \tag{2.1}$$

or in velocities as

$$\mathbf{F}_i = m_i \frac{d\mathbf{v}_i}{dt} , \tag{2.2}$$

or in atom positions as

$$\mathbf{F}_i = m_i \frac{d^2\mathbf{x}_i}{dt^2}, \tag{2.3}$$

where $\mathbf{F}_i$ is the force exerted on atom *i*, m_i is the mass of atom *i*, $\mathbf{a}_i$, $\mathbf{v}_i$ and $\mathbf{x}_i$ are the acceleration, velocity and position of atom *i*, respectively. For a system with a total number of *N* atoms, the force on atom *i* at a given time instant can be obtained from an inter atomic potential $u(\mathbf{x}_1, \mathbf{x}_2, ...\mathbf{x}_N)$, which is in general given as a function of the position vector **x** of all the atoms:

$$\mathbf{F}_i = -\nabla_i u(\mathbf{x}_1, \mathbf{x}_2, ...\mathbf{x}_N). \tag{2.4}$$

2.2.2 Force potential function

The force potential function is fundamentally important in a MD simulation since it determines the forces exerting on the atoms and hence the way in which the atomic system evolves in time. For this reason it is also termed as force field, or interaction potential. The choice of force potential is usually problem (material) dependent, and different types of functions may be available. One should consider both the accuracy and computational efficiency when making a choice. The available interaction potential can be categorized into two classes: *pair potential* and *multi-body potential* (Allen and Tildesley, 1987). The multi-body potential can be very computationally expensive, since it needs to consider the interaction between multiple atoms, resulting in multiple nested summations. The pair potential, considering the most important contributions, depends only on the magnitude of the spacing of the pair of atoms *i* and *j*, $r_{ij} = \left|\mathbf{x}_i - \mathbf{x}_j\right|$. The pair potential gives a remarkably good description of most of the problems simulated, and is widely used. Alternatively, averaged multi-body effects can be partially included by defining an effective pair potential.

When the pair potential is used, the total potential energy of the system of *N* atoms is then given in a form of

$$u(\mathbf{x}_1, \mathbf{x}_2, ...\mathbf{x}_N) = \sum_{i}^{N-1} \sum_{j>i}^{N} u(r_{ij}). \tag{2.5}$$

There are many kinds of pair potential, each with its advantages and disadvantages. Some simple examples are listed as follows.

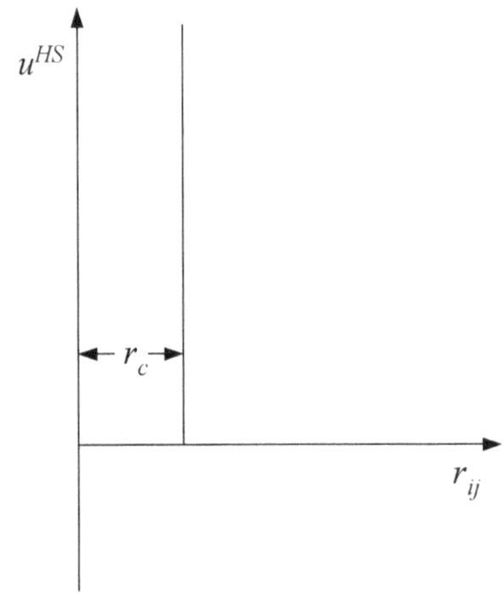

Figure 2.1 The hard sphere potential.

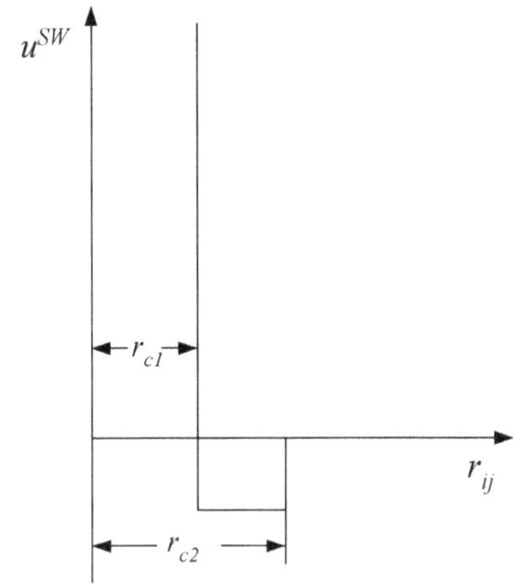

Figure 2.2 The square well potential.

The *hard sphere potential* (**Figure 2.1**) is given by

$$u^{HS}(r_{ij}) = \begin{cases} \infty, & r_{ij} < r_c \\ 0, & r_{ij} \geq r_c \end{cases}, \tag{2.6}$$

where r_c is a cutoff distance.

The *square well potential* (**Figure 2.2**) has the form of

$$u^{SW}(r_{ij}) = \begin{cases} \infty, & r_{ij} < r_{c1} \\ -\varepsilon, & r_{c1} \leq r_{ij} < r_{c2} \\ 0, & r_{ij} \geq r_{c2} \end{cases}, \tag{2.7}$$

where r_{c1} and r_{c2} are two cutoff distances.

The soft-sphere potential (**Figure 2.3**) with a repulsion parameter γ can be written as

$$u^{SS}(r_{ij}) = \varepsilon\left(\frac{\sigma}{r_{ij}}\right)^{\gamma}, \tag{2.8}$$

where ε governs the strength of the interaction; σ defines a length scale.

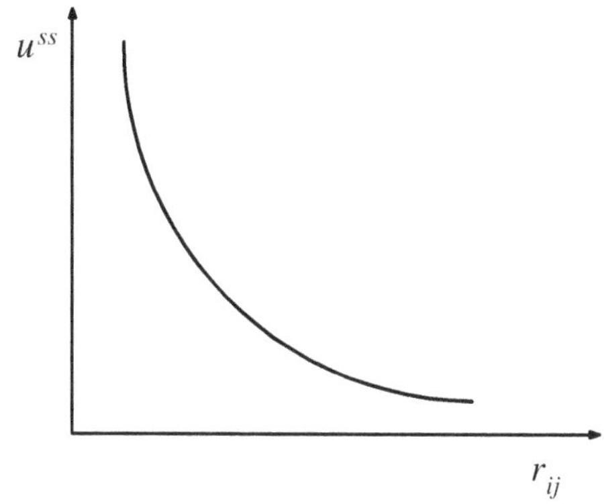

Figure 2.3 The soft sphere potential.

Some other examples include

- the ionic potential, which takes account of the Coulomb interaction of charges;
- the Morse potential, which is more suitable for cases when attractive interaction comes from the formation of some kind of chemical bond;
- the Bunkingham potential, which provides a better description of strong repulsion due to the overlap of the closed shell electron clouds.

Many other forms of potentials can be found in (Meyer and Pontikis, 1991). Xu and Liu developed an approach to establish the inter-atomic potentials by inversely fitting of experimental data using molecular dynamics simulations and the inter-generation projection genetic algorithm (Xu and Liu, 2003).

Among all the pair potentials, the most widely used is the *Lennard-Jones* (LJ) *potential* (**Figure 2.4**), which can be written as

$$u^{LJ}(r_{ij}) = 4\varepsilon\left[\left(\frac{\sigma}{r_{ij}}\right)^{12} - \left(\frac{\sigma}{r_{ij}}\right)^{6}\right], \quad r_{ij} \le r_c , \tag{2.9}$$

where r_c is a cutoff distance, ε governs the strength of the interaction, and σ defines a length scale.

The LJ potential has a long distance attractive tail (term $-1/r_{ij}^6$), which represents the van der Walls interaction due to electronic correlations, a negative well of depth ε, and a strongly repulsive core (term $1/r_{ij}^{12}$) arising from the non-bonded overlap between the electron clouds. MD simulations using the LJ potential can give a reasonably good agreement with the experimental properties of liquid argon if proper parameters are chosen. The LJ potential is popular in

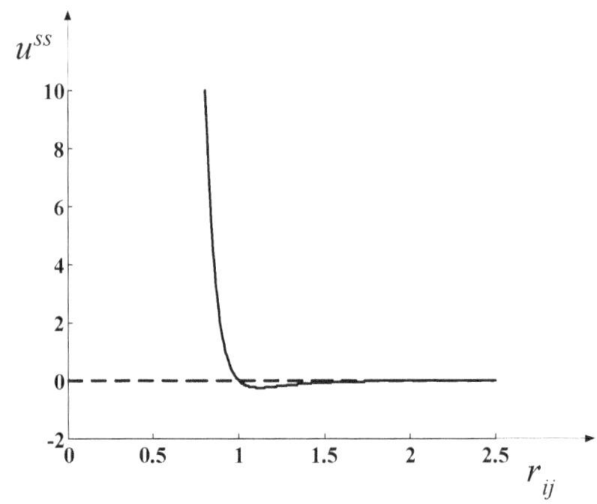

Figure 2.4 The LJ potential.

MD simulations when the objective is to model a general class of effects, which only require a reasonably accurate potential.

2.2.3 Time integration

After the potential function is determined, the Newton's equation of motion expressed in equations (2.1) and (2.3) can be numerically integrated in time to obtain the time-dependent atomic velocities and positions. There are numerical integration algorithms available for partial differential equations, e.g. Euler's scheme, Runge-Kutta scheme, leapfrog etc. A good time integration algorithm should conserve energy and momentum, be computationally efficient and has a time step Δt as large as possible. A most popular way of time integration is to march position and velocity with Taylor series expansion based on another time step and often with some transformations.

$$\mathbf{x}(t+\Delta t)=\mathbf{x}(t)+\mathbf{v}(t)\Delta t+\frac{1}{2}\mathbf{a}(t)\Delta t^2+\ldots, \tag{2.10}$$

$$\mathbf{v}(t+\Delta t)=\mathbf{v}(t)+\mathbf{a}(t)\Delta t+\ldots. \tag{2.11}$$

Applying forward and backward Taylor series expansion to position and then combining the resultant expressions together result in the following famous *Verlet algorithm*

$$\mathbf{x}(t+\Delta t)=\mathbf{x}(t-\Delta t)+2\mathbf{x}(t)+\mathbf{a}(t)\Delta t^2. \tag{2.12}$$

The velocities in the Verlet algorithm, which are necessary in computing kinetic energy so as to check the conservation of the total energy, do not appear in the algorithm explicitly. One variant of the Verlet algorithm, the *velocity Verlet algorithm* is to introduce velocities into the algorithm

$$\mathbf{x}(t+\Delta t)=\mathbf{x}(t)+\mathbf{v}(t)\Delta t+\frac{1}{2}\mathbf{a}(t)\Delta t^2, \tag{2.13}$$

$$\mathbf{v}(t+\Delta t)=\mathbf{v}(t)+\frac{1}{2}\Delta t[\mathbf{a}(t)+\mathbf{a}(t+\Delta t)]. \tag{2.14}$$

Another popular integration scheme is the well-known *leapfrog algorithm*, which staggers velocity and position at different instants (Liu and Liu, 2003).

2.2.4 Periodic boundary treatment

A typical molecular dynamics simulation can have a huge number of atoms. Because of the computer hardware, it is limited by the number of atoms. In order to simulate the bulk material properties with a "representative cell" of small number of atoms, it is necessary to use the *periodic boundary condition* (**Figure 2.5**), by which one assumes that the bulky material consists of an infinite assembly of the representative cells. The periodic boundary condition is implemented with a "wraparound" procedure: an atom that leaves the representative cell through a particular bounding face reenters immediately the region through the opposite face. An atom lying within the cutoff distance from a boundary interacts with atoms in an adjacent copy of the representative cell, or equivalently with atoms near the opposite boundary. This wraparound implementation of the periodic boundary condition should be performed in both the integration of the equations of motion when moving the atoms and the interaction computations between interacting atoms. As shown in **Figure 2.5**, in

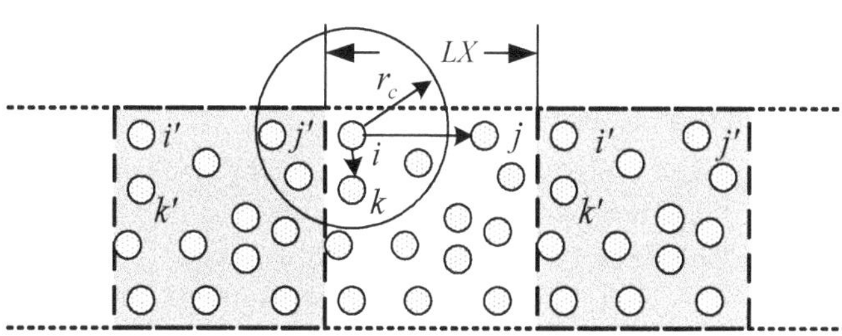

Figure 2.5 Periodic boundary condition in one direction for a molecular dynamics simulation. Only the middle portion is modeled in MD.

the integration process, it is necessary to check if an atom has moved out outside the region. If so, its coordinates must be adjusted to bring it back inside from the other side of the boundary. The checking and adjustment are: if $\mathbf{x}_i \geq LX/2$, replace $\mathbf{x}_i$ with $\mathbf{x}_i - LX$, and if $\mathbf{x}_i \leq -LX/2$, replace $\mathbf{x}_i$ with $\mathbf{x}_i + LX$. Similar checking and adjustment also apply to the interaction calculation process. The only difference is to replace the coordinate of a single atom with the position vector difference between two atoms.

It is an important aspect in molecular dynamics simulation to describe the interactions of atoms in the MD representative cell with surrounding environment. This interaction of MD atoms with surrounding environment involves more complex boundary conditions (BCs), e.g. free surface BC, rigid solid BC, non-slip BC, etc. Implementing these boundary conditions usually requires special methods for temperature and pressure control, which exchange heat and work between the MD computational cell and the environment. To sufficiently consider environmental effects on the motion of the MD atoms, the surrounding environment has been gradually incorporated into the MD simulation. The reality is, however, any MD simulation is always limited by the number of atoms that the computer can accommodate.

2.2.5 Classic MD simulation implementation

As shown in **Figure 2.6**, a typical molecular dynamics simulation consists of four sequential stages: initialization, equilibration, production and analysis state.

1. **Initialization:** For the first run of an MD simulation, it is necessary to initialize the coordinates of the atoms, their velocities and the target temperature for the simulation. Typically the atoms are initially placed in a regular lattice spaced to give the desired density. The initial velocities are assigned with random directions and a fixed magnitude. It is preferred to initialize the velocity with the appropriate *Maxwell-Boltzman distribution* for the specified temperature. However, the usual rapid equilibration renders the careful fabrication of a Maxwell-Boltzman distribution unnecessary. Initialization of atom velocities is subjected to a number of conditions:
    a) There is no overall momentum in any Cartesian direction;
    b) For a non-periodic calculation, there is no angular momentum with respect to any of the Cartesian axes;
    c) The total kinetic energy is appropriate to the temperature specified.

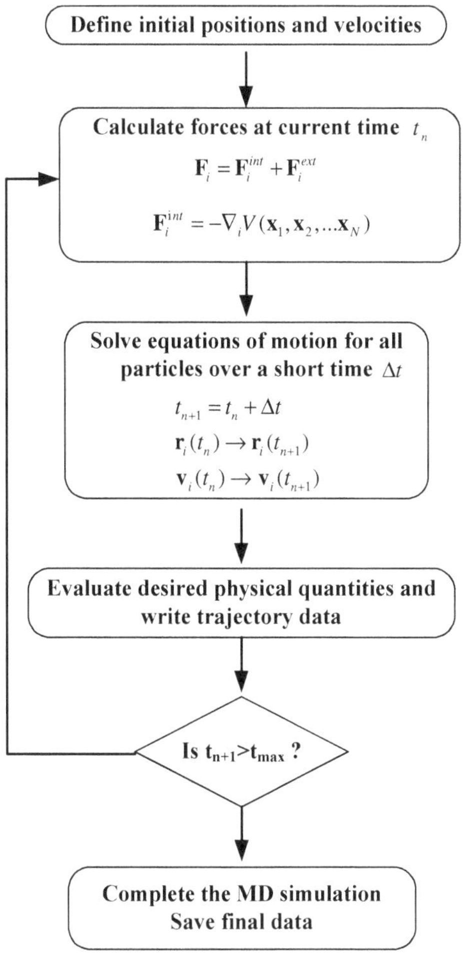

Figure 2.6 Computational procedure of a MD simulation. It is noted that $\mathbf{F}_i^{\text{int}}$ in this figure is the same as the force $(\mathbf{F})$ in equations (2.1) to (2.4).

2. **Equilibration:** After the initialization of the MD simulation, it should take some period of time for the system to achieve equilibrium before collecting data. This equilibration involves the achievement of the correct partitioning of energy between kinetic and potential energy, as well as attaining a Maxwell-Boltzmann velocity distribution corresponding to the concerned temperature. The time needed for equilibration to occur is variable as it depends on the nature and size of the system being run. During this stage, the velocities are normally scaled to maintain a proper temperature (Allen and Tildesley, 1987).

Equilibration of the system can be accelerated by first starting the simulation at a higher temperature and later cooling by rescaling the velocity.

3. **Production:** After the MD system reaches to the equilibrium state, it usually takes another period of time to collect data. In this production stage after equilibration, no velocity scaling for temperature control is involved, while the trajectories are written out in some interval to the external file for later analysis. The number of time steps in this production stage usually depends on the nature of the problem and the purpose of the simulation.
4. **Analysis:** This stage involves analyzing the information stored in the trajectory file in the production stage. The trajectory file usually contains the absolute Cartesian coordinates, the velocities and various observables of the system, such as the energy, temperature, pressure, etc. Depending on the different purpose of the simulation, the trajectory information can be extracted and employed to analyze either the material properties or other physical characteristics. The analysis of the trajectory information can be related to experimental observables.

2.2.6 MD simulation of Poiseuille flow

As mentioned above, molecular dynamics has been widely used to different areas with various applications. In fluid mechanics, molecular dynamics techniques combined with proper inter atomic potential function can be used either to predict the properties of fluids, or to model complex flow phenomena. Presented here is a molecular simulation of the Poiseuille flow, which is a classic benchmarking problem and important in engineering and sciences. The Poiseuille involves flow between two stationary plates, which is driven by some kind of force, and finally reaches to the equilibrium state after some time (**Figure 2.7**).

The classic molecular dynamics with LJ potential is used to simulate this Poiseuille flow of liquid argon. In the simulation, periodic boundary condition applies to the x and y direction, while reflection solid boundary condition (Rapaport, 2004) applies to the z direction. The MD simulation is carried out

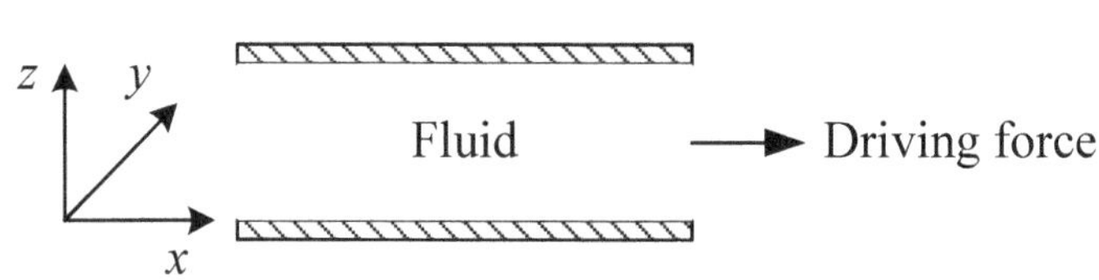

Figure 2.7 The Poiseuille flow in two parallel infinite plates driven by an external force.

using a leapfrog integrator with a time step of $0.005\tau_0$, where $\tau_0 = \sqrt{m\sigma^2/\varepsilon}$ is the characteristic time scale for molecular motion. The parameters used in the MD simulation are as follows. The characteristic length, energy and time scales of liquid argon are $\sigma = 3.4$ A, $\varepsilon/k_B = 120$ K, and $\tau_0 = 2.161\times10^{-12}$ s. The fluid temperature is constant at $T = 1.2\varepsilon/k_B$, where k_B is the Boltzmann constant. The fluid density is initially given as $\rho\sigma^3 = 0.80$. The flow is driven by a uniformly distributed external force of $g = 0.1\sigma/\tau_0^2$ in the x direction to maintain a low shear rate.

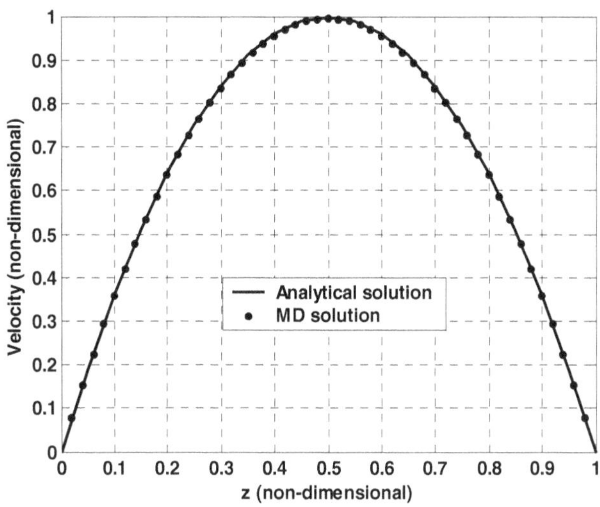

Figure 2.8 Velocity profile of the Poiseuille flow.

Figure 2.8 shows the velocity profile along the z coordinate. The results are interpolated from 50 layers in the z direction. The velocity magnitude and the z coordinate are nondimensionalized by the maximal velocity and the thickness of the entire problem domain in the z direction respectively. For the Poiseuille flow, there is analytical solution when using constant viscosity. The solid line in **Figure 2.8** corresponds to the analytical solution of the Poiseuille flow with a dynamic viscosity of $\mu = 2.2(m\varepsilon)^{1/2}/\sigma^2$. This analytical solution is very much close to the MD solution, in which no viscosity is directly used in the simulation. Therefore, matching the MD result with the analytical solution can give rise to the dynamic viscosity of the fluids in the MD simulation. This approach of matching the analytical solutions for the classic Poiseuille flow with the MD results (velocity and temperature) is usually used to predict the fluid transport properties such as viscosity and heat conductivity (Rapaport, 2004).

2.3 Coupling MD with macro scale methods

2.3.1 An overview

Recent development of micro and nano systems has been attracting the attention of researchers in many different areas. Flows in micro and nano mechanics systems are often very complex in nature, and usually involve multi-scales and multi-physics. One of the major outstanding challenges in the simulation of complex fluid flows is the necessity of a systematic frame, which bridges the gap between nano, micro, meso and macro scales for physics on multiple scales. *Coupling length scale* (CLS) is very important for such simulations and is usually implemented by combining different approaches to account for the different phenomena that dominate. One example is the flows in nano and microfluidic devices. Modeling the flows in micro devices with molecular dynamics is impractical since the usual atomistic MD simulations are limited to very small length scale over very short times. Application of the macro continuum numerical methods such as the finite element method (FEM), finite difference method (FDM) and finite volume method (FVM) is invalid for the atomistic regions due to the continuum assumptions. Coupling the atomistic molecular dynamics with the continuum methods tends to be a good approach for multiple scale computations. In the coupling practice, MD is employed for atomistic regions with inhomogeneities and complex features, and a continuum approach is used for other regions.

Coupling atomistic and continuum simulation for solids is well investigated and widely practiced (Broughton et al., 1999; Rudd and Broughton, 1999; Smirnova et al., 1999). It is especially popular in fracture mechanics for simulating micro crack initiation and propagation. In the simulations, MD is applied for the nano sized region, while some well developed numerical methods (e.g. FEM) in other regions with larger scales. The coupling is implemented by some kind of *handshaking* algorithm to treat the interface region.

For fluid flows, there are only limited cases in the atomistic and continuum coupling simulation. O'Connell and Thompson proposed a MD-continuum hybrid technique by constraining the dynamics of the fluid atoms in the vicinity of the MD-continuum interface (O'Connell and Thompson, 1995). The validity of the hybrid technique was demonstrated by the Couette flow problem using an overlap region mediating between a particle ensemble and an explicit FDM approximation for the incompressible Navier-Stokes equation. Hadjiconstantinou and Patera coupled the microscopic and macroscopic length scales by extracting the molecular solution in the vicinity of the contact line as the input of the boundary conditions for separate FEM computations (Hadjiconstantinou and Patera, 1997; Hadjiconstantinou, 1999a, b). This approach is generally used for simulating steady-state flows.

Garcia and his co-workers have developed a sophisticated adaptive mesh refinement (AMR) algorithm (Garcia et al., 1999), which embeds the Direct Simulation Monte Carlo (DSMC) (Bird, 1994; Pan et al., 2000) within a continuum method at the finest level of an AMR hierarchy. Due to the nature of DSMC, this approach is more suitable for rarefied systems. Flekkoy et al. have constructed another hybrid model for combined particle and continuum dynamics, which is symmetric in the sense that the fluxes of the conserved quantities are continuous across the particle-field interface (Flekkoy et al., 2000). Aktas and Aluru proposed a coupling approach, in which DSMC was coupled with a scattered point based finite cloud method for solving the Stokes equations for continuum fluids (Aktas and Aluru, 2002). The combination is implemented by an overlapped Schwarz alternating method with the Dirichlet-Dirichlet type boundary conditions. Qian et al. presented a combined molecular dynamics/ continuum approach with account of both the non-bonded and bonded interactions to model C_{60} in nanotubes (Qian et al., 2001).

In order to make the thermodynamics and transport properties across the interface between the two descriptions continuous, the hybrid models are either implemented by providing boundary conditions from one description to another, or through some kind of average to map the field variables from one description to another.

It is noted that there are different particle methods for different scales. For example, depending on the scale of the model, the particle may vary in size, from a single atom or molecule in the molecular dynamics method in atomistic scales, to a small cluster of atoms or molecules in the dissipative particle dynamics method in meso-scales, to an infinitesimal macroscopic region in the smoothed particle hydrodynamics (SPH) method in macro-scales. As MD, DPD and SPH are particle methods sharing similar features but for different scales, it is natural to couple MD, DPD and SPH for multiple scale simulation. Liu and Liu provided a combination of smoothed particle hydrodynamics method with MD for multi-scale fluid dynamic problems (Liu and Liu, 2003). SPH is applied to solve the Navier-Stokes equation in the continuum region. MD is employed in atomistic regions where complex flow features prevent a continuum description of the fluid. Since both SPH and MD are all practically particle methods, this hybrid particle-atom (SPH fluid particles and MD atoms) coupling approach is very flexible. The momentum exchange between SPH particles and MD atoms is realized through particle-atom interactions. Fedosov and Karniadakis developed triple-decker for interfacing atomistic-mesoscopic-continuum flow (Fedosov et al., 2010). In their work, MD (for atomistic scale), DPD (for mesoscopic scale) and spectral/hp element (Qiao and He, 2007) (for solving continuum flow with N-S equations) were coupled together. There are also some reports to couple the material point method with MD for multiple scale simulations (Chen et al., 2011; Liu et al., 2013).

2.3.2 Coupling MD with FEM

The *handshaking* algorithm is the most important part in the hybrid atomistic continuum combination. It determines the consistency of the transport properties and field variables around the handshaking interface area.

In the early approaches of coupling MD with FEM, the computational domain was divided into two parts without any overlap region (**Figure 2.9**). The atoms are usually treated as a node of the boundary elements. This treatment directly provides the boundary condition for both sides, but usually leads to property inconsistency in the boundary. Later, alternative MD-FEM coupling approach has been developed that employs a special region, in which the FEM elements and MD atoms are overlapped (Smirnova et al., 1999) (**Figure 2.10**). The implementation of boundary conditions for the FEM is to assign values on the overlapped nodes from the averaged properties over the overlapped atoms.

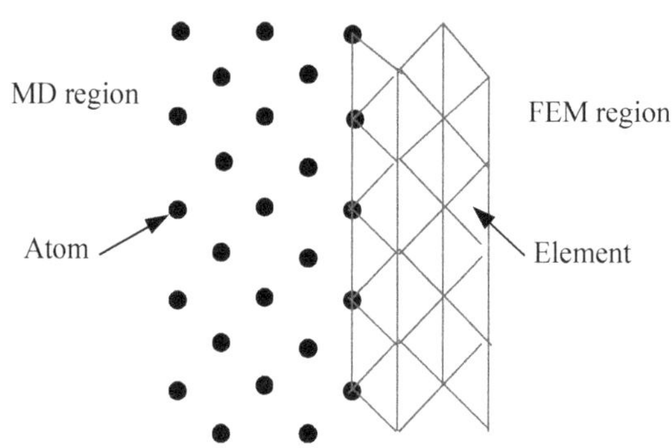

Figure 2.9 MD-FEM CLS simulation without overlap region.

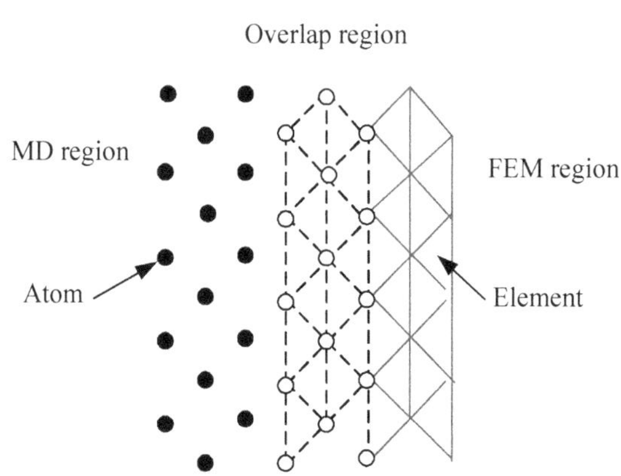

Figure 2.10 MD-FEM CLS simulation with overlap region.

The boundary conditions for the MD are to distribute the FEM force according to the Boltzman distribution to each atom.

Wu et al. have developed an atomistic simulation method combining molecular dynamics with finite element technique (Wu et al., 2006). In the combined method, the initial atomistic model is transformed to continuum model, and an approximate solution is first obtained with the finite element method for the system under the specified boundary conditions and external loadings. Then the deformed continuum model is transformed back to form a new atomistic model, and molecular dynamics simulation is performed to quickly reach the final stable equilibrium state. It was reported that this method can take advantages of both the efficiency of continuum mechanics method and the accuracy of atomistic simulation method.

2.3.3 Coupling MD with FDM

For the approaches of coupling MD with FDM, since MD is a moving particle method and FDM is based on the Eulerian grid, which is fixed in space. In a certain instants, there are atoms moving in or out of the FDM region. How to insert or remove atoms from the FDM domain is quite difficult. Similarly, two different handshaking treatment techniques have been used, one without overlap region (**Figure 2.11**), another with an overlap region (O'Connell and Thompson, 1995; Flekkoy et al., 2000) (**Figure 2.12**). If there is no overlap region, the fluid properties around the interface area are usually inconsistent. If with overlap region, since the atoms move in the overlap grid, it is not easy to calculate the mass, momentum flux in a grid cell.

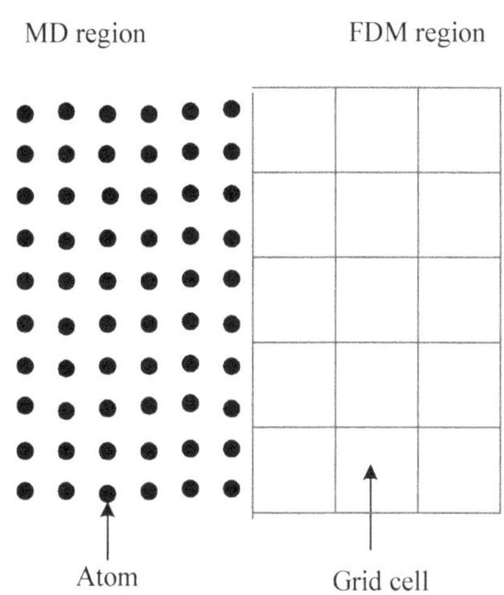

Figure 2.11 MD-FDM CLS simulation without overlap region.

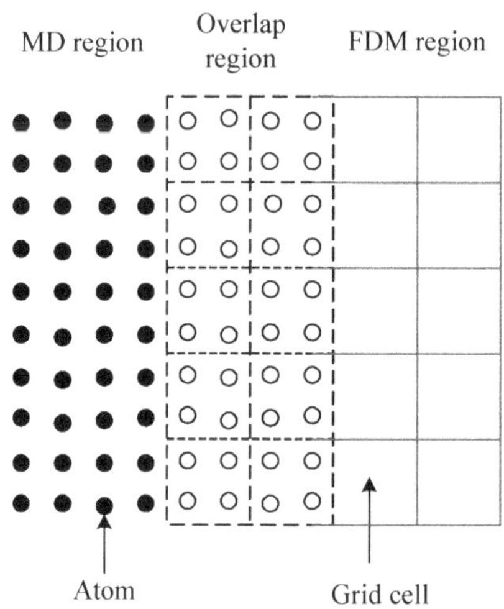

Figure 2.12 MD-FDM CLS with overlap region.

2.3.4 Coupling MD with SPH

As a meshfree Lagrangian particle method, SPH has much in common with the MD in the particle sense and therefore seems well suited for coupling with MD to simulate nano systems with multi-scale physics. The SPH method is used to give the continuum solution. For handshaking the interface, two different possible models can be employed, model I with overlap region, and model II without overlap region.

In model I, the domain is divided into three regions according to different characteristics, one region for MD simulation with a potential cutoff distance r_c for every atoms, another region for ordinary SPH particle simulation with a cutoff distance κh (where h is the smoothing length, and κ a scalar factor), and a layer of transitional SPH particles placed between the MD region and the ordinary SPH region as shown in **Figure 2.13**. Each SPH particle has its corresponding smoothing length representing the influencing area and length scale of the particle. For the ordinary SPH particle region, the length scale of the particles is graded down to the order of the atomic lattice size when approaching the interface area. The length scale of the transitional SPH particles is the same as the atomic lattice size. The transitional SPH particles interact with neighboring transitional and ordinary SPH particles that are within the influencing area of κh. They act also as virtual atoms which interact with the real and virtual atoms that are within the potential cutoff distance r_c. This dual role of the transitional SPH particle or virtual atoms acts as some kind of overlapping, or the layer of transitional SPH particles is overlapped with MD atoms. The influencing area of

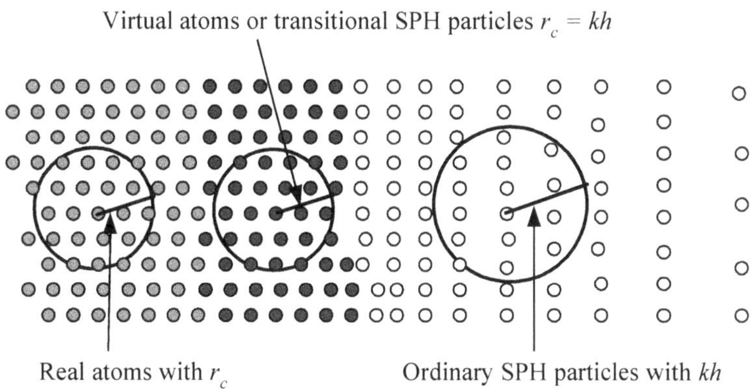

Figure 2.13 MD-SPH coupling: handshaking with overlap region.

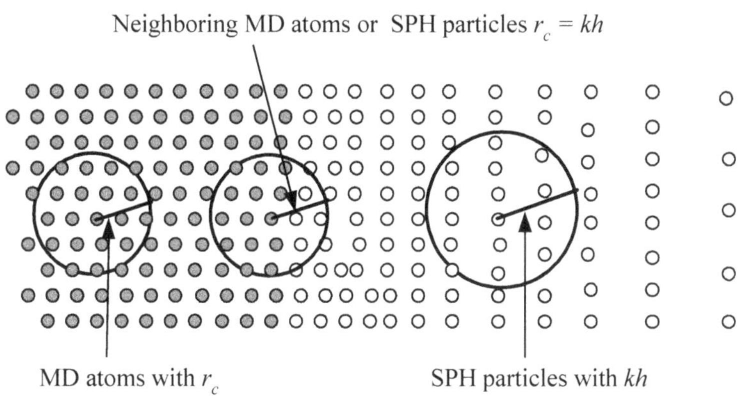

Figure 2.14 MD-SPH coupling: handshaking without overlap region.

κh of the transitional SPH particles is not necessarily equal to the potential cutoff distance r_c. The width of the transitional SPH region is around 4σ.

In model II, the computational domain consists of a MD region and a SPH region as shown in **Figure 2.14**. Similar to model I, there is also a layer of transitional SPH particles. The SPH particles in this transitional layer lie around the interface as the neighbors of the MD atoms, and are in the same length scale as the atoms. The influencing area of these neighboring SPH particles is the interaction potential cutoff distance of MD atoms. With the increasing distance from the MD region, the SPH particles are gradually coarse-grained from finer distribution so as to improve the computational efficiency. The handshaking is implemented by allowing interaction between neighboring MD atoms and SPH particles. For atoms near the interface, they not only feel the influence from other atoms, but also experience interaction with neighboring SPH particles. For SPH particles near the interface, they may also experience forces from the other SPH particles and neighboring atoms. The interactions between MD atoms and the interactions between SPH particles are treated traditionally. The interaction between atoms and neighboring SPH particles can be implemented using pairwise forces, which are exerting on the centerline of the neighboring pair of the MD atom and the SPH particle. The pairwise forces are equal in magnitude, but opposite in direction. It is convenient to take the pairwise force as some kind of potential force (e.g., LJ potential) within with a cutoff distance.

2.4 Molecular dynamics simulation of peptide-CNT interaction

2.4.1 General overview of CNTs

Carbon nanotubes (CNTs) are hollow cylindrical tubes consisting of webs of carbon atoms. Since their discovery in 1991 (Iijima, 1991), CNTs have stimulated ever-broader research activities in science and engineering devoted to production and application of various CNTs. The outstanding properties of CNTs such as high mechanical strength and remarkable electronic structure make CNTs special in applications in a vast variety of fields. A number of excellent reviews on general properties of CNTs are available (Dresselhaus et al., 1996; Harris, 1999; Dresselhaus and Avouris, 2001).

CNTs are normally classified into two categories: single-walled carbon nanotubes (SWCNTs) and multi-walled carbon nanotubes (MWCNTs). SWCNTs are made from a graphite sheet rolled into a cylinder, while MWCNTs are composed of multiple concentric graphite cylinders, as illustrated in **Figure 2.15**. Compared with MWCNTs, SWCNTs are more expensive and difficult to

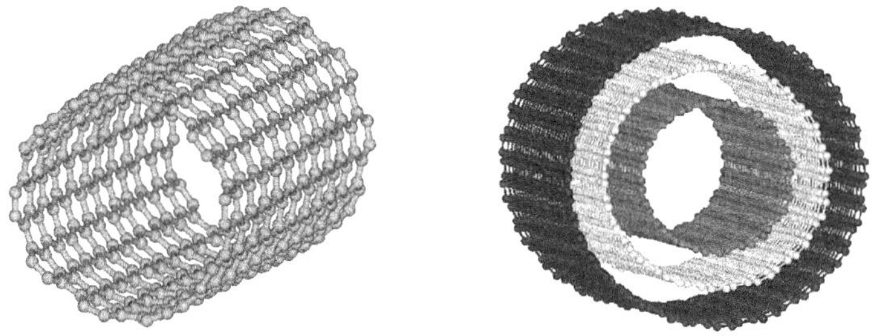

Figure 2.15 Structure of single-walled carbon nanotubes (SWCNT) (left) and multi-walled carbon nanotubes (MWCNT) (right).

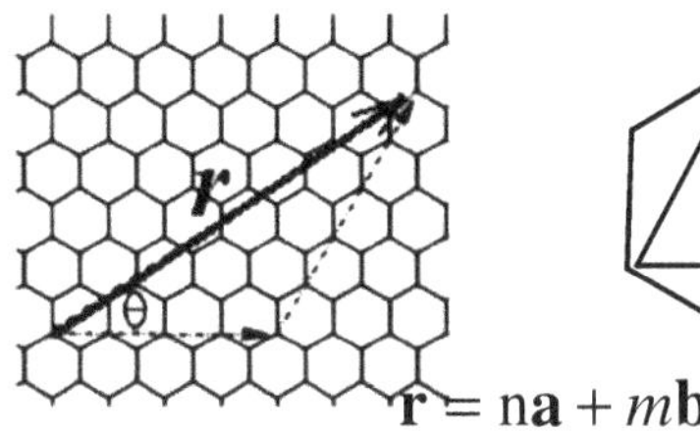

Figure 2.16 Definition of roll-up vector as linear combinations of base vectors **a** and **b**.

manufacture and clean, but they have been of great interest to researchers owing to their specific electronic, mechanical, and gas adsorption properties (Ebbesen, 1997).

CNTs can be considered as rolled-up graphite sheets. When carbon atoms geometrically combine together to form graphite, sp^2 hybridization occurs (Brown et al., 1999). Different types of CNTs can be characterized by a linear combination of base vectors $\mathbf{a}$ and $\mathbf{b}$ of the hexagon, or $\mathbf{r} = \mathrm{n}\mathbf{a} + m\mathbf{b}$, where n and m are integers of the vector equation (Thostenson et al., 2001; Qian et al., 2002) as shown in **Figure 2.16**. The values of n and m uniquely determine the chirality, or twist style of the nanotube. Three major categories of CNTs can be defined based on the value of n and m. If $n = m$, the CNT is armchair, if $n = 0$ or $m = 0$, the CNT is classified as zigzag. When $n \neq m$, the CNT is generally chiral. The chirality affects the conductance, the density, the lattice structure, and therefore affects other properties of the nanotube. A SWCNT is considered metallic if the value $n - m$ is divisible by three. Otherwise, the nanotube is semiconducting. Consequently, when tubes are formed with random values of n and m, it is expected that two-thirds of nanotubes would be semi-conducting, while the other third would be metallic, which happens to be the case. Representative configurations of the three types of CNTs are illustrated in **Figure 2.17**.

Given the chiral vector (n,m), the diameter d and the chiral angle θ of a carbon nanotube can be determined as

$$\begin{cases} \mathrm{d} = (\mathrm{n}^2 + \mathrm{m}^2 + \mathrm{nm})^{1/2} * 0.0783\ \mathrm{nm} \\ \theta = \sin^{-1}\left[\dfrac{\sqrt{3}m}{2(n^2 + nm + m^2)}\right] \end{cases} . \tag{2.15}$$

Many efforts have been made in order to investigate the mechanical properties of CNTs. For example, they were found to be bent mechanically by mechanical milling or embedding in a polymeric resin (Ajayan et al., 1994; Chopra et al., 1995; Ruoff and Lorents, 1995; Iijima et al., 1996). This flexibility

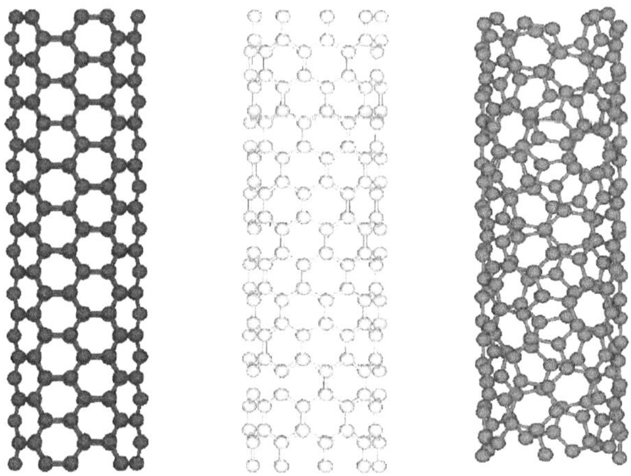

Figure 2.17 Three types of CNTs: armchair, zigzag, and chiral nanotubes.

property was also predicted through theoretical calculations (Robertson et al., 1992; Tersoff, 1992; Overney et al., 1993). High Young's modulus of CNTs was observed through these measurements. This high Young's modulus implies that CNTs are very strong material. On the other hand, the mechanical properties of composite materials containing CNTs are expected to be greatly enhanced, although those materials will not be as robust as individual nanotubes. CNTs also possess unique electrical properties. These properties are sensitive to the orientation of the hexagonal graphite lattice because it determines the density of electron states at the Fermi level (Gao and Kong, 2004). Furthermore, SWCNTs tend to self-assemble into bundles. The internal interactions of the tube may introduce small pseudogaps in bundles of nominally metallic nanotubes (Delaney et al., 1998; Kwon and Tomanek, 1998). The exceptional mechanical and electrical properties of CNTs facilitate their wide application in a number of fields in physics, chemistry, and material science including biosensors (Balavoine et al., 1999), atomic force microscopy (AFM) (Li et al., 1999; Jarvis et al., 2000) and fuel storage (Wang and Johnson, 1999; Lee and Lee, 2000).

2.4.2 General overview of proteins and peptides

Proteins are building blocks of a living cell, and they participate in essentially all cellular processes. One of the major functions of proteins is enzymatic catalysis of chemical conversions inside and around the cell. In addition, regulatory proteins control gene expression, and receptor proteins (which locate in the lipid membrane) accept intercellular signals that are often transmitted by hormones, which are proteins as well. Structural proteins form microfilaments and microtubules, as well as fibrils, hair, silk and other protective coverings. These proteins reinforce membranes and maintain the structure of cells and tissues. Some proteins provide the human body with entire bioenergetics, for example, light absorption, respiration, ATP production, and etc.

Proteins are polymers built of amino acids arranged in a linear chain and joined together by peptide bonds between the carboxyl and amino groups of adjacent amino acid residues. An $\alpha-$amino acid consists of a central carbon atom, called the α carbon, lined to an amino group, a carboxylic acid group, a hydrogen atom, and a distinctive R group. The R group is often referred to as the side chain. There are twenty kinds of amino acids, classified according to their side chains. The detailed structures for the individual amino acids can be found in references, for example in (Berg et al., 2002). The twenty types of side chains vary in size, shape, charge, hydrogen-bonding capacity, hydrophobic character, and chemical reactivity. All the proteins in all species are constructed from the same set of twenty amino acids. Owing to the diversity and versatility of these twenty building blocks, proteins are able to perform a wide range of functions.

Amino acids are often designated by a three-letter abbreviation or a one-letter symbol. Their essential properties such as the occurrence in proteins and the hydrophobicity scale of each amino acid are also listed. Hydrophilic molecules are in favor of interacting with water while hydrophobic ones tend to be nonpolar and thus prefer other neutral molecules and nonpolar solvent. The value of hydrophobicity is listed according to K-D method (Kyte and Doolittle, 1982), in which each amino acid has been assigned a value reflecting its relative hydrophilicity and hydrophobicity. A positive hydrophobicity value indicates that the amino acid is hydrophobic, and the negative value implies the hydrophilic property of the amino acid. The higher the hydrophobicity values, the more hydrophobic the amino acid is.

Protein structures can be described at four levels. The primary structure refers to the amino acid sequence. A series of amino acids joined by peptide bonds form a polypeptide chain, and each amino acid unit in a polypeptide is called a residue. The polymer chain consists of a chemically regular backbone called main chain and various side chains (R_1, R_2, …, R_M). The number M of residues in one protein could range from a few dozens to many thousands. This number is gene-encoded, and so are the positions of these amino acids in the protein chain. Most natural polypeptide chains contain between 50 and 2000 animo acid residues and are usually referred to as proteins. Polypeptides made of small number of amino acids are called oligopeptides or simply peptides.

Secondary structure refers to the conformation of the local regions of the polypeptide chain. Polypeptide chains can fold into regular structures such as the alpha helix, the beta sheet, and turns and loops. Although the turn or loop structures are not periodic, they are well defined and contribute together with alpha helices and beta sheets to form the final protein structure.

Tertiary structure describes the overall folding of the polypeptide chain. Finally, quaternary structure refers to the specific association of multiple polypeptide chains to form multisubunit complexes. A knowledge of the 3D structure of a protein is essential to understanding its function.

2.4.3 Setup of the MD simulation of peptide-CNT interaction

In spite of the fact that the marvelous properties of CNTs have triggered great interest of researchers to explore wide applications of CNTs, the mechanism of CNTs interacting with bimolecular still remains unclear. Furthermore, there have been only a few studies focusing on interactions between peptides and CNTs. Molecular simulation is a powerful tool which permits us to observe, examine and manipulate the detailed properties of the system in many ways beyond the abilities of experiments. A recent molecular dynamics simulation showed that SWCNTs could act as a hydrophobic channel for conduction of water molecules (Hummer et al., 2001). It was also shown that DNA oligonucleotides could be spontaneously inserted into SWCNTs in water solvent environment (Gao et al., 2003).

Recently, Liu, Chen and their team conducted a systematically research to investigate the interaction of peptide and CNT (Liu et al., 2005). In their work, MD simulations are performed using the software package Amber 7. Force field AMBER99 is used, which is suitable for general organic systems and drug-protein interaction simulation.

The selection of peptides is based on their specific biochemical properties such as hydrophobicity or their potential of being used as drugs. As shown in **Table 2.1**, some peptides are commonly used as therapeutic agents or extracted from disease-related proteins, others are designed to facilitate the identification of factors that affect peptides' insertion into CNTs. Specifically, oxytocin (Peptide 3) is a commonly used drug peptide. Angiotensin II (Peptide 6) is one of the famous peptide hormones23. Peptide 7 and 13 are extracted from *N*-terminal domain of mammalian PrPC 24 and yeast protein Sup35 (Gsponer et al., 2003).

Initial structures for simulation are generated using LEAP module in Amber 7 package (Case et al., 2002). Each simulated system consists of one SWCNT and one peptide solvated in water solvent. Initial structures of oxytocin and Angiotensin II are obtained from Protein Data Bank (PDB entry 1NPO and entry 1N9V, respectively). Other peptides adopt stretched initial conformations. The (12, 12) (diameter 16.1 Å) or (14, 14) (diameter 18.9 Å) types of SWCNTs are constructed through folding a graphite sheet of carbon rings to cylinder. Initially, the peptide is aligned along the nanotube axis with an appropriate initial distance between them. The peptide-SWCNT complex is then surrounded by TIP3P water molecules (Jorgensen et al., 1983).

Table 2.1 List of the simulated peptides, type of SWCNTs, number of surrounding water molecules as well as the initial distance between the most adjacent two atoms of the peptide and the SWCNT along the nanotube axis.

Assigned peptide name	Peptides sequences	Type of SWCNTs	Number of surrounding Water Molecules	Initial distances (Å)
pep1	VEAGG	(12, 12)	3172	2.5
pep2	AAFEL	(12, 12)	3072	4.1
pep3	GLPCNQIYC	(14, 14)	6373	2.7
pep4	LLLLLLLL	(12, 12)	4174	4.0
pep5	FLIGI	(12, 12)	3201	2.4
pep6	DRVYIHPF	(12, 12)	4985	5.6
pep7	PHGGGWGQ	(12, 12)	3750	4.0
pep8	EEEEEEEE	(12, 12)	4484	3.2
pep9	QQQQQQQQ	(12, 12)	4468	2.5
pep10	KKKKKKKK	(12, 12)	4628	3.2
pep11	SQNGNRE	(12, 12)	4058	3.5
pep12	DNNNRTEE	(12, 12)	3052	4.0
pep13	GNNQQNY	(12, 12)	4539	3.9
pep14	DDDDDDDD	(12, 12)	4488	3.2
pep15	DKNNRQE	(12, 12)	3993	4.4
pep16	RRRRRRRR	(12, 12)	4808	3.2
pep17	NNNNNNNN	(12, 12)	4451	4.0

Periodic boundary conditions are applied to the system. The periodic boundary conditions enable a simulation to be performed using a relatively small number of particles, in such a way that the particles experience forces as if they were in bulk fluid. The particles being simulated are enclosed in a box which is then replicated in all three dimensions to give a periodic array. During the simulation only one of the particles is represented, but the effects are reproduced over all the images. Each particle interacts not only with the other particles but also with its own images in neighboring boxes. The particles that leave one side of the box re-enter from the opposite side as their image. In this way the total number of particles in the central box remains constant.

The procedure of the simulations is as follows as shown in **Figure 2.18**. First, potential energy minimization is performed on each of the initial systems, then MD simulations are implemented on the energy-minimized systems, and the structure is saved every 5 ps. Constant volume and constant temperature ensemble (NVT) is run first for 100 ps to target the temperature of the system from initial value of 0 K to 300 K. Subsequently the simulation is performed at constant pressure (1 bar) and constant temperature (300 K) for 1900 ps. Newton's equations of motion are integrated with a step size of 1 fs, with all nonbonded cutoff distance of 8 Å being used. Bonds involving hydrogen atoms are constrained using the SHAKE algorithm and a relative tolerance of 0.00001. To make a fast implementation of the Ewald summation of the full electrostatic interactions of the unit cell (periodic box), the particle-mesh Ewald method (PME) algorithm with cubic-spline interpolation (1 Å grid width) is applied to calculate the full electrostatic energy of the simulated system.

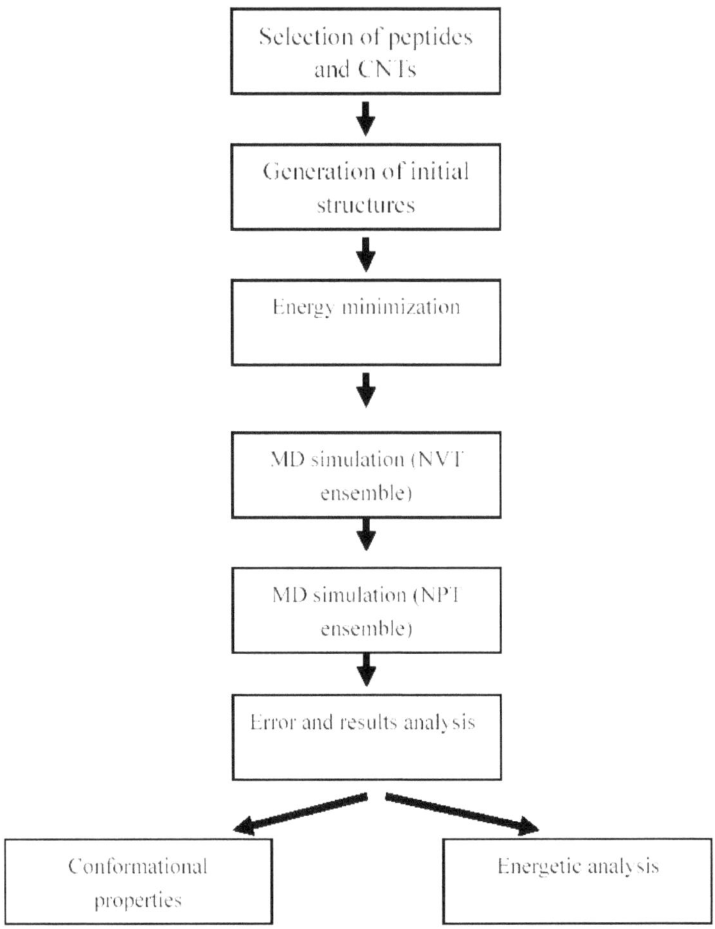

Figure 2.18 Strategies for implementation of simulation procedure.

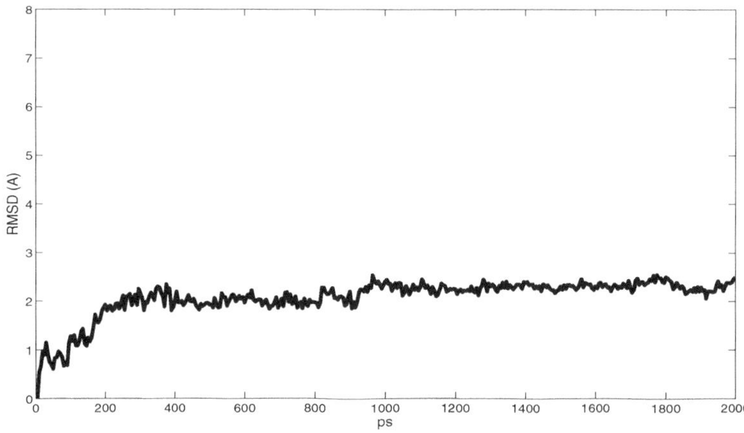

Figure 2.19 The RMSDs for the backbone atoms on pep3 against the simulation time.

2.4.4 Results and discussions

A. Diverse propensities

The stability of the conformational change of the peptides upon interacting with SWCNTs is tested through analysis of the root meant square deviations (RMSDs) of the backbone atoms on peptides. The RMSD trajectory of backbone atoms of one representative peptide during the 2 ns of simulation time is shown in **Figure 2.19**. It is observed that over the simulation time, the RMSDs are stable without unreasonable oscillations.

Through analyzing the simulation results, it is shown that some peptides are able to insert into SWCNTs while others are not. As an example of peptides that have strong affinities for SWCNTs, **Figure 2.20**(a)-(f) show the snapshots of structures of oxcytocin (pep3)-SWCNT system at different simulation time. Water molecules are stripped for a clearer visualization purpose. It is observed that the first residue of oxcytocin begins to enter the nanotubes at about 50 ps, and it has been encapsulated in the nanotube completely by the time up to 500 ps. Afterwards the peptide is trapped in the tube and does not escape, with only slight fluctuations in its relative position to the nanotube.

Significant differences in propensities are also observed. Five among the 17 simulated peptides, pep1 through pep5, could quickly self-insert into the SWCNTs. Pep12 through pep17 fail to insert into the SWCNTs, they either move away or approach the outer surface of the nanotube. Representative conformation of pep13-SWCNT interaction configuration at simulation time of 2 ns is shown in **Figure 2.21**. The remaining six peptides, pep6 through pep11 are able to enter or partly enter SWCNTs, but the insertion processes are very slow. Based on the above observation, these simulated peptides can be roughly classified into three classes as shown in **Table 2.2**: (i) Peptides that have strong affinities for SWCNTs, which could insert into SWCNTs within 800 ps simulation time (pep1 through pep5); (ii) Pep6 through pep11, which could completely or partly enter SWCNTs within 2 ns simulation time, but the insertion progresses are greatly hindered compared to those in class (i) (encapsulated completely at least after 1.2 ns); (iii) Peptides that move away from the SWCNTs (pep12 through pep17).

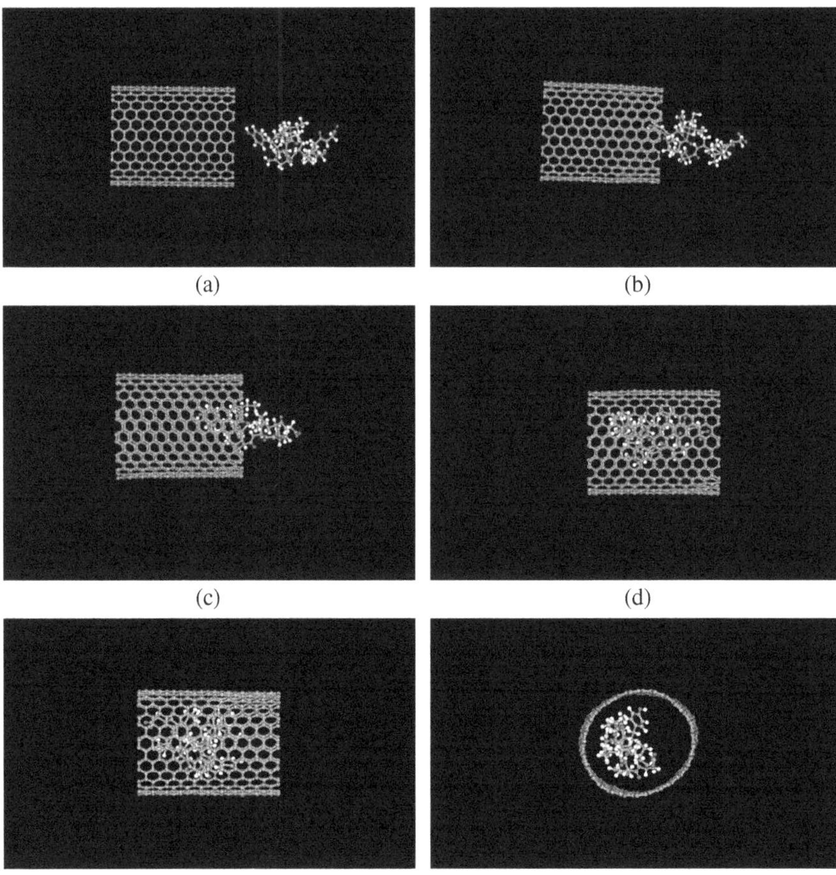

(a) (b) (c) (d) (e) (f)

Figure 2.20 The snapshots of the conformation of oxytocin (pep3) insertion into SWCNT at different simulation time: (a) initial structure, (b) 50ps, (c) 100ps, (d) 500ps, (e) 2ns. (f) shows the final structure (2ns) viewed along the axis of nanotube. The images are created with DS ViewerPro 5.0 software (Accelrys Inc., San Diego, CA).

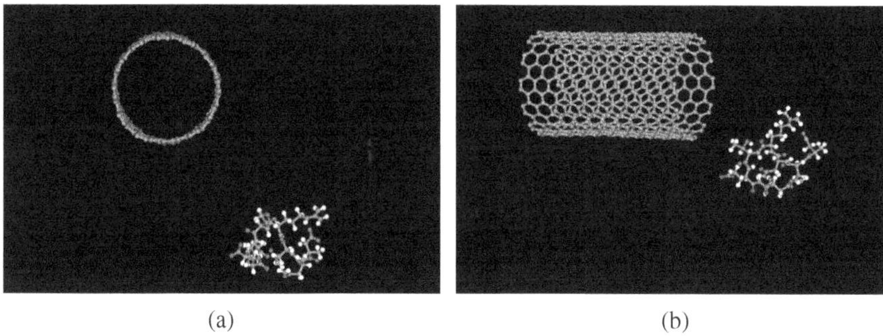

Figure 2.21 The snapshots of the final structure of pep13 interacting with SWCNT at simulation time of 2ns. The images are created with DS ViewerPro 5.0 software (Accelrys Inc., San Diego, CA).

Table 2.2 The list of the simulated peptides classified into three classes based on the insertion behaviors.

peptide name	Class
pep1-pep5	First class
pep6-pep11	Second class
pep12-pep17	Third class

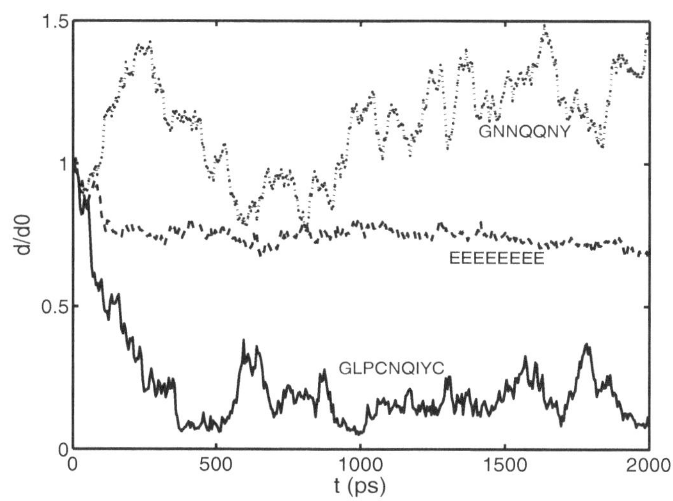

Figure 2.22 Normalized Center of Mass (COM) distances between the peptide and SWCNT as the function of MD simulation time. d0 is the initial COM distance between the peptide and the SWCNT, and d is the distance at the corresponding simulation time.

Different behaviors of peptides interacting with SWCNT could also be illustrated by tracing the relative distance between the two objects at instant simulation time. **Figure 2.22** shows the normalized center of mass (COM) distances as the function of simulation time of three representative peptides from the respective three categories. It is apparent that the peptide from the first class shows the sharpest reduction of COM distance to the CNT, while the one from the third class indicates the opposite tendency. The peptide from the second class takes the intermediate.

B. Energetics of peptide-CNT interaction

In order to examine the convergences and stabilities of MD simulations, the energetic trajectories of the simulation are also traced. The total potential energy trajectories and its components are analyzed with reference to simulation time. The data demonstrates that the energies converge with only small fluctuations.

The energetics of the process of peptides insertion into SWCNTs is further analyzed. For peptides that could insert into the nanotube, the potential energy of the whole system decreases with the reduction of COM distance between the peptide and the nanotube, as shown in **Figure 2.23**(a). The system potential energy reaches its minimum when the mass center of the peptide is close to the centre of the SWCNT, and subsequently remains at the minimum value. In contrast, in the case that the peptide could not insert into nanotube, the potential energy is not dependent on the distance between the peptide and the nanotube. For instance, the COM distance between pep13 and the nanotube never reaches low values (**Figure 2.24**(a)). The peptide tends to interact with the water molecules more than the nanotube.

Figure 2.23(b) demonstrates the relationship between the energy sum of van der Waals and electrostatic interactions (non-bonded interactions) and the COM distance. Comparing **Figure 2.23**(a) with **Figure 2.23**(b), it is found that the energy sum is dominant in driving the peptide into the potential well. Contributions of other interaction energies are shown in **Figure 2.23**(c), the difference between potential energy and non-bonded interactions energy is not dependent on the COM distance between the two objects in our simulation system. It is found that the electrostatic interactions and the van der Waals interactions among the peptide, the CNT and water play an important role in determining whether peptides could be encapsulated into SWCNTs. Although pep13 is not able to self-insert into the SWCNT, the consistent energetic contributions are also observed as shown in **Figure 2.24**(a)-(c).

In order to further examine the significance of van der Waals interaction on the insertion of the peptide into SWCNT, three systems were selected in which the peptides could insert into the nanotubes with original van der Waals parameters. The cross-section parameter σ_{cc} was cut by half (so that the van

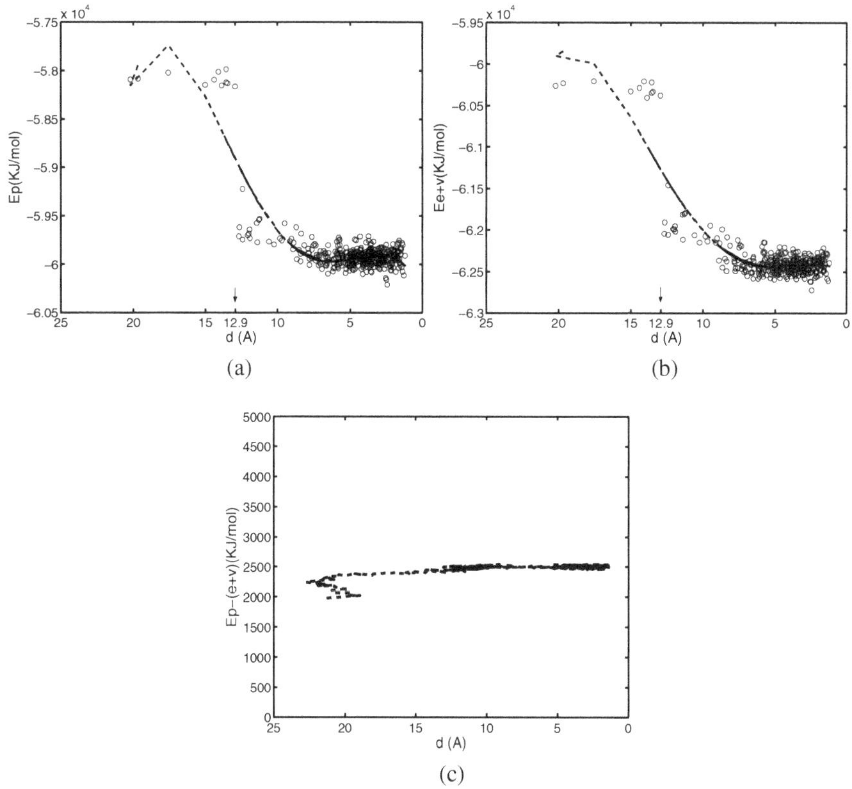

Figure 2.23 (a) Potential energy of the simulated oxytocin (pep3)-SWCNT system as the function of COM distance between SWCNT and pep3. (b) Energy sum of the van der Waals energy and the electrostatic energy (non-bonded interaction energy) as the function of COM for pep3-SWCNT system. (c) The difference between potential energy and non-bonded interaction energy as the function of COM distance between pep3 and SWCNT. The half length of the nanotube is 12.9 Å.

der Waals attraction is artificially reduced), and the simulation was repeated for 2 ns with the same initial conformations. Simulation results indicate that these peptides are not able to enter the nanotube with modified parameters. For the three selected systems with normal and modified van der Waals parameters, **Figure 2.25** compares the normalized COM distances between the peptide and nanotube as the function of simulation time. It is shown that the COM distances decrease rapidly within first 800ps for the normal van der Waals parameters cases; while the COM distances remained roughly unchanged with time for the systems with modified van der Waals parameters. This indicates that with the changed van der Waals interactions, the peptide only make random motion

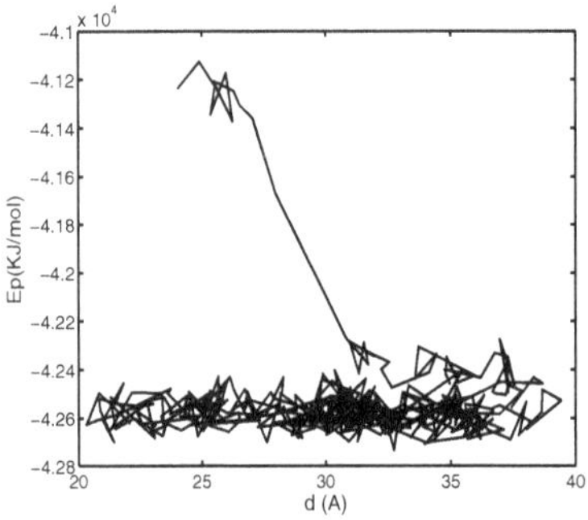

(a)

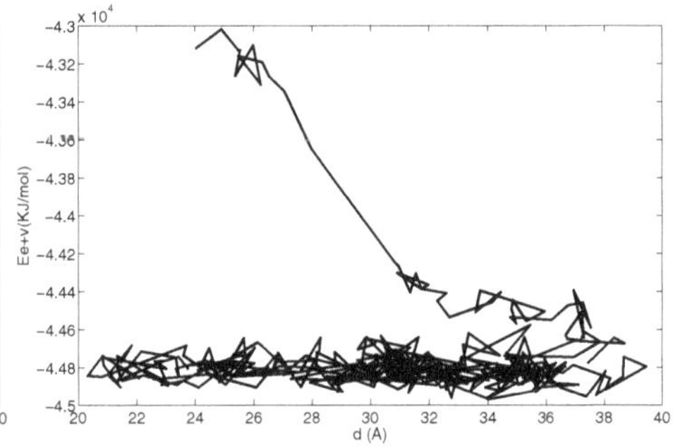

(b)

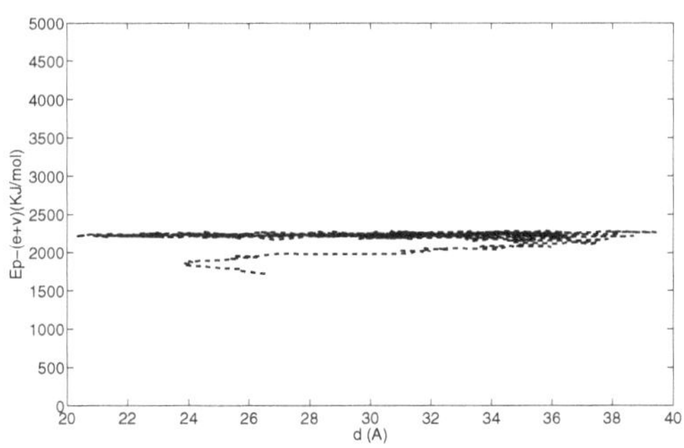

(c)

Figure 2.24 (a) Potential energy of the pep13-SWCNT system as the function of COM distance of SWCNT and pep13. (b) Energy sum of the van der Waals energy and the electrostatic energy (non-bonded interaction energy) as the function of COM for pep13-SWCNT system. (c) The difference between potential energy and non-bonded interaction energy as the function of COM distance between pep13 and SWCNT. The half length of the nanotube is 14.6 Å.

around the nanotube, and hence the insertion progress of peptide into nanotube is greatly hindered. These results imply that van der Waals interactions play an important role in the self-insertion of peptide into SWCNT.

For pep13 and pep14 (from the third class), the system was simulated with initial structure of peptide being positioned within the inner space of SWCNT, then observe its conformational change over 1 ns of simulation time. It is found that both peptides remain inside the SWCNT. It is speculated that if initially positioned inside the CNT, the peptide would be trapped by the potential well of the CNT. The peptide does not necessarily move out spontaneously due to the energy barrier, even for peptides exhibiting low propensities.

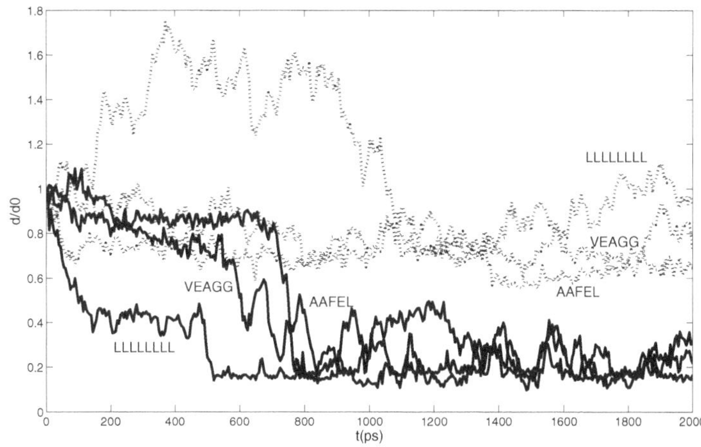

Figure 2.25 Normalized COM distances between the peptide and nanotube as the function of simulation time. Solid lines represent the cases with normal van der Waals parameters, dash lines are for the cases with the modified van der Waals parameters.

C. Impacts of CNT size

Other factors that may have effects on the propensities involve diameters of SWCNTs, lengths of SWCNTs, and etc. Taking into consideration that both the steric hindrance effect of CNTs and the van der Waals interactions between CNTs and peptides depend on the sizes of CNTs, the insertion process of peptides should be CNT-dimension dependent.

As shown in **Figure 2.26**, if the diameter of the nanotube is too small, some bulky residues (for example, those have aromatic rings) may be stucked at the entrance of the SWCNT even though the insertion process starts normally. Therefore peptides would not be able to fully enter the nanotube. By contrast, if

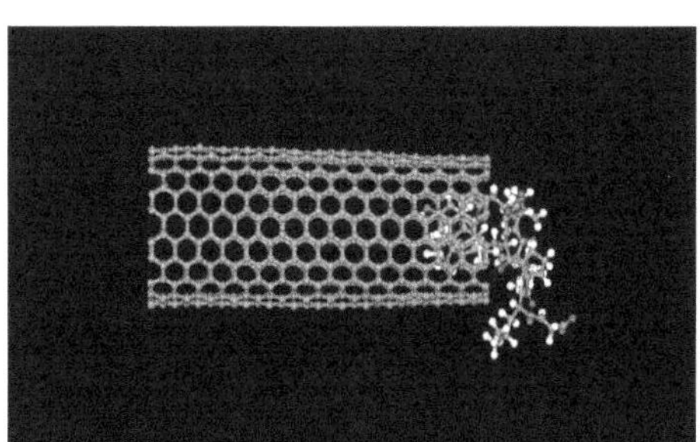

Figure 2.26 Snapshots of conformation of oxcytocin and (12, 12) type SWCNT at simulation time of 2ns. The diameter of the nanotube is 16.1 Å .

the diameter of the nanotube is large enough (see **Figure 2.20**), the same peptide can be encapsulated into the nanotube completely. As an example, the critical diameter for a SWCNT to intake pep3 is approximately 18.9 Å.

In order to clarify the effect of the SWCNT length on the propensity, two systems are tested, each composing of the same peptides but CNTs with different lengths. In both cases, peptides could self-insert into the CNTs, and their normalized Center of Mass (COM) distances are compared in **Figure 2.27**. It is observed that the COM distance between the longer CNT and the peptide does not decrease after the peptide is fully encapsulated into the CNT. This may be due to that for such cases, there exist a potential well with broadened at bottom, the system could stay anywhere at the flat bottom of the potential well, and thus the peptide does not necessarily reach the central of the CNT. Therefore lengths of SWCNTs are not critical factors as long as the tubes are long enough to encapsulate the peptides.

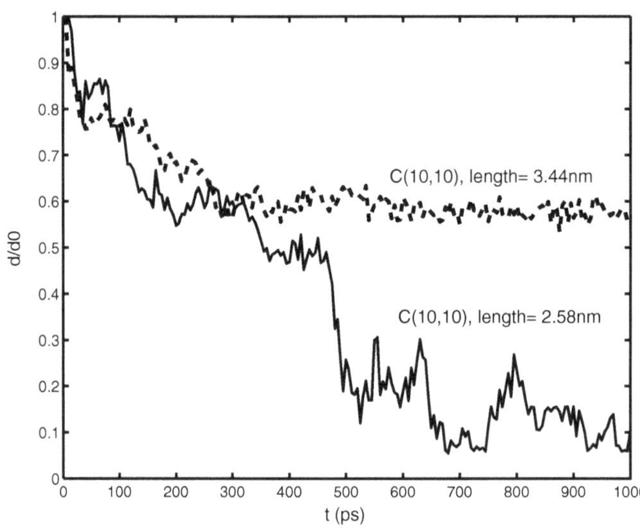

Figure 2.27 Normalized Center of Mass (COM) distances between the peptide and SWCNT as the function of MD simulation time for the same peptide inserting into SWCNTs of different length.

D. Correlations between hydrophobicities and propensities

It is well-established that hydrophobic interactions play an essential role on the interactions between CNTs and peptides. In the study, hydrophobic interactions are not calculated explicitly because the macroscopic statistical properties such as averaged free energy and the entropy are not easily accessible

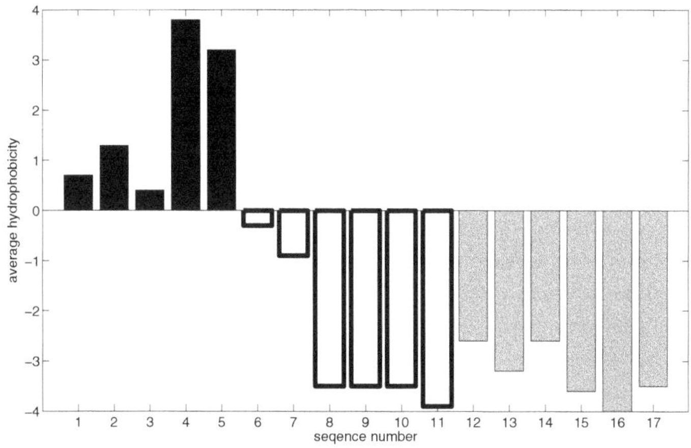

Figure 2.28 Average hydrophobicity for simulated peptides. Higher values of the average hydrophobicity imply that the peptides are more hydrophobic. Sequence numbers of peptides are in accordance as listed in Table 1. Pep1 through pep5 rapidly insert into the SWCNTs, pep6 through pep11 partially insert into SWCNTs or insert completely with slow speed, pep12 through pep17 fail to insert into SWCNTs.

in explicit solvent for MD system. Instead, The K-D method was used to assign each amino acid residue a hydrophobicity value. The average hydrophobicity for each peptide corresponding to **Table 2.1** was then calculated and plotted as bars in **Figure 2.28**. Comparing their average hydrophobicity values, it is found that peptides consisting of more hydrophobic amino acid residues tend to enter the SWCNTs more easily than those consisting of more hydrophilic polar residues. Such result is not surprising from the viewpoint of thermodynamics: hydrophobic solutes such as the SWCNT and hydrophobic peptide tend to aggregate to effectively reduce the hydrophobic surface exposed to polar solvent and consequently to reduce the overall free energy of the system. It is also observed that pep8 through pep17 have comparable average hydrophobicity value, while they are differentiated in their affinities for SWCNTs as mentioned above. Other characteristics of amino acid residues such as sizes, charges and aromaticities may have impact on the interactions between SWCNTs and peptides.

2.5 Concluding remarks

The molecular dynamics method is in principle capable of providing reliable results on all scales, but is yet restricted from practical applications due to the extremely small time scales (nanoseconds) and length scales (nanometers). There are basically three approaches to extend MD to large scale applications:

1) implementing MD with *high performance computing* (HPC) techniques. Parallelization of MD code using *message passing interface* (MPI) between the *central processing units* (CPUs) is very popular and has been implemented and described in many references. During the last decade, it is gradually found that the *graphic processing units* (GPUs) can be much more powerful and energy efficient than CPUs, with the peak performance of a modern GPU being 30 times as high as that of a modern CPU. Therefore it is more and more popular to develop parallelized codes using GPU for particle methods. It is possible to parallelize particle codes with co-design thinking for selecting, optimizing and developing numerical algorithms to meet the challenge of simulation at extreme scale (Tian, 2014).

2) coupling MD with macro scale computational methods. Some basic concepts of coupling MD with FEM, FDM and SPH have been briefed in this chapter. It is feasible to couple other macro or meso scale computational methods with MD for problems with larger applications both spatially and temporally.

3) coarse-graining of MD to obtain the main features of concerned physics, while ignores some trivial molecular details that do not affect the behavior at larger scales. The dissipative particle dynamics method is such a typical coarse-grained MD, and will be introduced in Chapter 3 and Chapter 4.

References

Ajayan, P. M., Stephan, O., Colliex, C. and Trauth, D. (1994) Aligned carbon nanotube arrays formed by cutting a polymer resin nanotube composite, *Science* **265**: 1212–1214.

Aktas, O. and Aluru, N. R. (2002) A combined continuum/dsmc technique for multiscale analysis of microfluidc filters, *Journal of Computational Physics* **178**: 342–372.

Alder, B. J. and Wainright, T. E. (1957) Phase transition for a hard sphere system, *Journal of Chemical Physics* **27**: 1208–1209.

Allen, M. P. and Tildesley, D. J. (1987) Computer simulation of liquids. Oxford University Press, Oxford.

Allen, M. P. and Tildesley, D. J. (1993) Computer simulation in chemical physics. Kluwer, Dordrecht.

Balavoine, F., Schultz, P., Richard, C., Mallouh, V., Ebbesen, T. W. and Mioskowski, C. (1999) Helical crystallization of proteins on carbon nanotubes: A first step towards the development of new biosensors, *Angewandte Chemie International Edition* **38(13/14)**: 1912–1915.

Berg, J. M., Tymoczko, J. L. and Stryer, L. (2002) Biochemistry. W.H. Freeman, New York.

Bird, G. A. (1994) Molecular gas dynamics and the direct simulation of gas flow. Oxford University Press, Oxford.

Broughton, J. Q., Abraham, F. F., Bernstein, N. and E., K. (1999) Concurrent coupling of length scale: Methodology and application, *Physical Reivew B* **60**(4): 2391–2403.

Brown, T. L. L., Bursten, B. E. and Lemay, H. E. (1999) Chemistry: The central science,. Prentice Hall PTR, Upper Saddle River.

Case, D. A., Pearlman, D. A., Caldwell, J. W., Cheatham, T. E., Wang, J., Ross, W. S., Simmerling, C. L., Darden, T. A., M., M. K., Stanton, R. V., Cheng, A. L., Vincent, J. J., Crowley, M., Tsui, V., Gohlke, H., Radmer, R. J., Duan, Y., Pitera, J., Massova, I., Seibel, G. L., Singh, U. C., Weiner, P. K. and Kollman, P. A. (2002) Assisted model building with energy refinement 7 (amber 7). University of California, San Francisco, CA.

Chen, H. W., I., H. and Tieu, A. K. (2011) A seamless coupling between molecular dynamics and material point method, *Japan Journal of Industrial and Applied Mathematics* **28**(1): 55–67.

Chopra, N. G., Benedict, L. X., Crespi, V. H., Cohen, M. L., Louie, S. G. and Zettl, A. (1995) Fully collapsed carbon nanotubes, *Nature* **377**: 135–137.

Ciccootti, G., Frenkel, D. and McDonald, I. R. (1987) Simulation of liquids and solids. North-Holland.

Delaney, P., Choi, H. J., Ihm, J., Louie, S. G. and Cohen, M. L. (1998) Broken symmetry and pseudogaps in ropes of carbon nanotubes, *Nature* **391**: 466–468.

Dresselhaus, M. S. and Avouris, P. (2001) Introduction to carbon materials research, *Carbon Nanotubes* **80**: 1–9.

Dresselhaus, M. S., Dresselhaus, G. and Eklund, P. C. (1996) Science of fullerenes and carbon nanotubes. Academic Press, San Diego, US.

Ebbesen, T. W. (1997) Carbon nanotubes: Preparation and properties. CRC Press, Baca Raton, Florida.

Fedosov, D. A., Caswell, B. and Karniadakis, G. E. (2010) Systematic coarse-graining of spectrin-level red blood cell models, *Computer Methods in Applied Mechanics and Engineering* **199**(29): 1937–1948.

Flekkoy, E. G., Wagner, G. and Feder, J. (2000) Hybrid model for combined particle and continuum dynamics, *Europhysics Letter* **52**(3): 271.

Fraga, S., Parher, J. M. R. and Pocock, J. M. (1995) Computer simulations of protein structures and interactions. Springer, Berlin.

Gao, H., Kong, Y., Cui, D. and Ozkan, C. S. (2003) Spontaneous insertion of DNA oligonucleotides into carbon nanotubes, *Nano Letters* **3**: 471–473.

Gao, H. J. and Kong, Y. (2004) Simulation of DNA-nanotube interactions, *Annual Review Of Materials Research* **34**: 123–150.

Garcia, A. L., Bell, J. B., Crutchfield, W. Y. and Alder, B. J. (1999) Adaptive mesh and algorithm refinement using direct simulation monte carlo, *Journal of Computational Physics* **154**: 121–134.

Gibson, J. B., Goland, A. N., Milgram, M. and Vineyard, G. H. (1960) Dynamics of radiation damage, *Physics Review* **120**: 1229–1253.

Gsponer, J., Haberthur, U. and Caisch, A. (2003) The role of side-chain interactions in the early steps of aggregation: Molecular dynamics simulations of an amyloid-forming peptide from the yeast prion sup35, *Proceedings of the National Academy of Sciences of the United States of America (PNAS)* **100**: 5154–5159.

Hadjiconstantinou, N. G. (1999a) Combining atomistic and continuum simulations of contact-line motion, *Physical Review E* **59**(2): 2475–2479.

Hadjiconstantinou, N. G. (1999b) Hybrid atomistic-continuum formulations and the moving contact line problem, *Journal of Computational Physics* **154**: 245–265.

Hadjiconstantinou, N. G. and Patera, A. T. (1997) Heterogneous atomistic-continuum representations for dense fluid systems, *International Journal of Modern Physics C* **8**(4): 967–976.

Harris, P. J. F. (1999) Carbon nanotube and related structures: New materials for the 21st century. Cambridge University Press, Cambridge.

Hoover, W. G. (1991) Computational statistical mechanics. Elsevier, Amsterdam.

Hummer, G., Rasalah, J. C. and Noworyta, J. P. (2001) Water conduction through the hydrophobic channel of a carbon nanotube, *Nature* **414**: 188.

Iijima, S. (1991) Helical microtubules of graphitic carbon, *Nature* (354): 56–58.

Iijima, S., Brabec, C., Maiti, A. and Bernholc, J. (1996) Structural flexibility of carbon nanotubes, *Journal of Chemical Physics* **104**: 2089–2092.

Jarvis, S. P., Uchihashi, T., Ishida, T., Tokumoto, H. and Nakayama, Y. (2000) Local solvation shell measurement in water using a carbon nanotube probe, *Journal of Physical Chemistry B* **104**(26): 6091–6094.

Jorgensen, W. J., Chandreskhar, J., Madura, J., Imprey, R. and Klein, M. (1983) Comparison of simple potential functions for simulating water, *Journal of Chemical Physics* **79**: 926–935.

Kwon, Y. K. and Tomanek, D. (1998) Electronic and structural properties of multiwall carbon nanotubes, *Physics Review B* **58**: R16001–R16004.

Kyte, J. and Doolittle, R. F. (1982) A simple method for displaying the hydrophobic character of a protein, *Journal of Molecular Biology* **157**: 105–132.

Lee, S. M. and Lee, Y. H. (2000) Hydrogen storage in single-walled carbon nanotubes, *Applied Physics Letters* **76**(20): 2877–2879.

Li, J., Cassell, A. M. and Dai, H. (1999) The carbon nanotube as afm tips: Measuring DNA molecules at the liquid/solid interface, *Surface and Interface Analysis* **28**: 8–11.

Liu, G. R., Cheng, Y., Mi, D. and Li, Z. R. (2005) A study on self-insertion of peptides into single-walled carbon nanotubes based on molecular dynamics simulation, *International Journal of Modern Physics C* **16**(8): 1239–1250.

Liu, G. R. and Han, X. (2003) Computational inverse techniques in nondestructive evaluation. CRC Press, Boca Raton.

Liu, G. R. and Liu, M. B. (2003) Smoothed particle hydrodynamics: A meshfree particle method. World Scientific, Singapore.

Liu, Y., Wang, H. K. and Zhang, X. (2013) A multiscale framework for high-velocity impact process with combined material point method and molecular dynamics, *International Journal of Mechanics and Materials in Design* **9**(2): 127–139.

Meyer, M. and Pontikis, V. (1991) Computer simulation in materials science, interatomic potentials, simulation techniques and applications. Kluwer, Dordrecht.

O'Connell, S. T. and Thompson, P. A. (1995) Molecular dynamics-continuum hybrid computations: A tool for studying complex fluid flows, *Physical Review E* **52**(6): 5792–5795.

Overney, G., Zhong, W. and Tomanek, D. (1993) Structural rigidity and low frequency vibrational modes of long carbon tubules, *Zeitschrift für Physik D Atoms, Molecules and Clusters* **27**: 93–96.

Pan, L. S., Liu, G. R., Khoo, B. C. and Song, B. (2000) A modified direct simulation monte carlo method for low-speed microflows, *Journal of Micromechanics and Microengineering* **10**: 21–27.

Qian, D., Liu, W. K. and Ruoff, R. S. (2001) Mechanics of c60 in nanotubes, *Journal Of Physical Chemistry B* **105**: 10753–10758.

Qian, D., Wagner, G. J., Liu, W. K., Yu, M. F. and Ruoff, R. S. (2002) Mechanics of carbon nanotubes, *Applied Mechanics Review* **55**: 495–533.

Qiao, R. and He, P. (2007) Simulation of heat conduction in nanocomposite using energy-conserving dissipative particle dynamics, *Molecular Simulation* **33**(8): 677–683.

Rahman, A. (1964) Correlations in the motion of atoms in liquid argon, *Physics Review* **136**: 405–411.

Rapaport, D. C. (2004) The art of molecular dynamics simulation. Cambridge University Press, Cambridge.

Robertson, D. H., Brenner, D. W. and Mintmire, J. W. (1992) Energetics of nanoscale graphitic tubules, *Physics Review B* **45**: 12592–12595.

Rudd, R. E. and Broughton, J. Q. (1999) Atomistic simulation of mems resonators through the coupling of length scales, *Journal of Modeling and Simulation of Microsystem* **1**(1): 29–38.

Ruoff, R. S. and Lorents, D. C. (1995) Mechanical and thermal properties of carbon nanotubes, *Carbon* **33**: 925–930.

Ruth, M. and Lynden, B. l. (1994) Computer simulations of fracture at the atomic level, *Science* **263**(5154): 1704–1705.

Smirnova, J. A., Zhigilei, L. V. and Garisson, B. J. (1999) A combined molecular dynamics and finite element method technique applied to laser induced pressure wave propagation, *Computer Physics Communication* **118**: 11–16.

Stillinger, F. H. and Rahman, A. (1974) Improved simulation of liquid water by molecular dynamics, *Journal of Chemical Physics* **60**: 1545–1557.

Tersoff, J. (1992) Energies of fullerenes, *Physics Review B* **46**: 15546–15549

Thostenson, E. T., Ren, Z. and Chou, T. W. (2001) Advances in the science and technology of carbon nanotubes and their composites: A review, *Composites Science and Technology* **61**: 1899–1912.

Tian, R. (2014) Simulation at extreme-scale: Co-design thinking and practices, *Archives of Computational Methods in Engineering* **21**(1): 39–58.

Wang, Q. and Johnson, J. K. J. (1999) Optimization of carbon nanotubes for hydrogen adsorption, *Physical Chemistry B* **103**: 4809–4813.

Wu, H. A., Liu, G. R., Han, X. and Wang, X. X. (2006) An atomistic simulation method combining molecular dynamics with finite element technique, *Chaos Solitons & Fractals* **30**(4): 791–796.

Xu, Y. G. and Liu, G. R. (2003) Fitting interatomatic potentials using molecular dynamics simulations and inter-generation projection genetic algorithm, *Journal of Micro-mechanics and Micro-engineering* **18**: 383–386.

Xu, Y. G., Liu, G. R., Behdinan, K. and Fawaz, Z. (2004) Stepwise-equilibrium and adaptive molecular dynamics simulation for fracture toughness of single crystals with defects, *Journal of Intelligent Material Systems and Structures* **15**(12): 933–939.

Yao, Z. H., Wang, J. S., Li, B. W. and Liu, G. R. (2005) Thermal conduction of carbon nanotubes using molecular dynamics, *Physical Review B* **71**(8): 085417.

Yao, Z. H., Wang, J. S., Liu, G. R. and Cheng, M. (2004) Improved neighbor list algorithm in molecular simulations using cell decomposition and data sorting method, *Computer Physics Communications* **161**(1–2): 27–35.

Chapter 3

Dissipative Particle Dynamics — Methodology

This chapter provides an overview on DPD in methodology including the basic concepts of DPD, the latest development of some important numerical aspects, and a number of validating examples to show the effectiveness of DPD in modeling complex physics and simulating the continuum hydrodynamic behavior.

The chapter is outlined as follows.

- In Section 3.2, the basic concepts of DPD are introduced, including governing equations, time integration algorithm, determination of DPD coefficients, and computational procedure of DPD simulations.
- In Section 3.3, some numerical aspects of DPD are described, including the assessment of dynamic properties, solid boundary treatment, conservative interaction potential and spring-bead chain models for simulating macromolecules. In particular, a generic algorithm for treating complex solid boundaries and a novel approach for constructing conservative interaction potential with short-rang repulsion and long-distance attraction are addressed in detail.
- In Section 3.4, the DPD method in modeling complex physics and simulating the continuum hydrodynamic behavior is demonstrated with a number of benchmark numerical examples.
- In Section 3.5, some remarks and conclusion are given.

3.1 Introduction

By integrating mechanical elements, sensors, actuators, and electronic components using the microfabrication technology, microelectromechanical systems (MEMS) can be made with fast response, high spatial resolution, and cost-effective (Gardner and Varadan, 2001; Hsu, 2008). Such MEMS technology has found its applications to chemical, biological and medical science. For example, MEMS for biomedical and biological applications (referred to as BioMEMS) are capable of delivering, processing and analyzing biochemical materials. They are also useful in disease diagnosis, clinical assays, drug screening and delivery, and even gene searching and sequencing. BioMEMS usually are more effective than traditional biomedical and biological techniques. Therefore, characterization of fluid flows in MEMS devices has increasingly becoming a very important topic. However, the fluidic behavior in MEMS is very different from what observed in daily life, presenting bigger challenges (Ho and Tai, 1998; Karniadakis et al., 2005).

One typical feature of fluid flows in MEMS devices is the size effects. For example, the delivery of drugs is usually conducted by the movement and suspension of macromolecules in micro channels, where the size of the drug agents (usually DNA molecules) and the size of the micro channel are important to understand the effects of the macromolecular conformation. If the *Knudsen number*, *Kn*, defined here as the ratio of the macromolecular length to the characteristic length of flow field, is much smaller than unity (e.g., $Kn << 1$), the movement and suspension of macromolecules (in macro channels) can be regarded as a continuum flow. If the Knudsen number is around (or even bigger than) unity, the movement and suspension of macromolecules (in micro channels) may not be regarded as a continuum flow. The suspension of DNA in micro channel is exactly the case with $Kn \approx O(1)$, because the length of a typical DNA molecule is usually in the same order as the size of a typical micro channel. For example, the size of a typical micro channel is about 9-40 μm (McAllister et al., 2000), and the uncoiled length of a λ-DNA is about 22 μm to 33 μm (Koplik and Banavar, 1995; Smith et al., 1999). Hence the standard rheological models developed from continuum assumptions for continuum applications may be misleading to describe such flows with $Kn \approx O(1)$. On the other hand, the molecular dynamics simulation is not feasible for modeling such flows, because MD deals problems with extremely small time scales (nanoseconds) and length scales (nanometers). Therefore the development of numerical methods at meso scale is required.

In general, *meso scale* denotes the (length and time) scale larger than *atomic scale*, but smaller than *macro scale*. The definition of meso scale cannot be too precise, and it can be different in computational material sciences, computational physics, computational biology, chemistry, and computational mechanics. For example, in computational material sciences and computational mechanics, meso scale usually involves a characteristic length ranging from 10^{-7} to 10^{-4} m and a characteristic time ranging from 10^{-9} to 10^{-3} s. This overlaps the micro scale (a characteristic length ranging from 10^{-8} to 10^{-6} m and a characteristic time ranging from 10^{-11} to 10^{-8} s) and macro scale (a characteristic length bigger than 10^{-4} m and a characteristic time bigger than 10^{-3} s) (**Figure 3.1**).

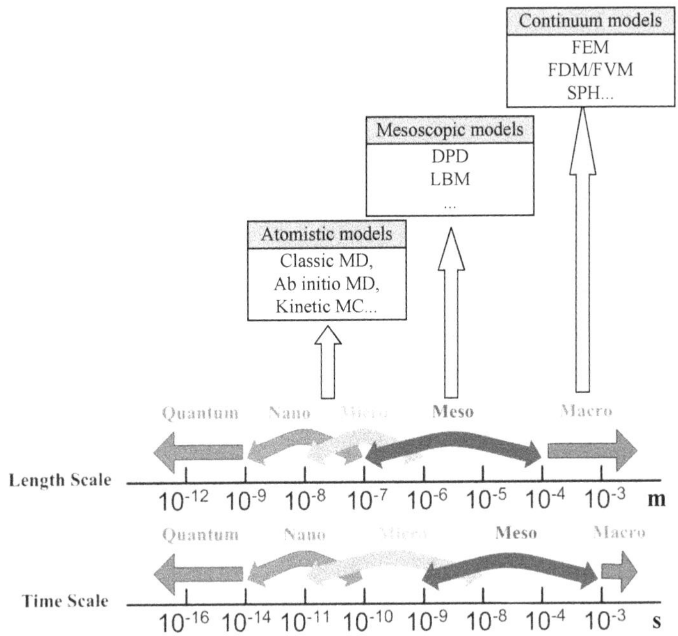

Figure 3.1 Different length and time scales and corresponding computational methods.

For problems at different scales, different computational models can be used respectively (Moeendarbary et al., 2009). For macro scale problems, computational models such as the finite element method (FEM) (Zienkiewicz and Taylor, 2000; Liu and Quek, 2003), the smoothed finite element method (S-FEM) (Liu and Nguyen, 2010), finite difference method (FDM) or finite volume method (FVM) (Peyret and Taylor, 1985; Chung, 2002) can be used. These macro scale computational models usually involve constitutive relations to solve a system of partial differential equations. When the length scale gradually reduces, the constitutive relations based on continuum assumptions may no longer be valid. For nano and micro scale problems, the atomistic models such as the classic molecular dynamics (MD) (Allen and Tildesley, 1987; Rapaport, 2004), Ab initio MD (Kresse and Hafner, 1993), and kinetic monte carlo (MC)

(Voter, 2007) can be used. The atomistic models provide a fundamental way of obtaining a better understanding of the behavior of fluid flow in micro channels. However, due to the very small length and time scales associated with these methods, they are computationally expensive, even for modern supercomputers, and they cannot be applied to many important scientific and practical problems.

Some mesoscopic or *coarse-grained methods* have been developed during the last two decades. The closely related *lattice-Boltzmann* (Ladd, 1994; He and Luo, 1997; Chen and Doolen, 1998) and *lattice-gas cellular automaton* (Pomeau and Frisch, 1986) models that are defined on a regular lattice or grid have been extensively investigated. Although lattice-Boltzmann and lattice-gas cellular automaton models have been extended to a wide range of applications such as colloidal systems and multiphase flows in porous media, they have some disadvantages that are associated with the restriction of the dynamics to the streaming of 'particles' between adjacent nodes on a regular lattice. Another approach is to use particle-based simulation methods, similar to molecular dynamics, in which the individual particles represent a volume of fluid that may vary in size, depending on the model, from a small cluster of atoms or molecules to macroscopic regions in a continuum solid or fluid. These off-lattice methods are manifestly *Galilean invariant* (unlike some lattice Boltzmann models). One of these methods, smoothed particle hydrodynamics (SPH), was originally invented to solve astrophysical problems (Gingold and Monaghan, 1977; Lucy, 1977), and it has been gradually modified for much smaller scale (Liu and Liu, 2003; Hu and Adams, 2006). In SPH, the fluid is represented by overlapping weight functions, or smoothing functions, centered on the particles. The particles move with the local velocity of the fluid, and the acceleration of each particle is calculated from the local pressure gradient and the fluid density. The density at any point can be calculated from the positions of the particles that are within the range of the weight function, and the corresponding pressure is obtained from an equation of state. Other forces, such as those due to viscosity, which act in concert with the forces associated with the pressure gradient to determine the particle accelerations, can be estimated using the positions and velocities of neighboring particles, the weight function and the derivatives of the weight function. The SPH method for mesoscopic applications is still under development, and quantitative relationships between model parameters and the macroscopic properties of the fluids that these models simulate are difficult to establish (Tartakovsky and Meakin, 2005).

Dissipative particle dynamics (DPD) (Hoogerbrugge and Koelman, 1992) is a relatively new meso scale technique that can be used to simulate the behavior of fluids. In DPD simulations, the particles represent clusters of molecules that interact via conservative (non-dissipative), dissipative and fluctuating forces. Español and Warren (Espanol and Warren, 1995) and Marsh (Marsh, 1998) established a sound theoretical basis for DPD based on statistical mechanics, and Groot and Warren obtained parameter ranges to achieve a satisfactory compromise between speed, stability, rate of temperature equilibration and compressibility (Groot and Warren, 1997). The applications of DPD have extended into many areas, especially in nano-fluidic devices for various purposes including DNA filtering. (Hoogerbrugge and Koelman, 1992; Espanol and Warren, 1995; Duong-Hong et al., 2008a; Duong-Hong et al., 2008b; Yan et al., 2012; Liu et al., 2014). Improvement efforts have also been made to DPD. Unlike traditional DPD methods that use a conservative pairwise force between particles that depends on their interparticle separation, the *multi-body DPD* (MDPD) model presented by Pagonabarraga and Frenkel assumes that the conservative force depends on the instantaneous local particle density, which in turn depends on the positions of many neighboring particles (Pagonabarraga and Frenkel, 2001). Therefore, the conservative interaction is a many-body interaction.

3.2 Basic concepts of dissipative particle dynamics

3.2.1 Coarse-graining

The classic molecular dynamics is a very important approach for investigating complex fluids such as polymers and macromolecules, and it is in principle capable of providing numerical results on all scales. As each particle in MD represents a true atom or molecule, MD can describe the dynamic behavior of a complex system with comprehensive details on every atom. The resolution of MD is essentially atoms. MD simulations thus are usually limited to extremely small time scales (nanoseconds) and length scales (nanometers) even if the state-of-art high performance computing techniques are used. However, many practical applications involve larger spatial and time scales. For example, polymers and other materials frequently show a hierarchy of length scales and associated time scales. This requires a very large number of atoms (and a very big number of degrees of freedom) if using molecular dynamics simulation. To reduce the number of degrees of freedom, coarse-grained molecular dynamic techniques have been developed (Nielsen et al., 2004; Bock et al., 2007; Knotts IV et al., 2007).

In the coarse-grained MD simulations, some trivial molecular details are ignored, while the main features of concerned physics can be effectively obtained. A general procedure in *coarse-graining* involves: 1) defining the goal and determining the degree of coarse-graining, 2) mapping atomistic model to coarse-grained model, 3) interaction between the coarse-grained particles, 4) reproducing target functions by the coarse-grained model, 5) optimizing parameters/functions in the coarse-grained model, and 6) conducting coarse-grained simulations. The goal and degree of coarse-graining are usually application-driven and they describe the number of atoms/molecules in a typical particle in the coarse-grained model. This is closely related to the minimal features of the atomistic model that should be retained to reproduce the desired properties in the coarse-grained model. Mapping atomistic model to coarse-grained model is very important in defining the positions of coarse-grained particles and it directly influences the parameterization of the coarse-grained force field. The interaction between the coarse-grained particles is usually conducted with analytical functions (e.g., LJ potential in classic MD) or numerical functions of the positions of the coarse-grained particles.

Dissipative particle dynamics is a coarse-grained molecular dynamics model, in which the particles represent clusters of molecules that interact via conservative (non-dissipative), dissipative and fluctuating forces. As a coarse-grained MD model, DPD follows the above-mentioned coarse-graining procedure.

3.2.2 Governing equations

In DPD models, a fluid system is simulated using a set of interacting particles. Each particle represents a cluster of small molecules instead of a single molecule. It is convenient to assume that all of the particles have equal masses, and use the mass of the particle as the unit of mass. Newton's second law governs the motion of each particle. The equation of motion for particle i can therefore be expressed as:

$$\frac{d\mathbf{r}_i}{dt} = \mathbf{v}_i, \quad \frac{d\mathbf{v}_i}{dt} = \mathbf{f}_i = \mathbf{f}_i^{int} + \mathbf{f}_i^{ext}, \tag{3.1}$$

where $\mathbf{r}_i$ and $\mathbf{v}_i$ are the position and velocity vectors, and $\mathbf{f}_i^{ext}$ is the external force including the effects of gravity. In equation (3.1), the inter-particle force acting on particle i, $\mathbf{f}_i^{int}$, is usually assumed to be pairwise additive and consists of three parts: a conservative (non-dissipative) force, $\mathbf{F}_{ij}^C$; a dissipative force, $\mathbf{F}_{ij}^D$; and a random force, $\mathbf{F}_{ij}^R$:

$$\mathbf{f}_i^{int} = \sum_{j\neq i} \mathbf{F}_{ij} = \sum_{j\neq i} \mathbf{F}_{ij}^C + \mathbf{F}_{ij}^D + \mathbf{F}_{ij}^R. \tag{3.2}$$

Here, $\mathbf{F}_{ij}$ is the force on particle i due to interaction with particle j, which is equal to $\mathbf{F}_{ji}$ in magnitude and opposite in direction. The symmetry of the interactions $\mathbf{F}_{ij} = -\mathbf{F}_{ji}$ ensures that momentum is rigorously conserved. The pairwise particle-particle interactions have a finite cutoff distance, r_c, which is usually taken as the unit of length in DPD models.

The *conservative force*, $\mathbf{F}_{ij}^{C}$, is a soft interaction acting along the line of particle centers, which is often given the form

$$\mathbf{F}_{ij}^{C} = a_{ij} w^{C}(r)\hat{\mathbf{r}}_{ij}$$
$$w^{C}(r) = \begin{cases} (1-r) & r < 1.0\,, \\ 0 & r \geq 1.0, \end{cases} \tag{3.3}$$

where a_{ij} is the maximum repulsion between particles i and j, $\mathbf{r}_{ij} = \mathbf{r}_i - \mathbf{r}_j$, $r = r_{ij} = |\mathbf{r}_{ij}|$ and $\hat{\mathbf{r}}_{ij} = \mathbf{r}_{ij} / r_{ij}$. Here, $w^{C}(r_{ij})$ is the weight function for the conservative force.

The *dissipative force*, $\mathbf{F}_{ij}^{D}$, represents the effects of viscosity, and it depends on both the relative positions and velocities of the particles. The form usually used for this interaction in DPD simulations is

$$\mathbf{F}_{ij}^{D} = -\gamma w^{D}\left(r_{ij}\right)\left(\hat{\mathbf{r}}_{ij} \cdot \mathbf{v}_{ij}\right)\hat{\mathbf{r}}_{ij}\,, \tag{3.4}$$

where γ is a coefficient, $\mathbf{v}_{ij} = \mathbf{v}_i - \mathbf{v}_j$ and $w^{D}\left(r_{ij}\right)$ is a distance-dependent weight function.

The *random force*, $\mathbf{F}_{ij}^{R}$, representing the effects of thermal fluctuations also depends on the relative positions of the particles, and it is defined as

$$\mathbf{F}_{ij}^{R} = \sigma w^{R}\left(r_{ij}\right)\xi_{ij}\hat{\mathbf{r}}_{ij}\,, \tag{3.5}$$

where σ is a coefficient, $w^{R}\left(r_{ij}\right)$ is a distance-dependent weight function, and ξ_{ij} is a random variable with a Gaussian distribution and unit variance. The dissipative force and random force also act along the line of particle centers and therefore also conserve linear and angular momentum.

As pointed by Español and Warren (Espanol and Warren, 1995), in order to recover the proper thermodynamic equilibrium for a DPD fluid at a prescribed temperatures T, the coefficients and the weight functions for the random force and the dissipative force are related by

$$w^{D}(r) = \left[w^{R}(r)\right]^{2}, \tag{3.6}$$

and

$$\gamma = \frac{\sigma^{2}}{2k_{\mathrm{B}}T}, \tag{3.7}$$

as required by the *fluctuation-dissipation theorem*. In equation (3.7), k_B is the Boltzmann constant. All of the interaction energies are expressed in units of k_BT, which is assigned a value of unity. One simple, straightforward and commonly used choice is

$$w^D(r) = \left[w^R(r)\right]^2 = \begin{cases} (1-r/r_c)^s & r < r_c \\ 0 & r \geq r_c \end{cases}, \tag{3.8}$$

where r_c is the cutoff distance of the dissipative and random force. In conventional DPD formulation, it usually takes the same value as the cutoff distance of the conservative force (unit value), but can vary to modify the dynamic properties in DPD simulation as will be shown later. s denotes the exponent of the weighting function. It was reported by Fan et al. that different s can lead to different dynamic behavior of a DPD system (Fan et al., 2006). For conventional DPD formulation, $s = 2$. $w^D(r)$ and its gradient are both continuous at $r/r_c = 1$. In contrast, if $s < 1$, though $w^D(r)$ is still continuous, its gradient is not continuous at $r/r_c = 1$.

The random fluctuation force, $\mathbf{F}_{ij}^R$, acts to heat up the system, whereas the dissipative force, $\mathbf{F}_{ij}^D$, acts to reduce the relative velocity of the particles, thus removing kinetic energy and cooling down the system. Consequently, the fluctuating and dissipative forces act together to maintain an essentially constant temperature with small fluctuations about the nominal temperature T. Therefore, dissipative particle dynamics simulations are essentially thermostatted molecular dynamics simulations with soft particle-particle interactions.

In summary, because the effective interactions between clusters of molecules are much softer than the interactions between individual molecules, much longer time steps can be taken relative to MD simulations. The longer time steps combined with the larger particle size makes it much more practical to simulate hydrodynamics using DPD than MD. DPD is particularly promising for the simulation of complex liquids, such as polymer suspensions, liquids with interfaces, colloids and gels. Because of the symmetry of the interactions between the particles in typical simulations, DPD rigorously conserves the total momentum of the system, and because the particle-particle interactions depend only on relative positions and velocities, the resulting model fluids are Galilean invariant. Mass is conserved because the same mass is associated with each of the particles, and the number of particles does not change. While DPD is not as computationally efficient as lattice Boltzmann simulations, it is a more flexible method that does not suffer from the numerical instability associated with many lattice Boltzmann applications. DPD facilitates the simulation of complex fluid systems on physically interesting and important length and time scales.

Comparing with MD, the DPD formulation has two additional force terms: dissipative force and the random force. These two forces are resulted from the "lumping" effects of the cluster of molecules. To ensure that the addition of these two forces does not affect the conservation of the total energy of the system, the fluctuation-dissipation theorem is imposed among these two forces. It is therefore clear that such a controlled addition of these two forces make it possible to lump a group of molecules together, reducing the number of particles for larger scale problems, while ensuring the total energy conservation.

3.2.3 Time integration

The time integration algorithm is very important in DPD. Poor integration algorithms lead to serious problems such as equilibrium properties that depend on the magnitude of the time step. Early implementations of equation (3.1) in DPD made use of the Euler scheme

$$\begin{aligned}
\mathbf{r}_i(t+\Delta t) &= \mathbf{r}_i(t) + \Delta t \mathbf{v}_i(t) \\
\mathbf{v}_i(t+\Delta t) &= \mathbf{v}_i(t) + \Delta t \mathbf{f}_i(t) \\
\mathbf{f}_i(t+\Delta t) &= \mathbf{f}_i(\mathbf{r}_i(t+\Delta t), \mathbf{v}_i(t+\Delta t)),
\end{aligned} \tag{3.9}$$

where Δt is the time step. The Euler scheme is not time reversible and it can lead to an energy drift in the system and hence it has been avoided in recent DPD research. Groot and Warren used a modified version of the velocity Verlet algorithm (Groot and Warren, 1997)

$$\begin{aligned}
\mathbf{r}_i(t+\Delta t) &= \mathbf{r}_i(t) + \Delta t \mathbf{v}_i(t) + \frac{1}{2}(\Delta t)^2 \mathbf{f}_i(t) \\
\tilde{\mathbf{v}}_i(t+\Delta t) &= \mathbf{v}_i(t) + \lambda \Delta t \mathbf{f}_i(t) \\
\mathbf{f}_i(t+\Delta t) &= \mathbf{f}_i(\mathbf{r}_i(t+\Delta t), \tilde{\mathbf{v}}_i(t+\Delta t)) \\
\mathbf{v}_i(t+\Delta t) &= \mathbf{v}_i(t) + \frac{1}{2}\Delta t(\mathbf{f}_i(t) + \mathbf{f}_i(t+\Delta t)),
\end{aligned} \tag{3.10}$$

where $\tilde{\mathbf{v}}_i(t+\Delta t)$ is the prediction of the velocity at time $t+\Delta t$ and λ is an empirically introduced parameter, which accounts for the effects of stochastic interactions. In this time integration algorithm, the velocity is first predicted to obtain the force and then corrected in the last step while the force is calculated only once during each integration step. It is found that for a velocity independent total force, the standard velocity Verlet algorithm can be recovered at $\lambda = 1/2$. Groot and Warren reported that when simulating an equilibrium system with $\rho = 3.0$ and $\sigma = 3.0$, the optimum value of λ is 0.65, which can lead to a considerable large time step to $\Delta t = 0.06$ without losing temperature balance (Groot and Warren, 1997).

Pagonabarraga et al. proposed a leap-frog scheme which is reported to be self-consistent and can recover the correct equilibrium properties but needs iteration at each time step (Pagonabarraga et al., 1998).

3.2.4 Stress tensor

After obtaining the positions, velocities and forces on all DPD particles, the stress tensor, **S**, is then calculated using the (Irving and Kirkwood, 1950) model expressed by

$$\mathbf{S} = -\frac{1}{V}\left[\sum_i m\mathbf{u}_i\mathbf{u}_i + \frac{1}{2}\sum_{i\neq j}\mathbf{r}_{ij}\mathbf{F}_{ij}\right], \tag{3.11}$$

where V is volume and it is the reciprocal of the number density (n) of particles, $\mathbf{u}_i = \mathbf{v}_i - \overline{\mathbf{v}}(\mathbf{r})$ is the peculiar velocity of particle i, $\overline{\mathbf{v}}(\mathbf{r})$ is the stream velocity at position $\mathbf{x}$. m is the mass of DPD particles, which is usually taken as unity. The first term in the brackets is the kinetic (ideal gas) contribution describing momentum transfer and the second term is the contribution from the particle-particle interactions (or inter-particle force). Just as expressed in equation (3.2), for simple DPD particles, the inter-particle force) is the summation of conservative, dissipative and random forces. For particles acting as a bead of molecular chains, the inter-particle force should include the total spring force on the particle.

The pressure, *p*, is obtained from the trace of the stress tensor

$$p = -\frac{1}{3}tr\mathbf{S}. \tag{3.12}$$

3.2.5 Determination of coefficients

The selection of coefficients in the DPD formulation directly influences the properties of the modeled DPD fluid (simulated properties). In order to match the simulated properties to the real properties and to maintain computational accuracy, parameters in DPD simulation need to be carefully chosen. Some coefficients can be determined by fitting the relevant data of the real fluid, some are selected to maintain the numerical accuracy in simulating simple cases with analytical solutions (e.g., Poiseuille flow). For complex system, just as pointed out by Fan et al., there is no solid physical basis to determine the coefficients characterizing interaction strengths between different components (Fan et al., 2006).

Coefficients of dissipative and random force: The coefficients of dissipative and random force (γ and σ) are co-related by fluctuation-dissipation theorem, as expressed in equation (3.7). Therefore there is only one independent coefficient, and also the coefficient is closely related to noise amplitude of system temperature. Groot and Warren ever tested the uniformly distributed random numbers and Gaussian distributed random numbers of the same variance and they found that there is no statistical difference between these two approaches (Groot and Warren, 1997). For temperature noise generated with uniformly distributed random numbers, increasing σ beyond 8 can lead to rapidly growing temperature and unstable simulation. Taking $\sigma = 3$ with suitable parameters in the time integration algorithm (e.g., for the modified version of the velocity-Verlet algorithm expressed in equation (3.10), $\lambda = 0.5$ and $\Delta t = 0.04$) is usually a recommended value to get a reasonable balance between fast temperature equilibration, a fast simulation and a stable, physically meaningful system.

Time step: It is found by Groot and Warren that, for the modified version of the velocity-Verlet algorithm expressed in equation (3.10), stable temperature control is obtained only when the term $\frac{1}{2}(\Delta t)^2 \mathbf{f}_i(t)$ is included in the position update (Groot and Warren, 1997). If this term is omitted, the simulation results are nearly as bad as the Euler algorithm. Empirically adjusting λ for a given system (with specific ρ and σ) can lead to a big time step without significant loss of temperature control. Groot and Warren reported that for a system with $\rho = 3$ and $\sigma = 3$ and an optimum value of $\lambda = 0.65$, the time step can be increased to $\Delta t = 0.06$ (Groot and Warren, 1997).

Repulsion parameter: The repulsion parameter (a) for the conservative force (see equation (3.3)) can be determined through matching the compressibility of the model fluid with real fluid. Groot and Warren found that for sufficiently high density ($\rho > 2$), a good approximation for pressure can be expressed as (Groot and Warren, 1997)

$$p = \rho k_B T + \alpha a \rho^2 , \tag{3.13}$$

where $\alpha = 0.101 \pm 0.001$.

As the compressibility for a fluid can be expressed as

$$\kappa^{-1} = \frac{1}{k_B T} \left(\frac{\partial p}{\partial \rho} \right)_T , \tag{3.14}$$

it can be further written as

$$\kappa^{-1} = 1 + \frac{2\alpha a\rho}{k_B T}. \tag{3.15}$$

As the known compressibility of water under room temperature is approximately 16, it is found that $a = 75k_B T/\rho$. Therefore for a given DPD system with specific temperature and density, the repulsion factor can be determined. For example, if $k_B T = 1$ and $\rho = 3$, the repulsion parameter (for DPD fluid mimicking the behavior of water) $a = 25$.

It should be noted that repulsion parameter a for particles from different fluids can be different. For example, for particle interactions from the same kind of fluid A or B, the repulsion parameter a_{AA} may or may not equal a_{BB}. Again for particle interactions from two different fluids A and B, a_{AB} (or a_{BA}, where $a_{BA} = a_{AB}$) may also be different from a_{AA} and a_{BB}, and in many cases, a_{AB} can be taken as $\sqrt{a_{AA}a_{BB}}$. The different repulsion parameter can lead to different behavior of two fluids as mixture or phase separation (Groot and Warren, 1997; Liu et al., 2006).

Also in DPD simulation, the interaction of fluid particles with particles from solid obstacles (solid particles) are necessary. However, there is no physical base on how the solid particles interact with each other, and interact with fluid particles. By taking a repulsion factor between solid particles (a_{ww} or a_w, where w means wall) different from that between fluid particles (a_{ff} or a_f, where f means wall), it is feasible to get different repulsion factor between fluid and solid particles, a_{wf}. The interaction behavior thus can be quite different. For example, when modeling two-phase flow in micro channels or fractures, it is found that gradually increasing the ratio of a_f to a_w from 0 can lead to different wetting behavior from strong non-wetting to moderate non-wetting, weak wetting, moderate wetting, strong wetting effects, and even film flows (Liu et al., 2007a).

3.2.6 Computational procedure

DPD method is a coarse-grained molecular dynamics method, and its computational implementation is also similar to that in the classic MD. **Figure 3.2** shows a typical computational procedure of a DPD simulation. As shown in **Figure 3.2**, there are basically sequential stages: initialization, force computation, time integration and data analysis.

1. **Initialization:** For the first run of a DPD simulation, it is necessary to initialize the coordinates of the DPD particles, their velocities and the target temperature ($2E_k/k_B N_{dof}$, where E_k is the kinetic energy, and N_{dof} is the total degree of freedom of the system) for the simulation.

Typically the DPD particles can be initially placed in a regular lattice space to give the desired density. They can also be injected into the computational domain according to a specific number density. The initial velocities are assigned with random directions and a fixed magnitude. It is preferred to initialize the velocity with the appropriate Maxwell-Boltzman distribution for the specified temperature. However, the usual rapid equilibration renders the careful fabrication of a Maxwell-Boltzman distribution unnecessary. Initialization of DPD particle velocities is subject to a number of conditions. For example, there is no overall momentum in any Cartesian direction, and the total kinetic energy is appropriate to the temperature specified.

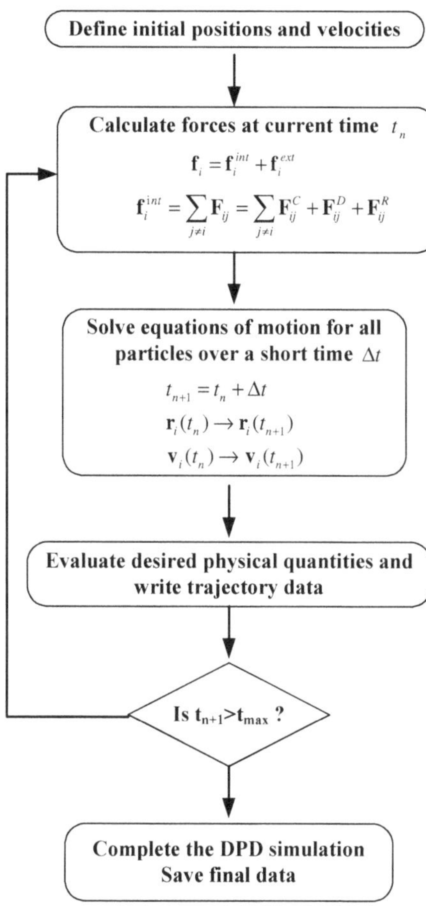

Figure 3.2 Computational procedure of a DPD simulation.

2. **Force computation:** In this stage, forces including the conservative force, dissipative force and random force are computed according to equations (3.1)-(3.5). External forces such as the gravitational force can also be computed according to the specific physics.
3. **Time integration:** After getting the forces, it is then possible to update the positions and velocities of all DPD particles according to a specific time integration algorithm.
4. **Data analysis:** In this stage, desired physical quantities such as stress can be evaluated, and the trajectory data is then saved.

3.3 Numerical aspects

3.3.1 Assessment of dynamic properties

Assume the radial pair distribution function, $g(r) \approx 1.0$, it is possible to derive the dynamic properties such as viscosity, diffusivity, and Schmidt number (Groot and Warren, 1997; Fan et al., 2006). For a dissipative particle system with weight function expressed in equation (3.8) for the dissipative and random force, the dissipative viscosity can be expressed as a function of s as follows

$$\eta^D = \frac{2\pi\gamma\rho^2 r_c^5}{15}\left(\frac{1}{s+1} - \frac{4}{s+2} + \frac{6}{s+3} - \frac{4}{s+4} + \frac{1}{s+5}\right). \tag{3.16}$$

It is noted that due to the soft interaction between DPD particles, the speed of momentum transfer is slow, and has the same order as the speed of particle diffusion. Therefore, the *Schmidt number* (*Sc*), defined as the rate of the speed of momentum transfer to the speed of particle diffusion, is about unity, much lower than $O(10^3)$ in a real fluid. For a typical DPD system, the dynamic viscosity is around 10^{-4} cP, which is also much lower than approximately 1 cP in real fluid. Therefore increasing the dynamic properties such as the Schmidt number and viscosity is usually necessary.

Figure 3.3 shows the influence of s on the dissipative viscosity. It is clear that reducing s can lead to considerably increasing viscosity. **Table 3.1** shows the dynamic properties for a DPD system with s = 1/2, s = 1.0 and s = 2.0. It is found that different s can lead to different dynamic properties. For example, for a given DPD system, the dynamic viscosity obtained with s = 1/2 is around eight times the dynamic viscosity obtained with s = 2.0, and the Schmidt number is increased around 35.5 times when reducing s from 2.0 to 1/2. Therefore reducing the exponential factor s is an effective way to improve dynamic properties of the system with the same computational requirement.

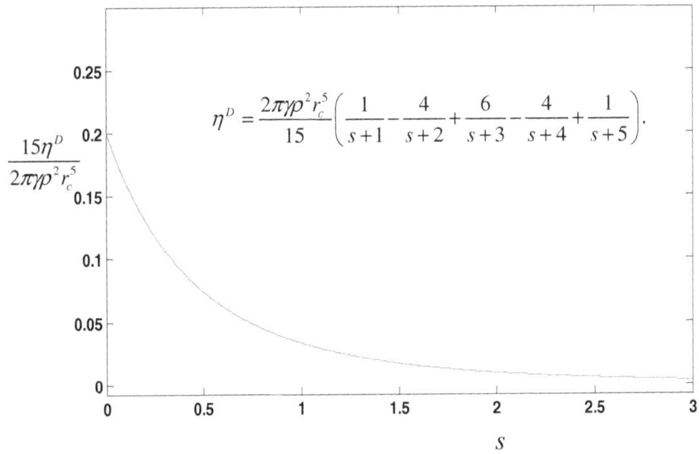

Figure 3.3 Viscosity as a function of s.

Table 3.1 Dynamic properties for DPD systems

Formulation	Conventional ($s = 2$)	Modified ($s = 1$)	Modified ($s = ½$)
Diffusivity, D	$\frac{45k_BT}{2\pi\gamma\rho r_c^3}$	$\frac{9k_BT}{\pi\gamma\rho r_c^3}$	$\frac{315k_BT}{64\pi\gamma\rho r_c^3}$
Viscosity, η	$\frac{\rho D}{2} + \frac{2\pi\gamma\rho^2 r_c^5}{1575}$	$\frac{\rho D}{2} + \frac{\pi\gamma\rho^2 r_c^5}{225}$	$\frac{\rho D}{2} + \frac{512\pi\gamma\rho^2 r_c^5}{51975}$
Schmidt number, Sc	$\frac{1}{2} + \frac{(2\pi\gamma\rho r_c^4)^2}{70875k_BT}$	$\frac{1}{2} + \frac{(\pi\gamma\rho r_c^4)^2}{2025k_BT}$	$\frac{1}{2} + \frac{(2\pi\gamma\rho r_c^4)^2}{1999k_BT}$

Another approach to modify the dynamic properties of a DPD system is to change r_c (cutoff distance for the dissipative, as expressed in equation (3.8)) and γ (strength coefficient for the dissipative force as expressed in equation (3.4)), as the dynamic properties is dependent on r_c and γ. Increasing γ can result in larger fluctuation of thermal energy and requires good control of system temperature. Increasing r_c is thus the most effective and easiest way to reduce the diffusivity and increase the dynamic viscosity and Schmidt number of the DPD system. However, increasing r_c means enlarged computational cost. Therefore combining the modified weighting function and moderately increasing the cutoff radius for dissipative weighting function can enhance the dynamic viscosity and Schmidt number with reasonable computation costs. For example, for a DPD system with $\gamma = 4.5$, $\rho = 4.0$ and $k_B T = 1.0$, the influence of r_c on the viscosity and Schmidt number for $s = 0.5$, 1 and 2 are shown on **Figure 3.4** and **Figure 3.5**. It is clear that increasing r_c can produce larger dynamic viscosity and Schmidt number. When $s = 0.5$ and $r_c = 1.88$, *Sc* can reach about 1000, which is of the same order as the Schmidt number of real fluid. In MD-like simulations, $r_c = 2.0 \sim 2.5$ is found to be satisfactory (Fan et al., 2006).

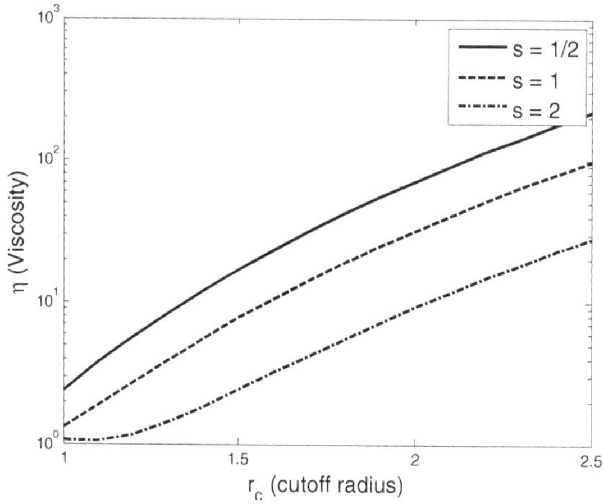

Figure 3.4 Viscosity as a function of r_c for different s.

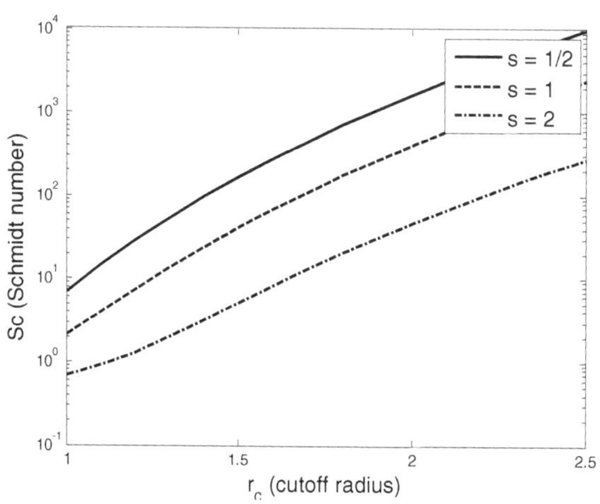

Figure 3.5 Schmidt number as a function of r_c for different s.

3.3.2 Solid boundary treatment

Just as in other CFD problems, solid boundary treatment is very important in DPD. To model the interaction between fluids and solid walls, both fluids and solid walls can be represented by DPD particles, which can be referred to as fluid particles and solid particles respectively. In DPD, a good solid boundary treatment algorithm should satisfy three requirements,

1) the fluid particles should not penetrate the solid walls unphysically,
2) there should not be large oscillation of physical variables in the boundary area, and
3) slip or no-slip boundary condition should be well implemented, either for fixed solid wall or moving solid obstacles.

3.3.2.1 Reflection

During the simulation, some of the mobile particles that are used to represent the fluid(s) may penetrate into the wall particles because of the soft interaction between the DPD particles. In order to avoid such unrealistic penetration, one possible solution is to use a higher particle density for the walls or a larger repulsive force between the wall particles and fluid particles. This may cause large density oscillation in the boundary area.

Another frequently used approach in preventing unphysical penetration is based on reflection, in addition to the interactions between fluid and wall particles. Revenga et al. (Revenga et al., 1998) investigated three different reflection models:

a) *specular reflection* in which only the normal velocity component is reversed and the tangential velocity keeps unchanged (and therefore leading to free slip condition),
b) *bounce-back reflection* in which all velocity components are reversed (same magnitude and opposite direction, and therefore leading to no-slip condition) and
c) *Maxwellian reflection* in which particles are reflected back into the system according to Maxwell distribution.

It is noted that when implementing the Maxwellian reflection, the velocities of particles that enter a thin layer next to the wall are selected randomly from the Maxwell distribution at temperature T (thermal condition), with a zero mean corresponding to the zero fluid velocity at the boundary (no-slip condition). The velocity components can be reversed if the velocity points outward from the bulk fluid (Fan et al., 2006).

The treatment of solid boundaries by using frozen boundary particles and a thin reflecting boundary layer was found to be an effective way of implementing no-slip boundary conditions (Fan et al., 2003; Fan et al., 2006). The thickness of the thin layer is selected to ensure that the probability of penetration is very low but the reflective layer occupies as little as possible of the fluid domain. In general, a thickness of 0.1 DPD unit is preferable for most applications. This thickness is small compared with the size of the fluid domain so it does not affect the bulk flow and it allows the fluid and wall particles to interact strongly enough to control the wetting behavior. On the other hand, it is large enough to prevent unphysical penetration. The implementation of no-slip boundary conditions with frozen wall particles and a thin boundary layer was found to be very flexible, especially for problems with complex geometries such as flow through porous media (Liu et al., 2007b).

3.3.2.2 Representation of solid grains

In DPD simulations, the effects of solid walls are usually simulated by using fixed particles to represent the solid matrix near to the solid-fluid interface. In the implementation, the entire computational domain can be discretized using a 'shadow' grid and grid cells are labeled "0" for regions occupied by pore spaces and "1" for solid filled regions (**Figure 3.6**a). This simple identification of fluid and solid cells can be used to represent any arbitrary pore geometries including those determined from high-resolution x-ray and NMR tomography. The unit vectors normal to the solid-fluid interfaces, which define the local orientation of the interface, can be obtained by simply calculating the surface gradient from the indicator numbers ("0" for liquid regions and "1" for solid regions). At the beginning of each DPD simulation, the particles are initialized and positioned randomly within the entire computational domain until a pre-defined particle number density is reached, and the system is then run to equilibrium using a DPD simulation with repulsive particle-particle interactions. The particles within

(a) (b) (c)

Figure 3.6 Illustration of the treatment of solid obstacles. (a) The cells in the entire computational domain are first labeled, "0" for fluid (void) cells and "1" for solid (obstacle) cells, (b) After equilibration, the DPD particles in the obstacle cells are frozen. (c) Only the frozen particles that are close to the fluid cells (within 1 DPD unit) are retained as boundary DPD particles (Liu et al., 2007b).

the solid cells (marked as "1") are then 'frozen' (their positions are fixed) to represent the solid grains (**Figure 3.6**b). The solid grains in porous media can occupy a considerable fraction of the entire computational domain, and hence the number of frozen particles representing the solids can be very large, particularly for low porosity media. Most of the frozen particles inside the solid grains are more than one DPD unit away from the adjacent fluid cells. These particles do not contribute to the solid-fluid interactions and consequently they have no influence on the movement of the mobile DPD particles within the fluid cells. Therefore, only the frozen particles that are within one DPD unit (or r_c) from the solid-fluid interface are retained as boundary DPD particles (**Figure 3.6**c), and the rest of the particles further inside the solid grains are removed from the model domain. **Figure 3.7** and **Figure 3.8** respectively show the representation of solid grains in a porous media and fracture network with

Figure 3.7 Representation of solid grains in a porous media with frozen DPD particles within one DPD unit away from the adjacent fluid cells.

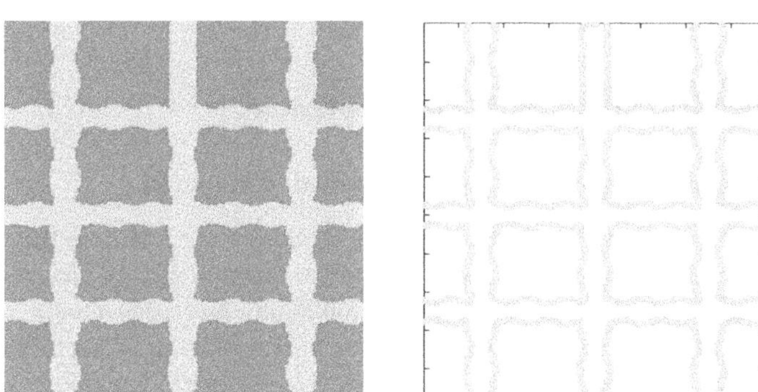

Figure 3.8 Representation of solid grains in a fracture network with frozen DPD particles within one DPD unit away from the adjacent fluid cells.

frozen DPD particles within one DPD unit away from the adjacent fluid cells. It is clear that this treatment of solid grains is convenient to implement and suitable for arbitrary complex geometries.

3.3.2.3 Implementing solid boundary condition

By using the above approach in representing solid grains and a suitable reflective model within a thin reflective boundary layer (see **Figure 3.9**), it is possible to implement solid boundary conditions, either non-slip or slip. It is noted that this treatment of solid boundaries with frozen DPD particles within one DPD unit away from the adjacent fluid cells, and a thin reflective boundary layer in the fluid domain is effective in modeling complex solid obstacles, either fixed or movable (Liu et al., 2007b).

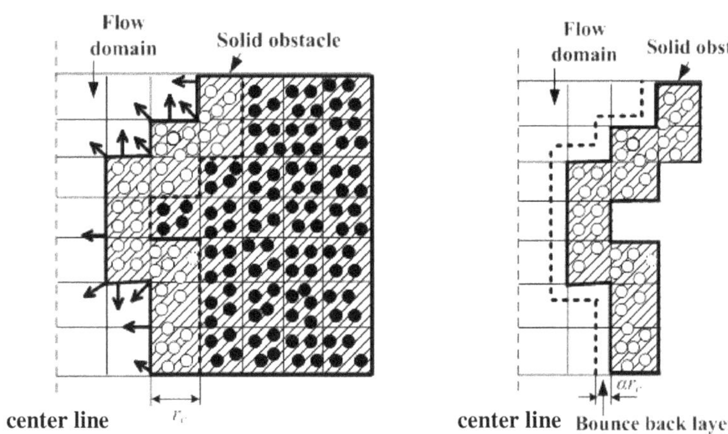

Figure 3.9 Illustration of implementing solid boundary condition.

3.3.3 Conservative interaction potential

3.3.3.1 Constructing conservative interaction potential

In conventional DPD implementations, a conservative force weight function in a simple form $w^C(r) = 1 - r$ with a cutoff distance of r_c (=1.0) has been used. Because the fluid generated by DPD simulations with this purely repulsive conservative force is a gas, it cannot be used to simulate the flow of liquids with free surfaces, the behavior of bubbly liquids, droplet dynamics and other important multiphase fluid flow processes. A direct solution of this problem is to include a long-range attractive component in $w^C(r)$. Like the repulsive component, the attractive component should also be a soft interaction to retain the advantages of the DPD method. And at short particle separations, the repulsive component should be strong enough, relative to the attractive component to prevent the particle density from becoming too high. Moreover, the magnitude of the conservative force weight function and the location of the transition point from repulsion to attraction should be easily adjustable to allow the behavior of different fluids to be simulated.

Based on such considerations, Warren developed a *many-body DPD* (MDPD) for modeling vapor-liquid co-existing problems (Warren, 2003). In MDPD, the conservative force can be expressed as

$$\mathbf{F}_{ij}^{C} = a_{ij}^{A} w^{A}(r)\hat{\mathbf{r}}_{ij} + b_{ij}^{R} w^{R}(r)\hat{\mathbf{r}}_{ij} \,, \tag{3.17}$$

where the first term in equation (3.17) is the attractive force between particles *i* and *j*, and the second term is the repulsive force between particles *i* and *j*. $w^{A}(r)$ and $w^{R}(r)$ stand for conservative weight functions with different cutoff distance r_A and r_R for the attractive and repulsive force between interacting DPD particles. a_{ij}^{A} and a_{ij}^{R} are the corresponding strength coefficients for attraction and repulsion.

It is possible to construct polynomials that include both short-range repulsion and long-range attraction with a single cutoff distance (Liu et al., 2003). Another approach is to combine commonly used SPH smoothing functions with different interaction strengths and cutoff distances to construct a particle-particle interaction potential. The most commonly used smoothing function in SPH is the cubic spline (Liu et al., 2003),

$$W(r) = W(r, r_c) = \begin{cases} 1 - \dfrac{3}{2}\left(\dfrac{2r}{r_c}\right)^2 + \dfrac{3}{4}\left(\dfrac{2r}{r_c}\right)^3 & 0 \le \dfrac{2r}{r_c} < 1 \\ \dfrac{1}{4}\left(2 - \left(\dfrac{2r}{r_c}\right)\right)^3 & 1 \le \dfrac{2r}{r_c} < 2 \,, \\ 0 & \dfrac{2r}{r_c} \ge 2 \end{cases} \tag{3.18}$$

where r_c is the cutoff distance (corresponding to the smoothing length, h, in SPH) of the smoothing function. For the cubic spline function, $r_c = 2h$. In SPH, the function $W(r)$ in equation (3.18) is multiplied by a coefficient, *C*, so that the normalization requirement $\int W(|\mathrm{r}|)d\mathrm{r} = 1$ is satisfied. The normalization coefficient, *C*, has values of $2/(3h)$, $10/(7\pi h^2)$ and $1/(\pi h^3)$ in the one-, two- and three-dimensional spaces. The cubic spline function defined in equation

(3.18) is a non-negative, monotonically decreasing function, and it is smooth at both the origin and the cutoff.

One way of obtaining particle-particle interactions with the required short-range repulsive and long-range attractive form is to use a sum of spline functions multiplied by an interaction strength coefficient a,

$$U(r)=a\left(AW_1(r)-BW_2(r)\right)=a\left(AW_1(r,r_{c1})-BW_2(r,r_{c2})\right), \tag{3.19}$$

to define the particle-particle interaction potentials, where $W_1(r)$ is a cubic spline with a cutoff length of r_{c1}, A is the coefficient for $W_1(r)$, $W_2(r)$ is a cubic spline with a cutoff length of r_{c2} and B is the coefficient for $W_2(r)$. $W_1(r)$ and $W_2(r)$ are non-normalized shape functions given in equation (3.18). The DPD conservative particle-particle interaction forces are given by

$$\mathbf{F}_{ij}^C=\frac{-dU(r)}{dr}\hat{\mathbf{r}}_{ij}. \tag{3.20}$$

In DPD simulations, all particle-particle interaction potentials can have the same shapes with different interaction strengths for different particle-particle interactions. For example for fluid 1, the particle-particle potentials are defined by

$$U_{11}(r)=a_{11}\left(AW_1(r,r_{c1})-BW_2(r,r_{c2})\right). \tag{3.21}$$

If a second fluid component is present, then the particle-particle interactions for that fluid are given by

$$U_{22}(r)=a_{22}\left(AW_1(r,r_{c1})-BW_2(r,r_{c2})\right), \tag{3.22}$$

and the interactions between pairs of particles representing different fluids are given by

$$U_{12}(r)=U_{21}(r)=a_{12}\left(AW_1(r,r_{c1})-BW_2(r,r_{c2})\right). \tag{3.23}$$

In equations (3.21)-(3.23), a_{ij} is the interaction strength between two particles representing component i and component j respectively.

A variety of functions can be obtained by using different combinations of A, r_{c1}, B, r_{c2}. For example, taking $A=2.0, r_{c1}=0.8$ and $B=1.0, r_{c2}=1.0$, and $a=1$ (see equation (3.19)), a function with positive and negative components is obtained. The positive and negative components, $AW_1(r)$, $-BW_2(r)$, and the resulting function $U(r)=AW_1(r)-BW_2(r)$ is shown in **Figure 3.10**. It shows

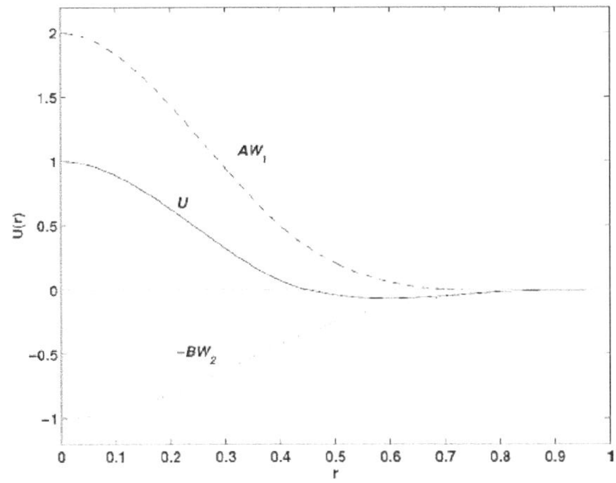

Figure 3.10 Construction of a particle-particle interaction potential, $U(r)$, that is repulsive at short distances, attractive at intermediate distances and zero at large particle separation, from two cubic spline functions, $AW_1(r)$ and $BW_2(r)$ (Liu et al., 2006).

that the function $U(r)$ is positive at the origin, gradually decreases, and then becomes negative at $r = 0.4529$. After reaching a minimum, $U(r)$ begins to increase until $U(r) = 0$ at $r = 1.0$. $U(r)$ is smooth at the origin and at the point $r = 1.0$. If $A = 1.0, r_{c1} = 1.0$ and $B = 0.0$, the resulting function $U(r)$ is the cubic spline expressed in equations (3.18), which is non-negative everywhere (see **Figure 3.11**).

The spline function $W(r)$ describes a purely repulsive interaction and its negative counterpart $-W(r)$ describes a purely attractive interaction. The parameters A and B can be regarded as the strengths of the repulsive and attractive interactions. Different interaction strengths with corresponding cutoff distances generate different potential functions, $U(r)$, and corresponding weight functions, w^C $(= -U'(r))$, which can be used to simulate different phenomena. The two SPH cubic spline potential functions $U(r) = 2W_1(r, 0.8) - W_2(r, 1.0)$ (obtained by using $A = 2.0$, $r_{c1} = 0.8$, $B = 1.0$ and $r_{c2} = 1.0$, and $a = 1.0$) and $U(r) = W_1(r, 1.0)$ (obtained by using $A = 1.0, r_{c1} = 1.0, B = 0.0$ and $a = 1.0$) as well as the conventional potential function $0.5 - (r - 0.5r^2)$ (corresponding to the conventional weight function $1 - r$) are shown in **Figure 3.11**. The corresponding conservative force weight functions (or shape functions) are shown in **Figure 3.12**.

Figure 3.12 shows that the conventional DPD conservative force weight function is non-negative and describes a purely repulsive interaction. Similarly the weight function obtained using $A = 1.0, r_{c1} = 1.0, B = 0.0$ is also a purely repulsive non-negative function. While the weight function resulting from using $A = 2.0, r_{c1} = 0.8, B = 1.0$ and $r_{c2} = 1.0$ is a function with positive and negative sections, which corresponds to an interaction with short-range repulsive and long-range attractive characteristics. The conventional DPD weight function is a monotonically decreasing function of the inter-particle separation with a constant negative (repulsive) slope whereas the new weight functions have regions with both positive (attractive) and negative slopes.

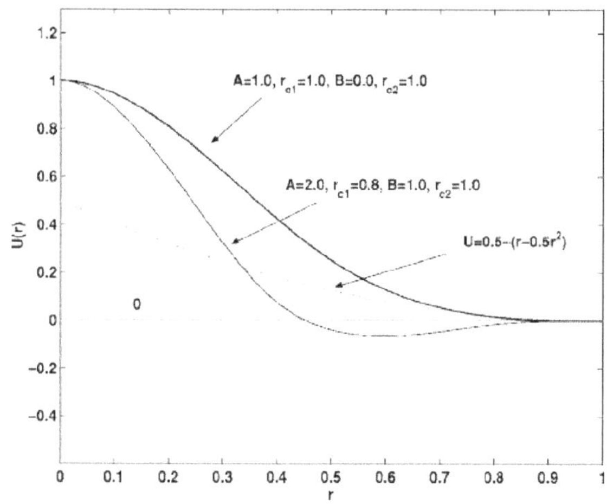

Figure 3.11 Cubic spline potential functions, $U(r)=W_1(r,1.0)$, $U(r)=2W_1(r,0.8)-W_2(r,1.0)$ and the conventional DPD potential function, $U(r)=0.5-(r-0.5r^2)$ (Liu et al., 2006).

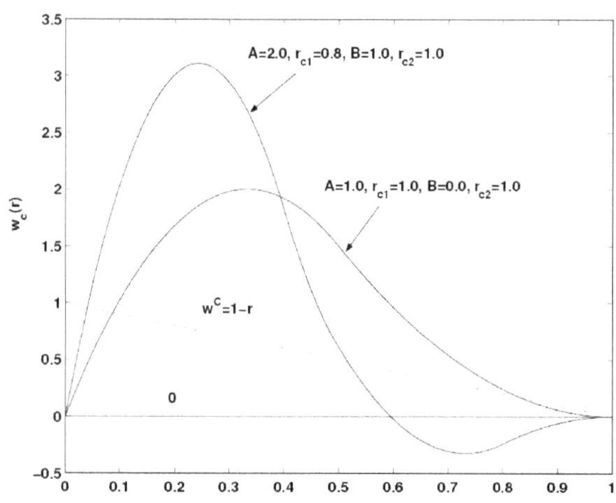

Figure 3.12 Cubic spline conservative force weight functions and the conventional DPD conservative force weight function (Liu et al., 2006).

Comparing equation (3.17) and equation (3.19), it is found that Liu's approach is actually is equivalent to MDPD. In both approaches, the conservative force is divided into two components, namely attraction and repulsion. The attractive force and repulsive force between two interacting particles are associated with different cutoff distances and different strength coefficients for modeling the properties of different fluids. Different from MDPD, in which the basic form of conservative weight function is the conventional form ($w^C(r)=1-r$), Liu's modified DPD approaches can use different polynomials (including the linear polynomial $1-r$) to construct the interaction potential (or weight function).

3.3.3.2 Pressure-density relation

The combination of the attractive and repulsive interactions in the cubic spline potential makes it possible to simulate systems with co-existing liquid and gas phases and liquid-gas phase transitions. For a DPD system with attractive and repulsive interactions, the pressure-density relation can be numerically calculated. The fluid pressure can be calculated as a function of density from the particle-particle interactions using the virial theorem to obtain a numerical equation of state (Allen and Tildesley, 1987; Rapaport, 2004). Because the random and dissipative forces have average values of zero, they do not contribute to the virial pressure (Groot and Warren, 1997), and the total pressure is given by

$$P = P_k + \frac{\rho}{3}\sum_{j<i}(\mathbf{r}_i - \mathbf{r}_j)\cdot \mathbf{F}_{ij}^C, \tag{3.24}$$

where P_k is the kinetic contribution ($P_k = \rho k_B T$, where ρ is the fluid density).

The van der Waals (vdW) equation of state can also be used to model co-existing liquid and gas phases and liquid-gas phase transitions. The formulation of the van der Waals equation was motivated by the idea that short range repulsive forces lead to an effective volume for the gas molecules, which reduces the average free volume per molecule from v to $v-\bar{b}$ and long range attractive forces reduces the pressure from $k_BT/(v-\bar{b})$ to $k_BT/(v-\bar{b})-\bar{a}/v^2$. The resulting equation, $(p+\bar{a}/v^2)(v-\bar{b})=k_BT$, provides a quantitative model for the phase behavior of simple fluids. In particular, for van der Waals fluids (model fluids described by the van der Waals equation) gas and liquid phases may coexist in a non-zero region of the (p,v,T) or (p,ρ,T) parameter space (depending on the coefficients $\bar{a}$ and $\bar{b}$) where ρ is the average fluid density. The van der Waals fluid is the classic example of a fluid with co-existing liquid and gas phases and liquid-gas phase transitions. The equation of state for a van der Waals fluid can be expressed in the form

$$p = \frac{\rho k_B T}{1 - \rho \bar{b}} - \bar{a}\rho^2, \tag{3.25}$$

where $\bar{a}$ controls the strength of the attractive force, and $\bar{b}$ is related to the size of the particle. This equation of state can be obtained from the macroscopic free energy density for interacting particles with short range repulsive interactions and long range attractive interactions in the mean field (infinite interaction range) limit (Lebowitz and Penrose, 1966). Giving $\bar{a}$ and $\bar{b}$, it is easy to plot the pressure-density relation for a constant temperature. **Figure 3.13** shows the pressure-density relations for a van der Waals fluid with $\bar{b} = 0.016$ and $\bar{a} = 1.9b$ while the temperatures are $k_B T = 1$ and $k_B T = 0.54$ respectively.

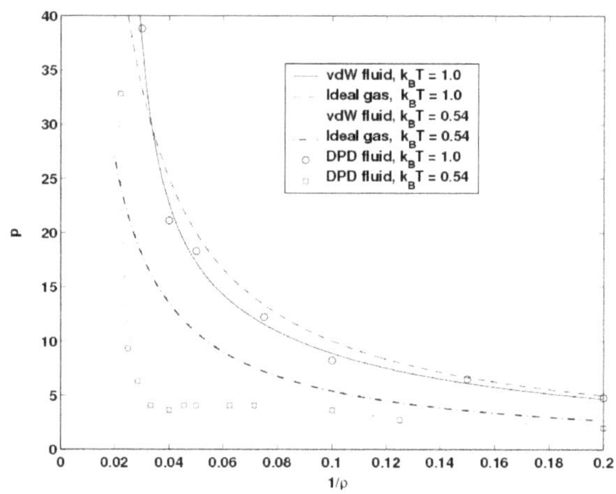

Figure 3.13 Pressure-density relations for a van der Waals fluid with $\bar{b} = 0.016$, $\bar{a} = 1.9b$, and a DPD fluid with $A = 2.0$, $r_{c1} = 0.8$, $B = 0.95$, $r_{c2} = 1.0$. The temperatures are $k_B T = 1$, and $k_B T = 0.54$ respectively (Liu et al., 2006).

The pressure-density-temperature relationship for a wide range of fluids can be represented quite well by a van der Waals equation of state over a limited part of the parameter space. By tuning the parameters $\bar{a}$ and $\bar{b}$, it is possible to obtain a van der Waals equation of state for DPD fluids. **Figure 3.13** shows the pressures calculated at a number of densities for a DPD fluid with $A = 2.0$, $r_{c1} = 0.8$, $B = 0.95$, $r_{c2} = 1.0$, and for $k_B T = 1$ and $k_B T = 0.54$. The simulations were implemented by placing different number of DPD particles into a box of size $10 \times 10 \times 10$ with periodic boundary conditions in all three directions to model the effects of different global densities, $\rho = n/1000$. The averaged total pressure was calculated using the virial theorem relationship given in equation (3.24). The pressure calculated for the DPD fluid at different densities can be represented well by the van der Waals equation, as **Figure 3.13** shows.

3.3.4 Spring-bead chain models

In the DPD model, a macromolecule chain can be represented by a chain of particles (beads) connected by springs in DPD system. Similar to fluid particles (for modeling simple fluids) that can be thought of as a small regions of fluid, the macromolecule beads can be thought of as polymeric chain segments consisting of number of monomeric units. The macromolecule beads exchange momentum with each other according to the spring force and other ordinary DPD interactions. Hydrodynamic and thermodynamic interactions between the macromolecule and solvent then emerge naturally in these simulations. Numerous simulations have verified that the DPD model can capture many essential physical phenomena of the macromolecule systems.

A number of spring-bead chain models have been used in polymer rheology as the coarse-grained models of macromolecules. Typical of them are the *worm-like chain* (WLC) model and *finitely extensible nonlinear elastic* (FENE) model. In the WLC model, the spring force law of a worm-like chain segment can be expressed as

$$\mathbf{F}_{ij}^{S} = -\frac{k_B T}{4\lambda_p^{eff}}\left[\left(1-\frac{r_{ij}}{l}\right)^{-2} + \frac{4r_{ij}}{l} - 1\right]\hat{\mathbf{r}}_{ij}, \tag{3.26}$$

where l is the maximum length of one chain segment and λ_p^{eff} is the effective persistence length of the chain. If the total length of the chain is L and the number of bead in the chain is N_b, $l = L/(N_b - 1)$. It was found that the mechanical properties of DNA molecules in an aqueous solution can realistically be modeled by the worm-like chains (Vologodskii, 1994; Larson et al., 1999).

The spring force law of a FENE chain segment can be expressed as the following equation

$$\mathbf{F}_{ij}^{S} = -\frac{H\mathbf{r}_{ij}}{1-\left(r_{ij}/r_m\right)^2}, \tag{3.27}$$

where H is the spring constant. r_m is the maximum length of one FENE chain segment. From equation (3.27), we can see that the spring force increases intensely and approaches infinity when r_{ij}/r_m approaches 1. As a result, the distance between two neighboring beads in FENE chain should be less than r_m.

3.4 Validation of the DPD method

The effectiveness of DPD in modeling complex physics and reproducing the continuum hydrodynamic behavior has been demonstrated in various applications (Espanol and Warren, 1995; Warren, 1998; Fan et al., 2003). Here we provide three examples including 1) binary mixture, 2) Poiseuille flow and 3) fully saturated flow through porous media.

3.4.1 Binary mixture

This example is similar to that used by Groot and Warren (Groot and Warren, 1997) in their investigation of phase separation in binary mixtures and polymers. In their work, they proposed a relationship between the purely repulsive particle-particle interactions used in their DPD model and the Flory-Huggins interaction parameter, χ, which is associated with the interaction energy between the two components of the binary mixture, 1 and 2. In the lattice-based Flory-Huggins model (Flory, 1942), the contribution of these interactions to the free energy per lattice site, F_{int}, is given by $F_{\text{int}} = k_B T \chi \phi_1 \phi_2$, where ϕ_1 and ϕ_2 are the volume fractions of components 1 and 2.

In the simulations, the size of the system was $8 \times 8 \times 20$. 3840 DPD particles, half of component 1 and half of component 2, were randomly injected into the box resulting in an average number density of 3. Depending on the inter-particle interactions, and the temperature, the system may remain mixed or undergo phase separation.

This binary mixture system was simulated using both the conventional weight function ($w^C = 1 - r$) and a cubic spline weight function with the parameters $A = 1.0, r_{c1} = 1.0$ and $B = 0.0$. The interaction strength coefficients were set to $a_{11} = a_{22} = 25.0$ and $a_{12} = 37.5$ respectively. Therefore, the particle-particle interaction potential function for fluid 1 was given by

$$U_{11}(r) = a_{11}\left(AW_1(r) - BW_2(r)\right) = 25.0W_1(r, 1.0)\,. \tag{3.28}$$

Similarly, the particle-particle interaction potential between particles representing fluid 2 was $U_{22}(r) = 25.0W_1(r, 1.0)$ and the particle-particle interaction between particles representing fluid 1 and particles representing fluid 2 was $U_{12}(r) = 37.5W_1(r, 1.0)$. Because the repulsive interactions between unlike particles are substantially larger than the repulsive interactions between like particles, the fluid is separated into phases that are rich in either component 1 or component 2.

The weight functions for the dissipative force and random force are defined as in the conventional DPD method and the parameters $\sigma = 1.5$ and $k_B T = 1.0$ ($\gamma = 1.125$). The leap-frog time integration algorithm was used for time evolution with a time step of 0.03.

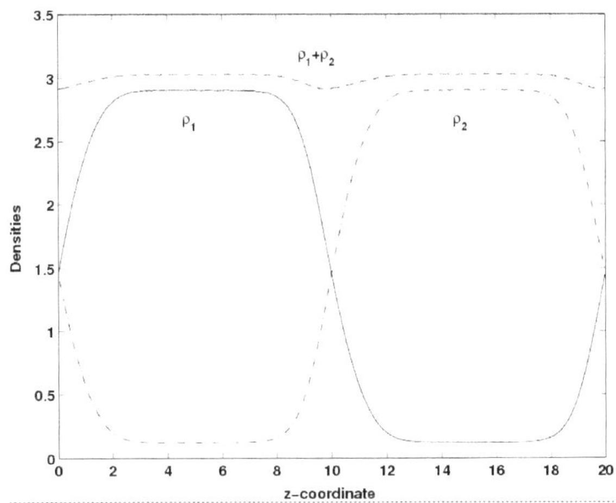

Figure 3.14 Density profiles obtained using the conventional weight function $w^C = 1 - r$ (Liu et al., 2006).

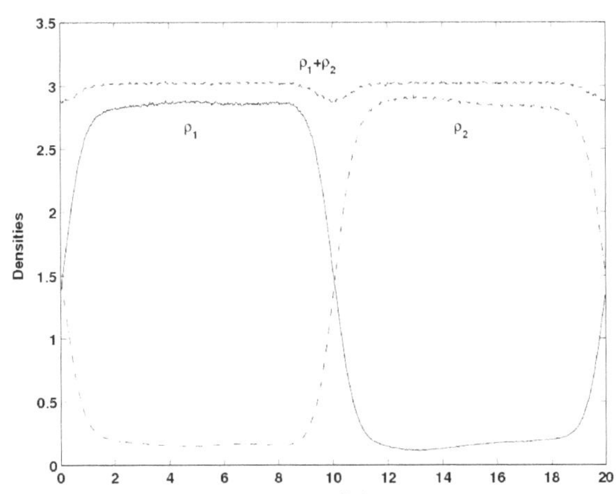

Figure 3.15 Density profiles obtained using the cubic spline weight function with the parameters $A = 1.0$, $r_{c1} = 1.0$ and $B = 0.0$ (Liu et al., 2006).

After the system had equilibrated (over around 10^5 time steps), the density profiles of type 1 and type 2 particles were sampled across the fluid-fluid interface and averaged over 10^5 time steps along the z direction. The averaged density profiles using the DPD simulations with both the conventional DPD conservative interaction and the cubic spline interaction potential are shown in **Figure 3.14** and **Figure 3.15**. A comparison of **Figure 3.14** and **Figure 3.15** shows that both simulations lead to the formation of narrow inter-phase regions with very similar density profiles. There is a small dip in the sum of densities $\rho_1 + \rho_2$ at the interface due to the relatively larger repulsive interactions between particles of different types. If the particle-particle interactions are purely repulsive, then this dip will increase in magnitude and become unrealistically large if the difference between the like and unlike particle-particle interactions becomes too large.

The radial distribution functions (RDF) $g(r)$ for bulk fluid calculated from simulations using the conventional DPD and DPD with cubic spline potential functions are shown in **Figure 3.16**. The radial distribution functions were obtained from binary mixtures rather than from single component systems. Hence, irrespective of the particle type, all particles are taken into account when the radial distribution functions were calculated. The different shapes of the radial distribution functions are a reflection of the differences between the inter-particle interactions (Landau and Lifshitz, 1980). The temporal fluctuations in the temperature calculated from the kinetic energy using the conventional DPD and DPD with cubic spline potential functions are shown in **Figure 3.17**.

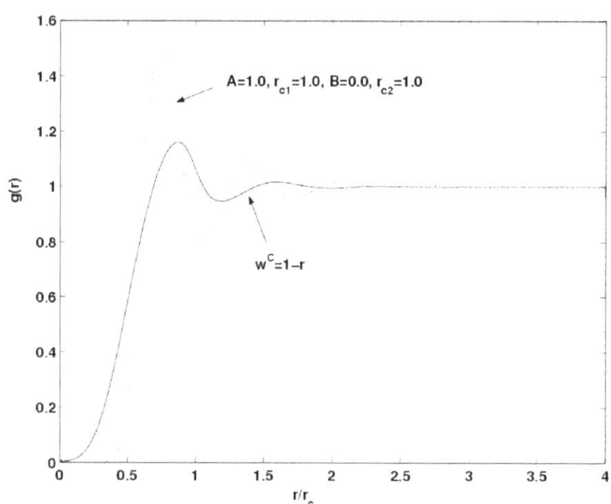

Figure 3.16 Radial distribution functions (RDF) for the conventional weight function $w^C = 1 - r$ and the cubic spline weight function with the parameters $A = 1.0$, $r_{c1} = 1.0$ and $B = 0.0$ (Liu et al., 2006).

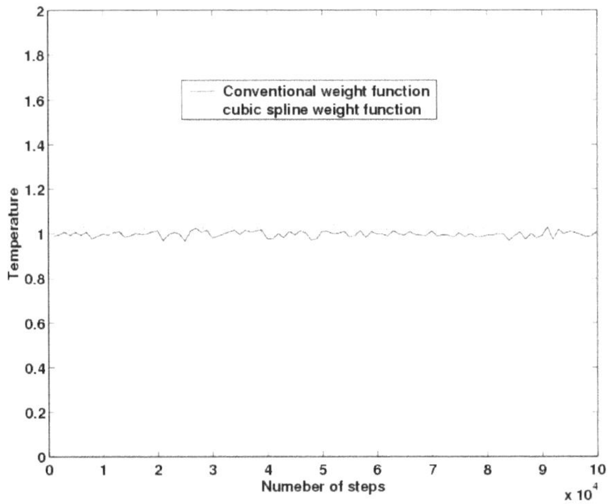

Figure 3.17 Kinetic temperature fluctuations obtained using the conventional weight function $w^C = 1 - r$ and the cubic spline weight function with the parameters $A = 1.0$, $r_{c1} = 1.0$ and $B = 0.0$ (Liu et al., 2006).

Both simulations maintained an approximately constant temperature with small oscillations about the fluctuation-dissipation temperature. If a DPD simulation results in a kinetic temperature (calculated from $T = \frac{1}{k_B d N_p} \sum_i \mathbf{v}_i^2$, where d is the number of dimensions and N_p is the number of fluid particles) that differs significantly from the nominal temperature, the simulation results will not be reliable. Comparison of these two temperatures provides a useful way of monitoring the accuracy of the simulations, and significant deviations indicate a problem with the numerical integration — such as a time step that is too large.

3.4.2 Poiseuille flow

DPD, as a mesoscopic method, can be used for studying the hydrodynamic behaviour of both simple and complex fluids. A DPD model should obey the Navier-Stokes (N-S) equation, at least when the time step in the time integration scheme tends to be zero (Espanol and Warren, 1995) and the system size is large enough for hydrodynamic (continuum) concepts and models to be valid. There were reports of DPD simulation of Poiseuille flow with conventional conservative interaction potential (Fan et al., 2003). Here the Poiseuille flow was simulated to evaluate how effectively DPD simulations with cubic spline interaction potentials reproduce hydrodynamic behaviours.

11520 particles of the same type were randomly injected into a system of size $40\times3\times24$, resulting in an average number density of 4.0. After equilibrium was reached, the particles located in the top and bottom regions within 2 DPD units of the boundary of the computational domain were fixed (2008 particles), while the rest of the particles (9512 particles) remained mobile in a domain of size $40\times3\times20$. The frozen particles were used to represent the channel walls and implement non-slip boundary conditions in the z direction. Moreover, we used a reflective boundary in addition to the interactions between fluid and wall particles. In our implementation, the velocities of particles that enter a thin layer next to the wall are selected randomly from the Maxwellian distribution at temperature T, with a zero mean corresponding to the zero fluid velocity at the boundary. The velocity components were reversed if the velocity points outward from the bulk fluid. This treatment of solid boundaries, using frozen boundary particles and a thin reflecting boundary layer, was found to be effective in yielding no slip boundary condition and obtaining good results.

Periodic boundary condition were used in x and y directions. The coefficients in the DPD model were assigned values of $a=18.75$, $\sigma=3.0$ and $k_BT=1.0$ ($\gamma=4.5$). The modified version of the velocity-Verlet algorithm was used to integrate the equation of motion of DPD particles with $\lambda=0.65$, and a time step of $\Delta t=0.02$. To drive the flow, a gravity force of $g=0.02$ (an acceleration of 0.02 for the particles of unit mass) was applied along the x direction, after the system reached equilibration. The parameters associated with the cubic spline interaction potential were $A=1.0, r_{c1}=1.0$ and $B=0.0$ ($U(r)=18.75W_1(r,1.0)$).

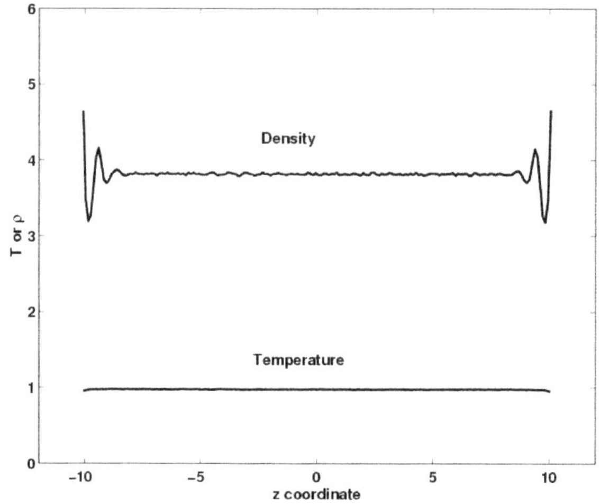

Figure 3.18 The density and temperature profiles along the z direction. The cubic spline potential is $U(r)=18.75W_1(r,1.0)$ (Liu et al., 2006).

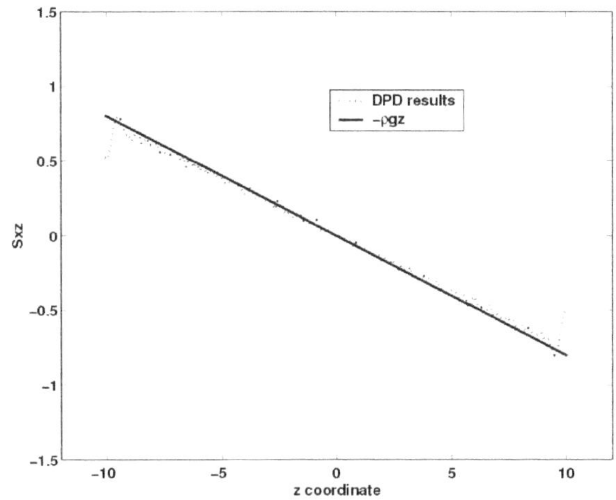

Figure 3.19 The shear stress (S_{xz}) distribution in the Poiseuille flow along the z direction. The cubic spline potential is $U(r)=18.75W_1(r,1.0)$ (Liu et al., 2006).

Figure 3.18 shows the density and temperature profiles along the z coordinate. The density is essentially uniform across the channel except in the boundary region near to the solid walls. Similarly, the temperature across the channel is also uniform, and remains almost equal to the initially specified temperature. **Figure 3.19** shows the profile of the xz component, S_{xz}, of the stress tensor. The DPD simulation results for S_{xz} agree well with the analytical solution $S_{xz}=-\rho gz$.

Figure 3.20 shows the steady state velocity profiles (velocity component in the x direction) across the channel. For a Newtonian fluid, the velocity profile from the Navier-Stokes equation is $v_x = V_{\max}\left(1-\left(\frac{z}{H}\right)^2\right)$, where $V_{\max}$ is the maximum velocity in the x direction, and H is half of the height of the channel (z direction). **Figure 3.20** shows that the velocity profile obtained from the DPD simulation agrees well with the analytical solution of the Navier-Stokes equation if the theoretical velocity profile is scaled so that its maximum is equal to the maximum of the velocity profile obtained from the simulation. The agreement between the simulation results and theory for the S_{xz} component of the stress tensor, and the shape of the velocity profile demonstrate that DPD simulations with cubic spline interaction potentials, like DPD simulations with the standard conservative force, reproduce the asymptotic (long length scale) hydrodynamic behavior embodied in the Navier-Stokes equation. The viscosity of the DPD fluid can be obtained by comparing the velocity profile shown in **Figure 3.20** with the velocity profile calculated using the Navier-Stokes equation.

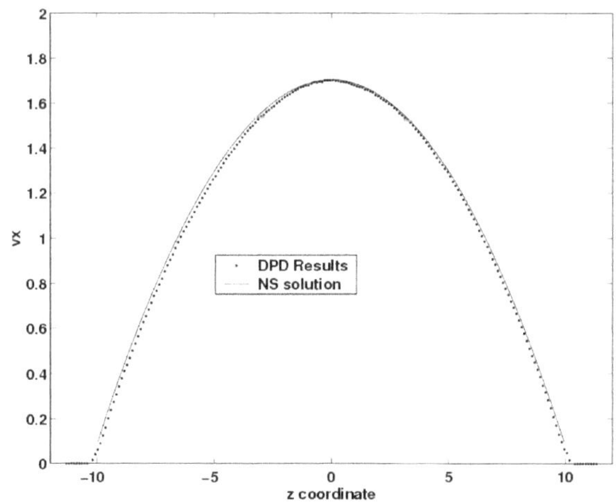

Figure 3.20 The steady state velocity profiles (velocity in the x direction) across the channel from DPD simulation results and the Navier-Stokes solution. The cubic spline potential is $U(r)=18.75W_1(r,1.0)$ (Liu et al., 2006).

3.4.3 Fully saturated flow through porous media

A DPD model should conform to the Navier-Stokes equations on scales that are large enough for hydrodynamic (continuum) concepts to be valid (on scales large enough for the effects of both the mean free path of discrete particles and their thermal fluctuations to be negligible), providing that the time step in the integration scheme is small enough to ensure accurate integration (Marsh, 1998). A number of previous investigations have shown that the results obtained from DPD simulations are in good agreement with the flow behaviour predicted by the Navier-Stokes equations for a variety of single phase fluid flows (Fan et al., 2003; Liu et al., 2006).

In this section, the DPD method is extended to model the fully saturated flow through porous media. The conventional conservative weight function is used in this simulation. The pore geometry can be viewed as a quasi two-dimensional granular porous medium with a micro fracture inside and it has an average total porosity of 0.443. The domain has a size of 160x160x2 (in units of the cutoff distance r_c, and 2 DPD units in thickness). A total of 204,800 particles of the same type were randomly positioned within the computational domain, in which 90,768 particles were used as fluid particles, 31,774 particles acted as wall particles, while the rest of the solid particles, which were more than 1 DPD unit away from the pore space, were removed. The number of particles corresponds to a density of 4.0 per unit DPD area (r_c^2) for both the fluid and wall particles. Periodic boundary conditions were used along all the external boundaries and a no-slip boundary condition was applied on the boundaries of the interior solid obstacles representing the grains in the granular porous medium. The parameters used in this DPD simulation include a = 18.75 and $k_B T = 1.0$ (hence $\gamma = 4.5$). These parameter values were chosen according to (Fan et al., 2003; Liu et al., 2006) so that the DPD fluid mimics the compressibility of water at room temperature if a simple interaction weight function ($w^C(r) = 1 - r$) is used. However, like SPH simulations, there are still no exact analytical expressions relating the fundamental fluid properties such as viscosity and surface tension to the parameters used in the particle-particle interaction models. The viscosity of a DPD fluid has two major contributors: the dissipative contribution due to the friction force acting on particles moving along different streamlines and the kinetic contribution due to particle diffusion across different streamlines. Quantifying these two terms in terms of particle interaction models requires integrating the weight function and velocity autocorrelation function that has an approximate analytical solution only for the most simple weight functions. By using the simple interaction weight function ($w^C(r) = 1 - r$), Groot and Warren derived approximate formulas for the viscosity of the DPD fluid. The important finding of their research is that the derived fluid viscosity is only within 10–30% of the measured results (Groot and Warren, 1997).

In this simulation, a modified version of the velocity Verlet algorithm (Groot and Warren, 1997) was used to simulate the particle dynamics, with $\lambda = 0.65$ and a time step of $\Delta t = 0.02$. A gravitational acceleration of g = 0.01 (a force of 0.01 for the particles of unit mass) was applied along the x direction to drive the flow. The system was divided into 160x160 bins along the x and z direction so that the temperature, velocity and density profiles, perpendicular to the direction of flow and the confining walls could be obtained by averaging the particle kinetic energies and particle velocities and determining the average particle density in each of these bins.

Figure 3.21a shows a contour map of the velocity magnitude field obtained using the DPD simulation. As expected, the majority of the fluid flow occurs within the fracture aperture inside the porous medium, and all peak values appear inside the fracture. The fluid flow within the pores is much less significant.

In order to validate the DPD simulation results, we also simulated the same problem by using a grid-based Navier-Stokes simulator that solves the dimensionless Navier-Stokes equations for incompressible fluids using a finite volume method (which reduces to a finite difference method when a regular grid is used). Different grid resolutions were used to test the convergence of the N-S simulation. A grid resolution of at least 3–5 grid cells per obstacle or per pore throat is necessary to guarantee the convergence of the N-S simulation results. **Figure 3.21**b shows the contour map of the simulated velocity magnitude field with a Reynolds number of 1 obtained by using a 320x320 uniform grid, which gives a grid resolution of at least 5–10 grid cells per obstacle. We tuned the *Re*

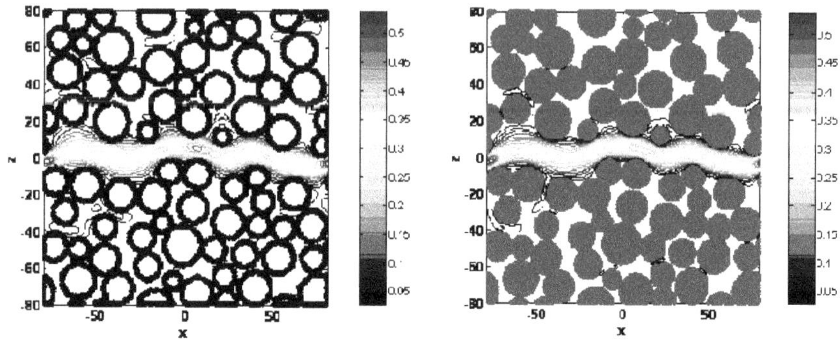

Figure 3.21 Velocity magnitude contours obtained using (a) the DPD simulation and (b) the grid-based Navier-Stokes simulator for fully saturated single-phase flow through a fractured porous medium (Liu et al., 2007b).

number so that both the DPD and N-S simulations gave the same averaged flux across the right boundary. An exact formula for the viscosity of a DPD fluid is not available, but the DPD model can be calibrated by comparing the DPD simulation results to experiments or other simulation techniques. The general flow patterns from the DPD and NS simulations agree very well with each other. In the middle section, which corresponds to the micro fracture and contains no obstacles, the fluid particles move smoothly and the velocity contours are aligned along the flow path. The aperture varies along the fracture, which results in several local velocity maxima along the fracture.

The velocity in the DPD simulation appears to be larger than that in N-S simulation in several disconnected pore spaces. The difference originates mainly from different representations of the solid obstacles in two approaches. In the N-S simulation, the solid obstacles are represented by staircase-like grid cells. In contrast, the DPD model uses randomly distributed particles to represent the solid obstacles, and those particles representation of solid obstacles give the surfaces of the obstacles a roughness that is not represented by a finite difference grid in the N-S simulation. This leads to some small discrepancies in the simulated velocity contour levels in some pore spaces. Nevertheless, the comparison indicates that the DPD model can effectively simulate hydrodynamics in porous media.

In order to more quantitatively compare the results of the two approaches, we also compare the simulated velocity profiles along the centerline of the fracture. **Figure 3.22** compares the velocity profiles. Except for some small discrepancies, the velocity profiles obtained using the DPD and N-S simulations are in a good agreement. The local velocity maxima from the two different approaches are also very similar.

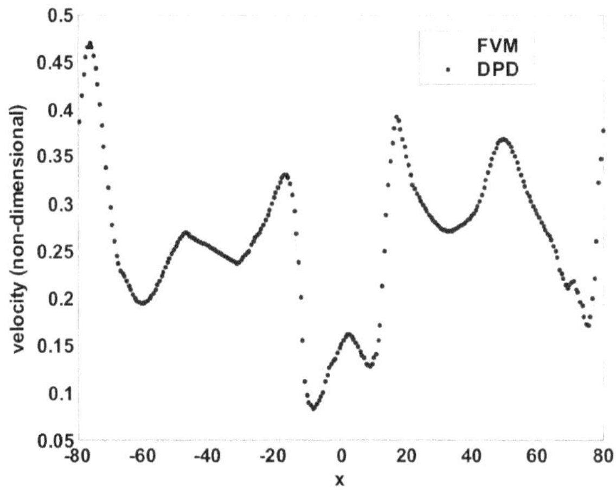

Figure 3.22 Velocity profiles along the centerline of the fracture (Liu et al., 2007b).

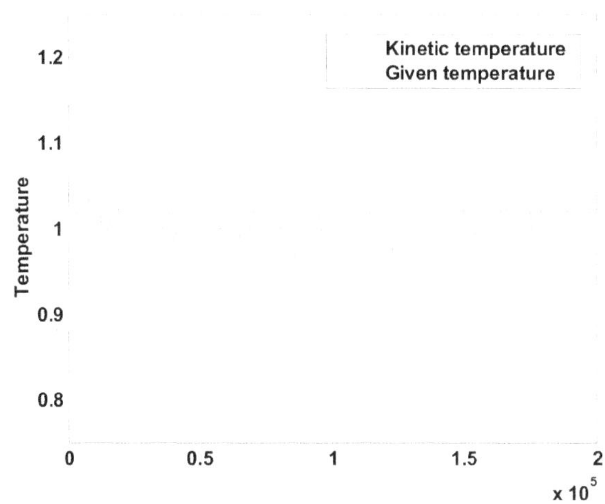

Figure 3.23 System kinetic temperature evolution for the simulation of fully saturated flow through a fractured porous media. The maximum discrepancy between the measured system kinetic temperature and the fluctuation-dissipation theorem temperature is less than 3.5% (Liu et al., 2007b).

Figure 3.23 shows the kinetic temperature evolution of the system for fully saturated flow through the porous media. The kinetic temperature calculated from the kinetic energy was uniform, and remained almost equal to the initially specified temperature. The maximum discrepancy between the measured kinetic temperature and the temperature defined through the fluctuation-dissipation theorem was less than 3.5%. The temperature variation is a direct measure of the accuracy of a DPD simulation. If a DPD simulation results in a kinetic temperature that differs significantly from the nominal temperature, the simulation results will not be reliable.

3.5 Concluding remarks

Dissipative particle dynamics is a coarse-grained molecular dynamics method, in which a particle contains a cluster of atoms/molecules. As the DPD method is associated with bigger particle size and soft conservative interaction potential, it should be able to model problems with larger time and length scale. As a meso scale method, DPD is suitable for modeling complex physics and it is able to reproduce continuum hydrodynamic when the time step in the time integration scheme approaches zero and the system size is large enough for hydrodynamic (continuum) concepts and models to be valid.

As a coarse-grained molecular dynamics, DPD has many similarities with MD. Both DPD and MD are deterministic Lagrangian particle methods (on meso and atomic scales respectively). The computational procedures of DPD and MD are also very much similar including particle initialization, system equilibration, neighbor particle searching, pairwise force computation, physical variable evaluation and even pre- and post-processing. The solid boundary treatment in DPD is also similar to that in MD, and usually involves the use of frozen particles and the inclusion of reflection models.

However, DPD is different from MD in many aspects. One primary feature of DPD is that DPD uses bigger particles and a soft conservative interaction potential, and it is therefore can be applied to larger length and time scale than those in MD. This makes the numerical simulation of the movement and suspension of macromolecules such as DNA (with an uncoiled length of $O(10\mu m)$) feasible. It is also possible to model the movement and deformation of a single cell (with a diameter of $O(10\mu m)$) or even a cluster of cells by using DPD method. The application of DPD to modeling the movement and suspension of macromolecules in micro channels and to modeling cell mechanics problems will be shown in the next chapter.

It is possible to couple DPD with smoothed particle hydrodynamics for multiple scale simulations or develop a variety of 'hybrid' models that combine DPD and SPH concepts. Español (Español, 1997) described a *fluid particle dynamics* (FPD) model that is a synthesis of dissipative particle dynamics and smoothed particle hydrodynamics, and Español and Revenga (Espanol and Revenga, 2003) combined features from DPD and SPH to develop the *smoothed dissipative particle dynamics* (SDPD) model in which the Navier-Stokes (N-S) equation governing the system is discretized using SPH approximations while thermal fluctuations are included in a consistent way. Therefore, SDPD is a modified SPH model that has little in common with the original DPD method, except for the random forces representing the thermal fluctuations, which are an essential component of DPD simulations.

The other main feature of the DPD method is the inclusion of the dissipative force and random force, which act together to maintain an essentially constant temperature with small fluctuations around the nominal temperature. The conservative force in DPD is similar to that in MD, but different in the interaction potential. The interaction potential is important in MD as it determines the pattern of interaction between particles. The interaction potential has been well investigated in MD, and different interaction potentials have been used for different materials/fluids. However, there are very few investigations on the conservative interaction potentials in DPD. The conventional DPD uses a simple conservative force weighting function ($w^C(r) = 1 - r$), which is repulsive and is effective in modeling fluids behaving like gas. It is not able to simulate the flow of liquids with free surfaces, the behavior of bubbly liquids, droplet dynamics and other important multiphase fluid flow processes. By combining two SPH smoothing functions (of polynomial form), it is possible to construct conservative interaction potentials with short-range repulsion and long-distance attraction. The corresponding strength coefficients and cutoff distances of the repulsive part and attractive part determines the exact form of the interaction potential and further determines the physical properties of the modeling DPD fluid. By using this new interaction potential, it is possible to model complex systems with co-existing liquid-gas-solid phases.

In classic molecular dynamics, dimensional, primal variables can be used in computer implementation, and the modeling parameters can be correlated to the physical properties of real materials/fluids. In contrast, DPD method is generally implemented in a non-dimensional form. In order to match the modeling parameters with the physical properties of real materials/fluids, it should be careful in choosing coefficients in DPD simulation. In general, some coefficients can be determined by fitting the relevant data of the real fluid, some are selected to maintain the numerical accuracy in simulating simple cases with analytical solutions. Another point is that due to the soft interaction between DPD particles, the Schmidt number and dynamic viscosity obtained from a conventional DPD simulation are usually much lower than those for a real fluid. This drawback can be remedied by increasing the cutoff distance or reducing the exponential factor of the weight function of the dissipative (and random) force.

References

Allen M. P. and Tildesley, D. J. (1987) Computer Simulation of Liquids. Oxford University Press, Oxford.

Bock H., Gubbins, K. E. and Klapp, S. H. L. (2007) Coarse graining of nonbonded degrees of freedom, *Physical Review Letters* **98**(26).

Chen S. Y. and Doolen, G. D. (1998) Lattice Boltzmann method for fluid flows, *Annual Review of Fluid Mechanics* **30**: 329–364.

Chung T. J. (2002) Computational fluid dynamics. Cambridge University Press.

Duong-Hong D., Han, J., Wang, J., Hadjiconstantinou, N. G., Chen, Y. Z. and Liu, G. R. (2008a) Realistic simulations of combined DNA electrophoretic flow and EOF in nano-fluidic devices, *Electrophoresis* **29**(24): 4880–4886.

Duong-Hong D., Wang, J. S., Liu, G. R., Chen, Y. Z., Han, J. Y. and Hadjiconstantinou, N. G. (2008b) Dissipative particle dynamics simulations of electroosmotic flow in nano-fluidic devices, *Microfluidics and Nanofluidics* **4**(3): 219–225.

Español P. (1997) Fluid particle dynamics: A synthesis of dissipative particle dynamics and smoothed particle dynamics, *Europhysics Letters* **39**(6): 605.

Espanol P. and Revenga, M. (2003) Smoothed dissipative particle dynamics, *Physical Review E* **67**(2): 26705.

Espanol P. and Warren, P. (1995) Statistical mechanics of dissipative particle dynamics, *Europhysics Letters* **30**(4): 191–196.

Fan X. J., Phan-Thien, N., Chen, S., Wu, X. H. and Ng, T. Y. (2006) Simulating flow of DNA suspension using dissipative particle dynamics, *Physics of Fluids* **18**(6).

Fan X. J., Phan-Thien, N., Yong, N. T., Wu, X. and Xu, D. (2003) Microchannel flow of a macromolecular suspension, *Physics of Fluids* **15**(1).

Flory P. J. (1942) Thermodynamics of high polymer solutions, *The Journal of Chemical Physics* **10**: 51.

Gardner J. W. and Varadan, V. K. (2001) Microsensors, MEMS and smart devices. John Wiley & Sons, Inc., Hoboken.

Gingold R. A. and Monaghan, J. J. (1977) Smoothed particle hydrodynamics-Theory and application to non-spherical stars, *Monthly Notices of the Royal Astronomical Society* **181**: 375–389.

Groot R. D. and Warren, P. B. (1997) Dissipative particle dynamics: Bridging the gap between atomistic and mesoscopic simulation, *Journal of Chemical Physics* **107**(11): 4423.

He X. Y. and Luo, L. S. (1997) Theory of the lattice Boltzmann method: From the Boltzmann equation to the lattice Boltzmann equation, *Physical Review E* **56**(6): 6811.

Ho C. M. and Tai, Y. C. (1998) Micro-electro-mechanical-systems (MEMS) and fluid flows, *Annual Review of Fluid Mechanics* **30**(1): 579–612.

Hoogerbrugge P. J. and Koelman, J. (1992) Simulating microscopic hydrodynamic phenomena with dissipative particle dynamics, *Europhysics Letters* **19**: 155.

Hsu T.-R. (2008) MEMS & Microsystems: Design, Manufacture, and Nanoscale Engineering. John Wiley & Sons, Hoboken.

Hu X. and Adams, N. (2006) A multi-phase SPH method for macroscopic and mesoscopic flows, *Journal of Computational Physics* **213**(2): 844–861.

Irving J. and Kirkwood, J. G. (1950) The statistical mechanical theory of transport processes. IV. The equations of hydrodynamics, *The Journal of Chemical Physics* **18**: 817.

Karniadakis G. E., Beðskèok, A. and Aluru, N. R. (2005) Microflows and nanoflows. Springer, Berlin.

Knotts IV T. A., Rathore, N., Schwartz, D. C. and de Pablo, J. J. (2007) A coarse grain model for DNA, *Journal of Chemical Physics* **126**(8): 084901.

Koplik J. and Banavar, J. R. (1995) Continuum deductions from molecular hydrodynamics, *Annual Review of Fluid Mechanics* **27**(1): 257–292.

Kresse G. and Hafner, J. (1993) Ab initio molecular dynamics for liquid metals, *Physical Review B* **47**(1): 558.

Ladd A. J. (1994) Numerical simulations of particulate suspensions via a discretized Boltzmann equation. Part 2. Numerical results, *Journal of Fluid Mechanics* **271**(1): 311–339.

Landau L. and Lifshitz, E. (1980) Statistical Physics, Vol. 1, *Oxford: Pergamon* **24**: 31–57.

Larson R., Perkins, T., Smith, D. and Chu, S. (1999). The hydrodynamics of a DNA molecule in a flow field, Flexible Polymer Chains in Elongational Flow, Springer: 259–282.

Lebowitz J. L. and Penrose, O. (1966) Rigorous Treatment of the Van Der Waals–Maxwell Theory of the Liquid–Vapor Transition, *Journal of Mathematical Physics* **7**: 98.

Liu G., R. and Nguyen, T. T. (2010) Smoothed Finite Element Methods. CRC Press, Boca Raton.

Liu G. R. and Liu, M. B. (2003) Smoothed particle hydrodynamics: a meshfree particle method. World Scientific, Singapore.

Liu G. R. and Quek, S. S. (2003) Finite element method: a practical course. Butterworth-Heinemann, Oxford.

Liu M. B., Liu, G. R. and Lam, K. Y. (2003) Constructing smoothing functions in smoothed particle hydrodynamics with applications, *Journal of Computational And Applied Mathematics* **155**(2): 263–284.

Liu M. B., Liu, G. R., Zhou, L. V. and Chang, J. Z. (2014) Dissipative particle dynamics (DPD): an overview and recent developments, *Archives of Computational Methods in Engineering*.

Liu M. B., Meakin, P. and Huang, H. (2006) Dissipative particle dynamics with attractive and repulsive particle-particle interactions, *Physics of Fluids* **18**(1): 017101.

Liu M. B., Meakin, P. and Huang, H. (2007a) Dissipative particle dynamics simulation of fluid motion through an unsaturated fracture and fracture junction, *Journal of Computational Physics* **222**(1): 110–130.

Liu M. B., Meakin, P. and Huang, H. (2007b) Dissipative particle dynamics simulation of pore-scale flow, *Water Resources Research* **43**.

Lucy L. B. (1977) A numerical approach to the testing of the fission hypothesis, *Astronomical Journal* **82**(12): 1013–1024.

Marsh C. (1998). Theoretical Aspects of Dissipative Particle Dynamics, University of Oxford Press, Oxford.

McAllister D. V., Allen, M. G. and Prausnitz, M. R. (2000) Microfabricated microneedles for gene and drug delivery, *Annual Review of Biomedical Engineering* **2**(1): 289–313.

Moeendarbary E., Ng, T. Y. and Zangeneh, M. (2009) Dissipative particle dynamics: introduction, methodology and complex fluid applications—a review, *International Journal of Applied Mechanics* **1**(04): 737–763.

Nielsen S. O., Lopez, C. F., Srinivas, G. and Klein, M. L. (2004) Coarse grain models and the computer simulation of soft materials, *Journal of Physics-Condensed Matter* **16**(15): R481.

Pagonabarraga I. and Frenkel, D. (2001) Dissipative particle dynamics for interacting systems, *Journal of Chemical Physics* **115**: 5015.

Pagonabarraga I., Hagen, M. H. J. and Frenkel, D. (1998) Self-consistent dissipative particle dynamics algorithm, *Europhysics Letters* **42**(4): 377-382.

Peyret R. and Taylor, T. D. (1985) Computational methods for fluid flow, Springer-Verlag, New York.

Pomeau B. H. Y. and Frisch, U. (1986) Lattice-gas automata for the Navier-Stokes equation, *Physical Review Letters* **56**(14): 1505.

Rapaport D. C. (2004) The art of molecular dynamics simulation. Cambridge University Press.

Revenga M., Zuniga, I., Espanol, P. and Pagonabarraga, I. (1998) Boundary models in DPD, *International Journal of Modern Physics C* **9**(08): 1319–1328.

Smith D. E., Babcock, H. P. and Chu, S. (1999) Single-polymer dynamics in steady shear flow, *Science* **283**(5408): 1724–1727.

Tartakovsky A. M. and Meakin, P. (2005) A smoothed particle hydrodynamics model for miscible flow in three-dimensional fractures and the two-dimensional Rayleigh Taylor instability, *Journal of Computational Physics* **207**(2): 610–624.

Vologodskii A. (1994) DNA extension under the action of an external force, *Macromolecules* **27**(20): 5623–5625.

Voter A. F. (2007). Introduction to the kinetic Monte Carlo method, Radiation Effects in Solids, Springer, Berlin.

Warren P. (2003) Vapour-liquid coexistence in many-body dissipative particle dynamics, *arXiv preprint cond-mat/0306027*.

Warren P. B. (1998) Dissipative particle dynamics: Dynamic aspects of colloids and interfaces, *Current Opinion in Colloid & Interface Science* **3**(6): 620–624.

Yan K., Chen, Y. Z., Han, J. Y., Liu, G. R., Wang, J. S. and Hadjiconstantinou, N. G. (2012) Dissipative particle dynamics simulation of field-dependent DNA mobility in nanoslits, *Microfluidics and Nanofluidics* **12**(1–4): 157–163.

Zienkiewicz O. C. and Taylor, R. L. (2000) The finite element method. Butterworth-Heinemann, Oxford.

Chapter 4

Dissipative Particle Dynamics — Applications

This chapter presents some typical DPD applications. Special interests are focused on

- micro drop dynamics including DPD modelling of the formation of drop with co-existing liquid-vapor, large-amplitude oscillation of a liquid drop and flow transition in controlled drug delivery (Section 4.2),
- multiphase flows in pore-scale fracture network and porous media (Section 4.3),
- movement and suspension of macromolecules in micro channels (Section 4.4), and
- movement and deformation of a single cell (Section 4.5).

4.1 Introduction

Molecular dynamics (MD) and *ab initio* quantum mechanics are the most fundamental ways of obtaining a better understanding of the behavior of solids and fluids. However, due to the high resolution in length and time scales associated with these methods, they are computationally expensive when the model size gets big, even for modern supercomputers. Hence, they cannot be applied to many important scientific and practical problems. The behavior of a wide range of materials such as gels, colloids, polymer solutions, proteins and DNA molecules can be understood in terms of meso scale interactions between components with effective sizes ranging from nanometers to micrometers. To reach the characteristic time and length scales associated with these mesoscopic materials, new approaches that discard some of the excessive short time and length scale details associated with quantum mechanical and classical molecular dynamical methods are needed. These mesoscopic methods have been developed to simulate the properties and behaviors of systems on super-molecular length scales and the associated time scales, with reasonable computer resources

As a relatively new meso scale technique that can be used to simulate the behavior of complex fluids, dissipative particle dynamics (DPD) (Hoogerbrugge and Koelman, 1992) uses particles to represent the state of a modeling system while a particle in DPD simulation include a cluster of molecules that interact via conservative (non-dissipative), dissipative and fluctuating forces. Because the effective interactions between clusters of molecules are much softer than the interactions between individual molecules, much longer time steps can be taken relative to MD models. The longer time steps combined with the larger particle size makes DPD much more practical to simulate hydrodynamics in micro-fluidic devices. DPD is also particularly promising for the simulation of complex liquids, such as polymer suspensions, liquids with interfaces, colloids and gels. Because of the symmetry of the interactions between the particles in typical simulations, DPD rigorously conserves the total momentum of the system, and because the particle-particle interactions depend only on relative positions and velocities, the resulting model fluids are Galilean invariant. Mass is conserved because the same mass is associated with each of the particles, and the number of particles does not change. While DPD is not as computationally efficient as lattice Boltzmann simulations, it is a more flexible method that does not suffer from the numerical instability associated with many lattice Boltzmann applications. DPD facilitates the simulation of complex fluid systems on physically interesting and important length and time scales.

As the smoothed particle hydrodynamics (SPH) is also a meshfree, Lagrangian particle method for macroscopic applications, it is possible to couple DPD with SPH for multiple scale simulations or develop a variety of 'hybrid' models that combine DPD and SPH concepts. Español described a fluid particle dynamics (FPM) model that is a synthesis of dissipative particle dynamics and smoothed particle hydrodynamics (Español, 1997). Español and Revenga combined features from DPD and SPH to develop the smoothed dissipative particle dynamics (SDPD) model in which the Navier-Stokes equation governing the system is discretized using SPH approximations while thermal fluctuations are included in a consistent way (Espanol and Revenga, 2003). Therefore, SDPD is a modified SPH model that has little in common with the original DPD method, except for the random forces representing the thermal fluctuations, which are an essential component of DPD simulations.

Like other Lagrangian particle based methods such as SPH, DPD models have special advantages over the traditional grid based methods in modeling multiphase flow in domains with complex solid boundaries. They do not require explicit and complicated interface tracking algorithms, and thus there is no need to explicitly track the material interfaces, and processes such as fluid fragmentation and coalescence can be handled without difficulty.

As a coarse-grained molecular dynamics method, DPD is attractive in modeling the hydrodynamic behavior of mesoscopic complex fluids. Therefore, since its invention, the DPD method has been extended to many applications including colloidal suspensions (Koelman and Hoogerbrugge, 1993), surfactants (Groot, 2000), dilute polymer solutions (Schlijper et al., 1995), biological membranes (Groot and Rabone, 2001), macromolecular dynamics (Fan et al., 2003; Pan et al., 2010) and many others (Moeendarbary et al., 2009). Most of the earlier applications focus on the equilibration process of complex fluids including the aggregation of polymer and surfactant, the mixture or phase separation and morphology evolution of complex fluids with multi-components or multi-phases. Recently DPD method is popular in modeling the dynamic flow process of mesoscopic complex fluids including liquid drop dynamics (drop formation, oscillation, coalescence, collision, impacting, and spreading) and the saturated or unsaturated flows in mesoscopic structures (micro channels, fractures and porous media).

Here we emphasize on the following areas of applications well suited to DPD method.

- multiphase drop dynamics
- multiphase flow in micro channels and fractures
- movement and suspension of macromolecules
- movement and deformation of a single cell

4.2 Micro drop dynamics

Characterization of fluid flows in microfluidic devices has increasingly becoming a very important topic since the fluidic behavior in MEMS is very different from what observed in daily life. Flows in microfluidic devices usually involve small or ignorable inertial force, but dominant viscous, electro-kinetic and surface effects especially when the surface-to-volume ratio increases (Karniadakis et al., 2005). Analytical or semi-analytical solutions for microfluidics are generally limited to a very few simple cases, whereas experimental studies are usually expensive. Numerical simulation of flows in microfluidic devices, as an effective alternate, has been attracting more and more researchers. However, simulation of microfluidic devices is not easy due to the involved complex features including movable boundaries (free surfaces and moving interfaces), large surface-to-volume ratio, and phenomena due to small scale physics. Numerical studies with reliable models are needed to develop a better understanding of the temporal and spatial dynamics of multiphase flows in microfluidic devices.

On the other hand, drop formation and break-up in micro/nano scales are fundamentally important to diverse practical engineering applications such as ink-jet printing, DNA and protein micro-/nano-arraying, and fabrication of particles and capsules for controlled release of medicines. Numerical studies provide an effective tool to improve better understanding of the inherent physical dynamics of drop formation and breakup. Computational models for drop formation and breakup in micro/nano-scales must be able to handle movable boundaries such as free surfaces and moving interfaces, large density ratios, and large viscosity ratios. These requirements together with micro scale phenomena and possible complex boundaries (fluid-fluid-solid contact line dynamics and fluid-fluid interface dynamics) in microfluidic devices present severe challenges to conventional Eulerian-grid-based numerical methods such as finite difference methods and finite volume methods which require special algorithms to treat and track the interfaces. Algorithms based on Lagrangian-grid-based methods such as finite element methods have been shown to agree quantitatively with experimental measurements, but they are only capable of modeling the dynamics of formation of a single drop or the dynamics until the occurrence of the first singularity.

Researchers have simulated drop dynamics for multiple component systems by using the DPD method with the conventional weight function for the conservative force (Clark et al., 2000). However, the conventional conservative force is able to simulate liquid-liquid and liquid-solid interfaces for multiple component systems, but is not able to simulate liquid-gas interfaces for single component systems. Recently the DPD method is modified to model the solid-liquid-gas co-existing systems, either by using the many body DPD approach developed by Warren (Warren, 2003) or by using a new conservative interaction potential with long-distance attraction and short-range repulsion proposed by Liu et al. (Liu et al., 2006), or by other approaches which are able to describe the attraction and repulsion between interacting DPD particles. For example, Li et al. investigated the 3D flow structures in a moving droplet on substrate by using the many body DPD (Li et al., 2013), Zhang et al. studied the movement of a droplet in a grooved channel by using Liu's conservative interaction potential (Zhang et al., 2012). Merabia and Pagonabarraga developed a mesoscopic model for simulating the dynamics of a non-volatile liquid on a solid substrate and they analyzed the kinetics of spreading of a liquid drop wetting a solid substrate and the dewetting of a liquid fill on a hydrophobic substrate (Merabia and Pagonabarraga, 2006). Based on mean-field theory, Tiwari and Abraham proposed a DPD model for two-phase flows involving liquid and vapor phases (Tiwari and Abraham, 2006). The DPD model is validated by a number of numerical examples including the small- and large-amplitude oscillations of liquid drops.

4.2.1 Formation of drop with co-existing liquid-vapor

Revisiting the interaction potential with long-distance attraction and short-range repulsion (as introduced in Chapter 3), different parameter sets, A, r_{c1}, B and r_{c2}, determine the shape of the interaction potential, and consequently the behaviour of the DPD fluid. In this study, the parameters selected for the interaction potential were $A = 2.0, r_{c1} = 0.8$ and $r_{c2} = 1.0$ with several values for B to investigate different DPD fluid behaviours resulted from different attractive effects. The coefficients associated with the fluctuating and dissipative forces were $a = 18.75$, $\sigma = 3.0$, and $k_B T = 1.0$ $(\gamma = 4.5)$. The particle-particle interaction potentials were given by $U(r) = 18.75(2W_1(r,0.8) - BW_2(r,1.0))$, and were shown in **Figure 4.1**.

In the simulations, 4000 particles of the same type were first randomly placed in a system of dimensions $20\times20\times20$. A modified version of the velocity-Verlet algorithm (Groot and Warren, 1997) was used to integrate the equation of motion, with $\lambda = 0.65$ and a time step of $\Delta t = 0.02$. Periodic boundary conditions were used in all three directions. The system was divided into $40\times40\times80$ bins (40×40 columns of bins along the z direction, each with 80 bins) along the x, y, and z direction so that the temperature, velocity and density profiles, perpendicular to the direction of flow and the confining walls

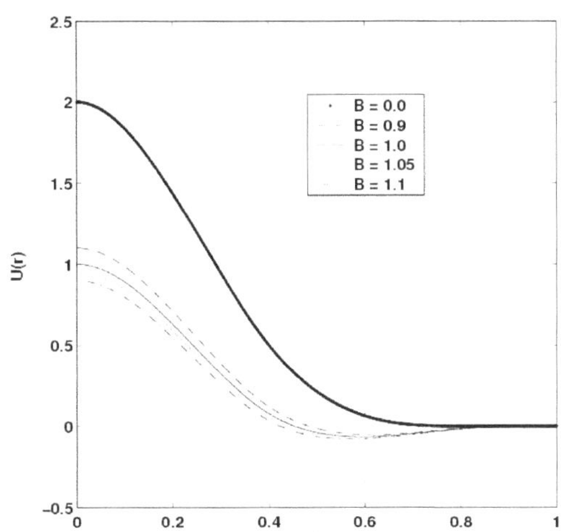

Figure 4.1 Cubic spline interaction potential functions, $U(r) = 2W_1(r,0.8) - BW_2(r,1.0)$ with different coefficients (Liu et al., 2006).

could be obtained by averaging the particle kinetic energies and particle velocities and determining the average particle density in each of the bins. The DPD system was first run to reach equilibrium by using the conventional, repulsive interaction potential, with a global density of 0.5. The new cubic spline potentials were then applied to the resultant DPD system to investigate the behaviors of the DPD fluids. Therefore the simulations started from single gas phase due to the repulsive conventional interaction potential.

The potential function represents a purely repulsive interaction when $B = 0$. **Figure 4.2** shows the particle distribution at 10^5 steps, and **Figure 4.3** shows the corresponding density profile along the centre column of bins (the bins whose bin numbers are 20 and 20 at x and y direction) in the z direction. Since there is no attractive component in the particle-particle interaction, the DPD particles did not separate into liquid and gas phases, and they did not form a liquid drop (or drops). Instead, the DPD particles were distributed randomly with local densities (dotted line in **Figure 4.3**) roughly equal to the global density (0.5, solid line in **Figure 4.3**). The density profile obtained using this purely repulsive potential agrees well with the density profile obtained using the conventional potential function (dash-dotted line in **Figure 4.3**).

A small value of B, corresponding to weak long-range attraction between the DPD particles is not sufficient to induce phase separation. When the critical value for B is reached, at a particular temperature, large density fluctuations will occur, and an additional small increase in B will lead to slow phase separation. **Figure 4.4** shows the particle distribution after 20000 steps in a simulation with $B = 0.9$. The fluid forms a spherical liquid drop surrounded by dense gas particles. **Figure 4.5** shows the density profiles at $t = 160$, 240, and 320 in a column of width 80 along the z direction, through the centre of the drop. The density of the liquid drop is around 6.1 while the density of the surrounding gas is around 0.3 (a ratio of about 20 compared with a ratio of about 1000 for water and air at standard temperature and pressure). The size of the spherical liquid drop remained approximately constant, while the location of the drop centre shifted a little. This is expected since collisions between 'gas' particles and the drop will induce the drop to undergo Brownian motion, and combined evaporation and condensation will also result in random motion of the centre of mass). The interfacial width is proportional to the interaction range (one DPD unit). The particles had sufficient time for the interfacial width to equilibrate, and the liquid drop was stabilized. The density fluctuations in the dense liquid phase were comparatively smaller than those in the gas phase, and this can be expected on the basis of the compressibility of the fluids. The formation of coexisting of liquid-gas phases was not dictated by the initial conditions, but was determined by the particle density and the cubic spline potential with attractive force.

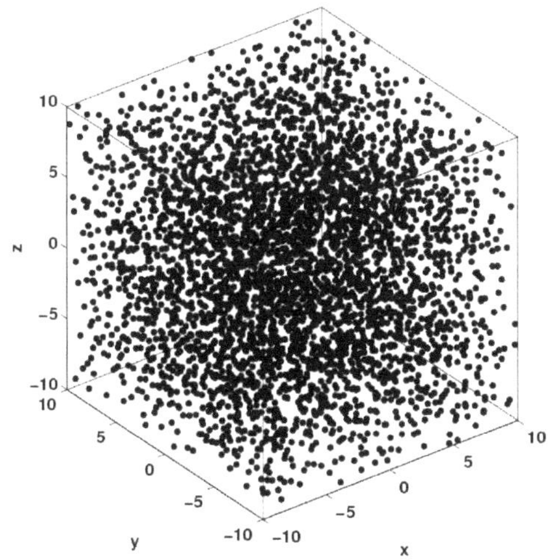

Figure 4.2 Particle distribution obtained using the cubic spline potential $U(r) = 37.5W_1(r, 0.8)$ (Liu et al., 2006).

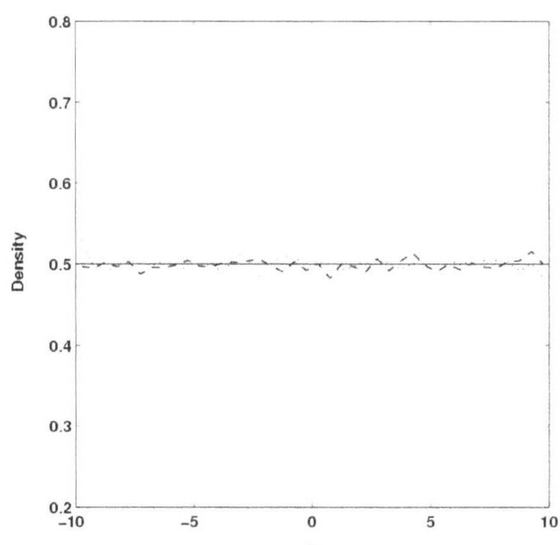

Figure 4.3 Density profiles along the centre column bins in z direction obtained using the cubic spline potential $U(r) = 37.5W_1(r, 0.8)$ (dotted line), the conventional potential (dash-dotted line), and the global density (solid line). (Liu et al., 2006).

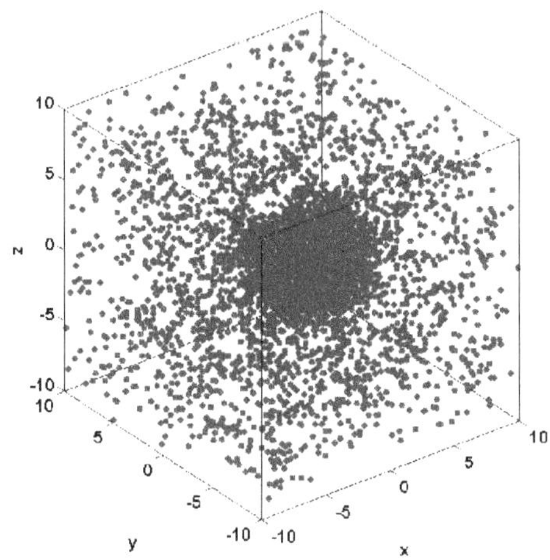

Figure 4.4 Particle distribution obtained using the cubic spline potential $U(r) = 18.75(2W_1(r,0.8) - 0.9W_2(r,1.0))$ (Liu et al., 2006).

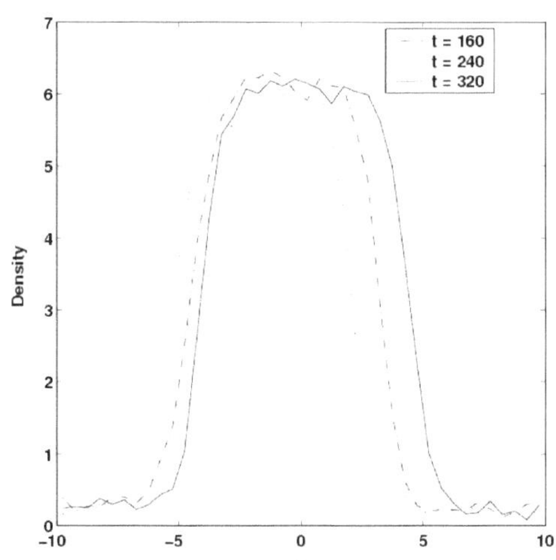

Figure 4.5 Density profile along the centre column bins in z direction obtained using the cubic spline potential $U(r) = 18.75(2W_1(r,0.8) - 0.9W_2(r,1.0))$ (Liu et al., 2006).

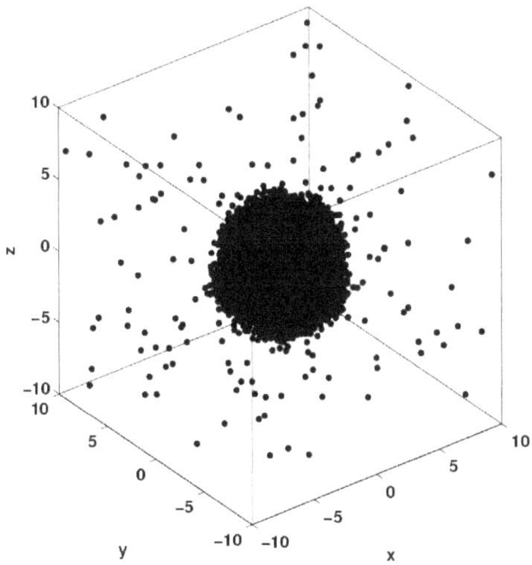

Figure 4.6 Particle distribution obtained using the cubic spline potential $U(r) = 18.75(2W_1(r,0.8) - 1.0W_2(r,1.0))$ (Liu et al., 2006).

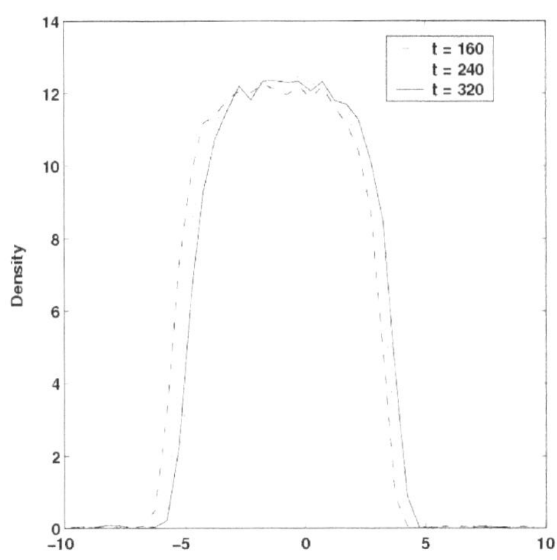

Figure 4.7 Density profile along the centre column bins in z direction obtained using the cubic spline potential $U(r) = 18.75(2W_1(r,0.8) - 1.0W_2(r,1.0))$ (Liu et al., 2006).

Figure 4.6 shows the particle distribution after 20000 steps, and **Figure 4.7** shows the density profiles at t = 160, 240, and 320 using B = 1.0. In this case, the fluid also forms a spherical liquid drop with sparse gas particles surrounding it. The liquid/gas density ratio is greater than 600. The shape of the liquid drop was stable, the drop size was almost constant, and the interface width was roughly equal to the interaction range. The number of surrounding gas particles was much smaller than that in the case illustrated in **Figure 4.4**, which was reflected by the very small density profiles in the gas phase in **Figure 4.7**.

Further increases in B result in stronger attractive effects in the interaction. **Figure 4.8** shows the particle distribution after 20000 steps from a simulation with B = 1.05, and **Figure 4.9** shows the density profiles at t = 160, 240, and 320. The bulk fluid forms a stable spherical liquid drop. In this case the density of the gas is very small, the number of 'gas' particles fluctuates strongly and at some time there may be no gas particles at all in the relatively small volume used in this simulation. This behaviour can be expected if the attractive part of the interaction potential is large enough compared with k_BT. Moving along a straight line from the exterior of the drop to its centre, the density profile started from zero, increased gradually and then reached a maximum value in the central region. This is quite different from the density profiles obtained with smaller values of B (weaker attractive interactions) for which the density started from a small value rather than from zero, and reached a plateau near the centre. The stronger attraction resulted in a thinner liquid-gas interface, which is smaller than the interaction range.

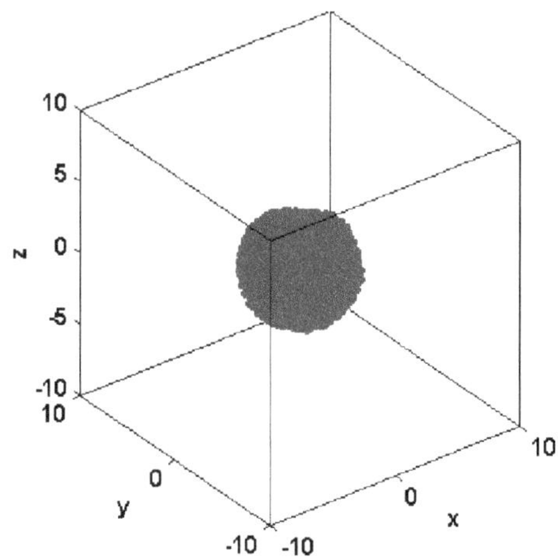

Figure 4.8 Particle distribution obtained using the cubic spline potential $U(r) = 18.75(2W_1(r,0.8) - 1.05W_2(r,1.0))$ (Liu et al., 2006).

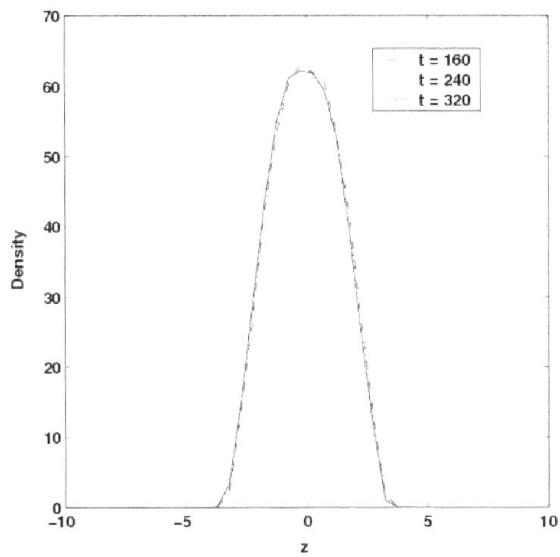

Figure 4.9 Density profile along the centre column bins in z direction obtained using the cubic spline potential $U(r) = 18.75(2W_1(r,0.8) - 1.05W_2(r,1.0))$ (Liu et al., 2006).

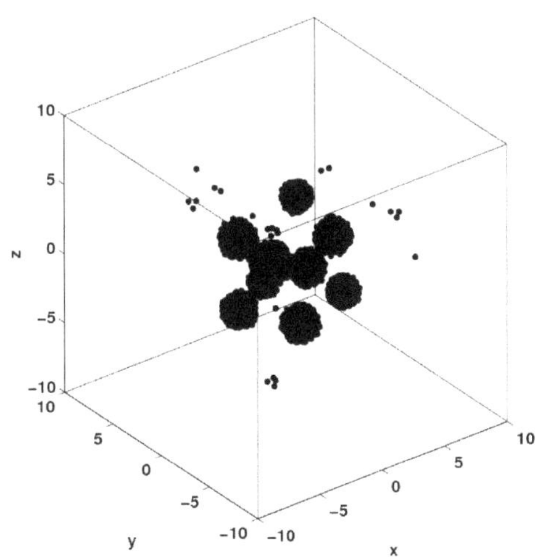

Figure 4.10 Particle distribution obtained using the cubic spline potential $U(r) = 18.75(2W_1(r,0.8) - 1.1W_2(r,1.0))$ (Liu et al., 2006).

Additional increases in B can result in different behaviour. **Figure 4.10** shows the particle distribution after 20000 steps using B = 1.1. The fluid underwent a phase transition but instead of forming a single liquid drop surrounding with gas particles, a number of small droplets formed. This can be expected due to the stronger particle-particle attraction, which resulted in a number of small droplets. Eventually a single drop should be formed by a process similar to Ostwald ripening (Boistelle and Astier, 1988) in which particles evaporate more rapidly from small droplets and condense more rapidly on large ones. However, as B is increased, the time scale for this coarsening process will become very long. The random relative motion of droplets pairs may also bring them into contact leading to coalescence.

4.2.2 Large-amplitude oscillation of a liquid drop

To further verify the validity of the DPD method with long-range attractive and short-range repulsive particle-particle interactions, the large-amplitude oscillations of an initially oblate liquid drop were studied. The initially oblate liquid drop with an aspect ratio of 5 was taken from a well-equilibrated circular drop which was obtained by injecting 20000 DPD particles into a $20\times1\times20$ box and running a DPD simulation using $A=2.0$, $r_{c1}=0.8$, $B=1.09$, $r_{c2}=1.0$, and $k_BT=1.0$ until the system had equilibrated. **Figure 4.11** shows snapshots of the DPD simulation at 8 typical stages. An SPH simulation was also conducted for an initially oblate liquid drop with an aspect ratio of 5 using the van der Waals equation of state with $\bar{a}=2.0$ and $\bar{b}=0.5$. **Figure 4.12** shows snapshots of the SPH simulation at equivalent stages. The DPD simulation agrees well with the SPH simulation. The shapes of the DPD drops were not as smooth as the corresponding SPH drops, as can be anticipated from the inclusion of the random forces in the DPD simulations. In both DPD and SPH simulations, the liquid drop underwent oscillations that closely resemble the oscillations of a large ball of water under micro gravity conditions observed experimentally in the space shuttle Columbia (Apfel et al., 1997).

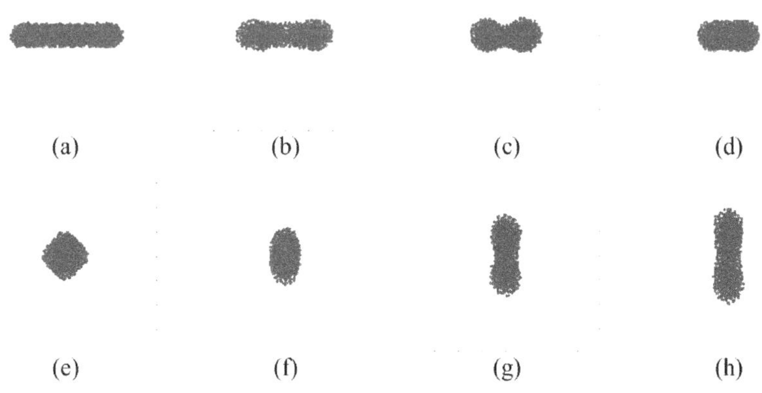

Figure 4.11 DPD simulation of the large-amplitude oscillations of a DPD fluid drop with an initial aspect ratio of 5 at 8 stages. The parameters for the DPD fluid are $A=2.0$, $r_{c1}=0.8$, $B=1.09$, $r_{c2}=1.0$ (Liu et al., 2006).

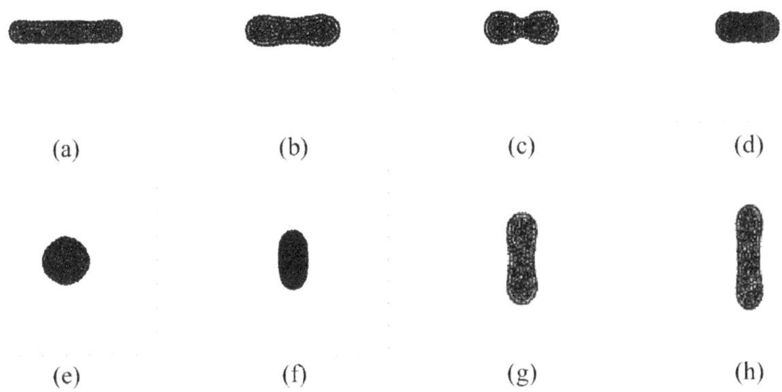

Figure 4.12 SPH simulation of the large-amplitude oscillations of a vdW fluid drop with an initial aspect ratio of 5 at 8 corresponding stages (Liu et al., 2006).

4.2.3 Controlled drug delivery

The DPD method with short range repulsion and long distance attraction is also applied to modeling of injection flow of drug agents for controlled drug delivery, which involves releasing the drug agents at the right time in a safe and reproducible way (Orive et al., 2003). As the modified DPD method is able to simulate multiphase systems, continuous flows, single and multiple drops can be effectively modeled, depending on the average particle density, the temperature and the details of the particle-particle interactions (Liu et al., 2007a). The material interfaces are naturally obtained from the DPD particle positions, and no complex interface tracking algorithm is required.

DPD particles were used to represent drug agents. The computational domain is $20\times3\times100$. DPD particles were randomly injected into the first row of cells in the z direction at the top of the computational domain until an average particle density of 4 was reached. The injected particle equilibrated with the particles that had previously entered the first row of cells, and the injected fluid particles move downward further, as the density of the injected particles and the concomitant pressure increased. A pressure drop is produced due to the particle injection. The surface tension of the fluid is determined by the interplay between the attractive and repulsive components of the interaction between the fluid particles. The interaction between the possible wall particles and the fluid particles can be different from that between the fluid particles, and these interactions can be tuned to give different wetting behaviours and capillary forces. The pressure drop, surface tension, and wetting behaviour or capillary force, together with possible external forces, govern the fluid flow in the micro channel, which may exhibit a variety of flow regimes. In delivering drug agents, if the specific drug agents (DNA, protein etc.) and the specific micro biomedical devices are fixed, the material behaviour of the drug agents and the interaction of the drug agents with the solid wall of the micro channels of the biomedical devices are determined. Hence the inject rate of the drug agents is the sole factor to influence the efficiency and accuracy of drug delivery. The influence of the injection rate on drug delivery was thus investigated.

In the simulation, periodic boundary condition was used along the *x* and *y* direction, and DPD particles that reached the bottom end of the system, at the minimum value of *z*, were removed (inflow and outflow boundary conditions). The coefficients used in the DPD model were $\sigma = 3.0$ and $k_B T = 1.0$ (and therefore $\gamma = 4.5$). The interaction strength between the fluid particles was $a = 18.75$. A gravity force of $g = 0.05$ per particle was applied along the *z* direction, after the system reached equilibrium. The parameters for the interaction potential as shown in Equation (8) were $A = 2.0, r_{c1} = 0.8$ and $B = 1.0, r_{c2} = 1.0$.

Figure 4.13 shows the sequential snapshots for a DPD simulation of the injection of drug agents with a particle injection rate of 500 particles per 100 steps. It is seen that few particles evaporated from the bulk fluid, and there forms a gas-liquid co-existing system. Along the entire channel (in *z* direction), the injected drug agents maintained a continuous flow mode without discontinuity. **Figure 4.14** shows the snapshots of DPD simulation with an injection rate of 100 particles per 100 steps. The initially continuous flow of the drug agent gradually necked down and a series of micro drops formed. Further reducing the injection rate to 50 particles per 100 time steps resulted in earlier necking down of drug agents and more micro drops (**Figure 4.15**). This numerical investigation thus verified that at different injection rate, the flow of drug agents can be different from continuous flow to discontinuous flow with micro drops.

Figure 4.13 DPD simulation snapshots of drug agent injection with an injection of 500 DPD particles per 100 time steps. The injected drug agents maintained a continuous flow mode without discontinuity (Liu et al., 2007a).

Figure 4.14 DPD simulation snapshots of drug agent injection with an injection of 100 DPD particles per 100 time steps. The initially continuous flow of the drug agent gradually necked down with a series of micro drops (Liu et al., 2007a).

Figure 4.15 DPD simulation snapshots of drug agent injection with an injection of 50 DPD particles per 100 time steps. The continuous injection of the drug agent quickly necks down with more micro drops (Liu et al., 2007a).

4.3 Multiphase flows in pore-scale fracture network and porous media

Pore-scale, multiphase fluids in contact with solid surfaces are important in almost all areas of science and technology including nuclear reactor heat exchangers, lubricated pipeline transport, manufacturing of multilayer films and fibers, chemical reactors and separators, coating systems, enhanced oil and gas production, and environmental remediation (De Gennes, 1985; Bracke et al., 1989). They involve complex physics of fluid-fluid-solid contact line dynamics and wetting behaviors which are closely related to the inter particle and intra molecular hydrodynamic interactions of the concerned multiple phase system. For example, unsaturated fractures in the vadose zone are very important for groundwater recharge, fluid motion and contaminant transport, and flow through fractures can lead to exceptionally rapid movement of liquids and associated contaminants (Nativ et al., 1995; Scanlon et al., 1997). The physics of fluid flows in unsaturated fractures is still poorly understood due to the complexity of multiple phase flow dynamics. Experimental studies of fluid flow in fractures are limited (Dragila and Weisbrod, 2004). In computer simulations it is usually difficult to take into account the fracture surface properties and microscopic roughness. Predictive numerical models can be divided into two general classes: volume-averaged continuum models (such as those based on Richard's equation) (Kwicklis and Healy, 1993) and discrete mechanistic models (Persoff and Pruess, 1995). Knowledge of the physical properties of the fluids and the geometry of the fracture apertures is required in both classes. Volume-averaged continuum models are more suitable for large-scale systems, and they usually involve the representation of fractures as porous media with porosity and permeability parameters adjusted to mimic flow within fractures. However, volume-averaged continuum models are unable to describe the details of flow dynamics in fractures, they do not reproduce the spatio-temporal complexity of multiphase fluid flow in fractures, and they often fail to predict the rapid fluid motion and contaminant transport observed in the fractured vadose zone. Small-scale studies with discrete mechanistic models are needed to develop a better understanding of the temporal and spatial dynamics of fracture flows. However, the complexity of fracture flow dynamics makes it difficult to develop successful numerical models for fluid flows in fracture networks. A broadly applicable model must be able to simulate a variety of phenomena including film flow with free surfaces, stable rivulets, snapping rivulets, fluid fragmentation and coalescence (including coalescence/fragmentation cascades), droplet migration and the formation of isolated single-phase islands trapped due to aperture variability.

Realistic mechanistic models for multiphase fracture flows must be able to handle moving interfaces, large density ratios (e.g., ≈1000:1 for water and air), and large viscosity ratios (e.g., ≈100:1 for water and air). These requirements combined with the complex geometries of natural fractures present severe challenges to mechanistic models. Grid-based numerical methods such as finite difference and finite volume methods and Eulerian finite element (FE) methods require special algorithms to treat and track the interface between different phases. These algorithms are usually complicated and fall into two general groups, *interface tracking* and *interface capturing*. Interface tracking algorithms generally use marker particles within grid cells intersected by the interface to identity the locations of interfaces (Harlow, 1964; Unverdi and Tryggvason, 1992). The particles are then advected with the flow, and the positions of the interfaces can be determined from the particle positions. This approach is computationally expensive, especially for three-dimensional simulations, and often requires additional interface repairing techniques when the interface topology changes. Interface capturing algorithms are usually based on an 'indicator' field function with different values for different phases. The location of the interface can be determined from the indicator function, $f(\mathbf{x})$ where $\mathbf{x}$ is the position in the D-dimensional computational domain, which may have a specific value at the interface, or a range of values with a large gradient near the interface. The evolution of the moving interface can be obtained from the evolution of the indicator function. The *volume of fluid* (VOF) approach (Hirt and Nichols, 1981) is based on an indicator function that specifies how much fluid of each phase is contained in each of the grid cells. In the *level-set* (LS) function approach (Sussman et al., 1994), the interface is a D-dimensional cut (contour) at $f = f^{O}$, through the D+1-dimensional surface $f(\mathbf{x})$. In most implementations, for two phase systems, $f(\mathbf{x})$, is positive in regions occupied by one phase, negative in regions occupied by the other, and $f^{O} = 0$. The VOF approach is robust and the mass loss/gain during a simulation is usually well controlled. But the captured interface usually spans several grid cells. In the LS approach, the interface is more sharply defined, but the loss/gain of mass during a simulation is larger.

There are a number of works in using the DPD method to model the multiphase flow in micro channels or fractures with surface tension and wetting effects (Liu et al., 2007b, c; Cupelli et al., 2008; Zhang et al., 2012). In this section, the DPD method with short range repulsion and long distance attraction is applied to modeling multiphase fluid flow in pore-scale channels, fracture networks and porous media.

4.3.1 Multiphase flows in micro channel and fractures

4.3.1.1 Unsaturated flow between parallel plates

In this example, DPD particles were randomly injected into a straight fracture at a constant injection rate. After equilibration, the DPD particles move into the aperture where they are influenced by each other, stationary wall particles and possibly external forces. If a purely repulsive interaction is used between the fluid particles, the injected particles will expand and occupy the whole fracture (the DPD fluid is a gas). If particle-particle interactions with short-range repulsion and long-range attraction are used, it is possible to simulate fluid flow with free surfaces, and flows with co-existing liquid and gas phases.

The simulations were carried out using walls that have a disordered internal structure. The DPD particles were randomly injected into a computational domain of 40×3×8 until an average particle density of 4 was reached. After equilibration, the particles at the bottom and top edges, within one DPD unit of the boundaries, were frozen and became the stationary wall particles. The fluid particles were then randomly injected into the fracture at a pre-selected rate. The fluid particles were injected into the first column of DPD cells in the x direction at the left hand side (**Figure 4.16**), and particles that reached the opposite end of the system, at the maximum value of x, were removed. The injected particle equilibrated with the particles that had previously entered the aperture and the wall particles, and the injected fluid particles move to the right, further into the aperture, as the density of the injected particles and the concomitant pressure increased. A pressure drop along the fracture is produced due to the particle injection. The surface tension of the fluid is determined by the interplay between the attractive and repulsive components of the interaction between the fluid particles. The interaction between the wall particles and the fluid particles can be different from that between the fluid particles, and these interactions can be tuned to give different wetting behaviours and capillary forces. The pressure drop, surface tension, and wetting behaviour or capillary force, together with possible external forces, govern the fluid flow in the fracture, which may exhibit a variety of flow regimes.

To simulate injection of fluid into the unsaturated fracture, periodic boundary condition was used along the y direction, and no-slip boundary conditions were applied on the fracture walls. The coefficients used in the DPD model were $\sigma = 3.0$ and $k_B T = 1.0\,(\gamma = 4.5)$. The interaction strength between the fluid particles was $a_f = 18.75$, while the interaction strength between the fluid and wall particles a_w was varied to mimic different wetting conditions. In the modified velocity-Verlet time integration algorithm, λ was 0.65, and a time step of $\Delta t = 0.02$ was used. A gravity force of $g = 0.02$ per particle was applied along the x direction, after the system reached equilibrium. The parameters for the SPH like potential and weight function were $A = 2.0, r_{c1} = 0.8$ and $B = 1.0, r_{c2} = 1.0$. Therefore the particle-particle interactions were given by $U(r) = a_f\left(2W_1(r,0.8) - W_2(r,1.0)\right)$ for fluid-fluid particle interactions and $U(r) = a_w\left(2W_1(r,0.8) - W_2(r,1.0)\right)$ for fluid-wall particle interactions.

The ratio between the interaction strengths, a_w/a_f, and the injection rate into the unsaturated fracture have a strong influence on the flow behavior. **Figure 4.16** illustrates a simulation with an injection rate of 10 particles per 100 steps, and a gravitational force of $g = 0.02$. The interaction ratio is 10 (the interaction between the wall and fluid particles is much larger than that between the fluid particles) and this generates strongly wetting behavior. The

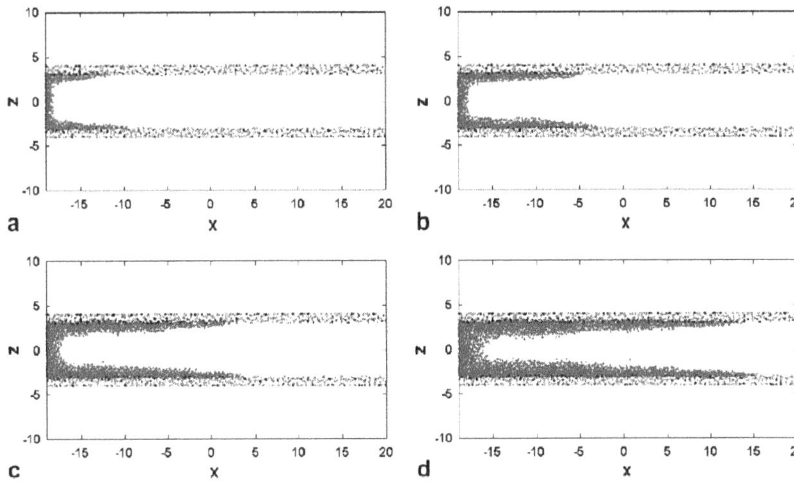

Figure 4.16 Particle distributions during injection of strongly wetting fluid into the straight fracture at (a) 10000, (b) 20000, (c) 32000, and (d) 55000 steps. The injection rate was 10 particles per 100 steps, $a_w/a_f = 10.0$ and $g = 0.02$ (Liu et al., 2007b).

figure shows that the particles near the walls move into the aperture much faster than those far from the walls, and few particles evaporate from the bulk fluid. Because the positions of the frozen wall particles are disordered and the interactions between the fluid particles and other particles have a random component, the distribution of fluid particles is only roughly rather than perfectly symmetric. The contact angle is very small. In contrast to grid-based methods in which the contact angle is exactly imposed on the fluid, the contact angle in DPD studies is approximately estimated from the position of the wall and liquid particles. Further investigation revealed that a smaller injection rate or/and a larger interaction strength between the wall and fluid particles resulted in a smaller effective contact angle. This velocity dependent contact angle behavior is observed in real systems (Hoffman, 1975; De Gennes, 1985; Cox, 1998).

Figure 4.17 shows a simulation of the injection of fluid into the unsaturated fracture with a particle injection rate of 100 particles per 100 steps and a gravitational force of $g = 0.02$. The interaction strength between the wall and fluid particles was five times the interaction strength between fluid particles. In this simulation the fluid propagates into the aperture with an approximately constant contact angle, which can be calculated from the shape of the advancing particle distributions. Again very few particles evaporated from the bulk flow. A larger injection rate or/and a smaller a_w ($a_w \geq a_f$) leads to a larger contact angle.

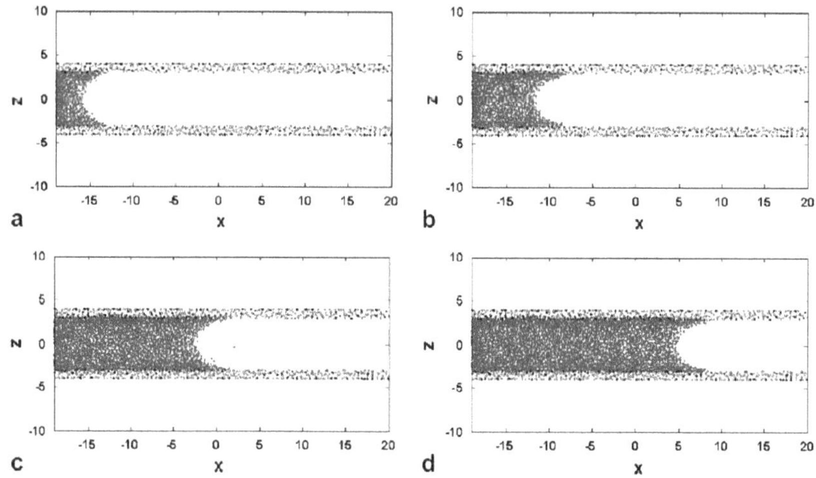

Figure 4.17 Particle distributions during injection of wetting fluid into a straight fracture at (a) 1000, (b) 2100, (c) 4500, and (d) 6500 steps. The injection rate was 100 particles per 100 steps, $a_w/a_f = 5.0$ and $g = 0.02$ (Liu et al., 2007b).

A convergence study was conducted for the two-phase injection flow through the unsaturated fracture. The convergence study is a little different from those in conventional grid-based methods and smoothed particle hydrodynamics. In grid-based methods and SPH, the convergence study is conducted by refining the mesh or increasing the overall number of particles so as to reduce the mass contained in each control volume (a mesh element or a particle). In contrast, DPD simulation employs a DPD unit system in which all of the particles have equal mass and the mass of the particles is used as the unit of mass. It is not convenient to reduce the mass of each particle (away from unity) and therefore to increase the number of particles for an unchanged computational lattice system to examine the numerical convergence. Instead, since the length unit in a DPD simulation is also non-dimensional, it is convenient to enlarge the lattice system in DPD simulation (increase the number of DPD unit in each direction) and to reduce the ratio of DPD unit to the size of the real geometry so as to increase computational accuracy. Therefore the convergence study in DPD simulations can be implemented by using a larger lattice system with a larger number of DPD particles.

Two scenarios with lattice systems of 40×8 and 80×16 respectively in x and z direction were simulated. In the two simulations, a gravitational force and interaction ratio were $g = 0.02$ and $a_w/a_f = 0.25$ respectively. The particle injection rate is 100 particles per 100 steps for the case with a lattice system of 40×8 in x and z direction and 400 particles per 100 steps for the case with enlarged lattice system. **Figure 4.18** shows the flow patterns for the two

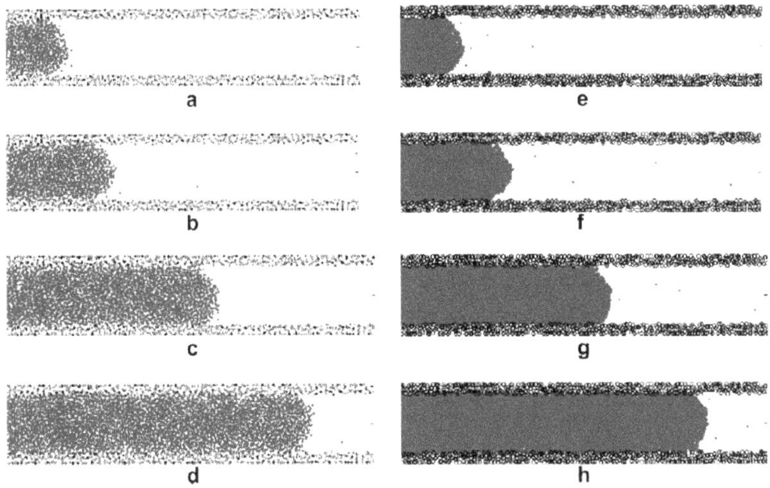

Figure 4.18 Flow patterns of the injection of a non-wetting fluid into the straight fracture. The 4 subplots in the left column (a-d) show the DPD simulation snapshots with a lattice system of 40×8 in x and z direction and the 4 subplots in the right column (e-h) show the results with a lattice system of 80×16 at 1100, 2000, 4200, and 6200 steps (Liu et al., 2007b).

Table 4.1 Position of the bulk fluid front in the channel. The positions of the bulk fluid front in the channel are normalized by the channel length (Liu et al., 2007b).

Time step	**1100**	**2000**	**4200**	**6200**
40×8	0.195	0.329	0.599	0.838
80×16	0.199	0.335	0.607	0.848

scenarios at 4 stages in a normalized non-dimensional configuration. The flow patterns obtained from two scenarios are very alike. Since the interaction strength between fluid particles is stronger than that between the fluid particles and wall particles, the contact angle is relatively large. Both scenarios demonstrate this non-wetting effect. The positions of the bulk fluid front obtained from the two scenarios are nearly the same. This shows that the obtained DPD simulation results are convergent (see **Table 4.1**).

Figure 4.16–Figure 4.18 show that the DPD method with cubic spline interaction potential functions can produce film flow (strong wetting flow), wetting flow and non-wetting flow at different injection rates and different ratios of the interaction strengths.

4.3.1.2 Fluid motion in an inverted Y shape micro channel junction

Micro channels with *Y* shape or *T* shape junction is fundamental components in MEMS devices and simulating the multiphase fluid dynamics in micro channels with *Y* shape or T shape junctions is therefore important to the understanding of dynamic physics in MEMS devices (Karniadakis and Beskok, 2002). In this section, fluid motion through a microchannel junction is investigated using the developed DPD model, and compared with the results obtained using VOF model (Huang et al., 2005b). The micro channel junction has an inverted Y shape consisting of a vertical channel that is divided into two branch channels with the same aperture (**Figure 4.19**). The fluid is injected into the micro channel junction at the top of the vertical branch.

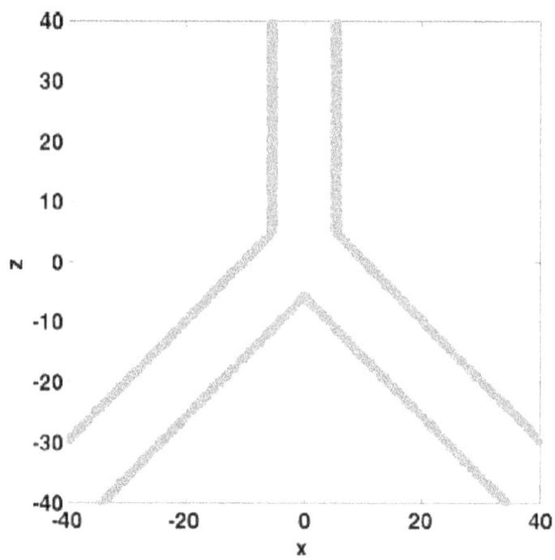

Figure 4.19 Geometry of an inverted *Y* shape micro channel junction (Liu et al., 2007c).

In the DPD model, the size of the computational domain was 80×3×80. The walls of the branched channel were represented by 2645 stationary particles using essentially the same procedure that was used for the simple parallel walled channel. No-slip boundary conditions were used along the channel walls, and particles were injected near the top of the vertical channel. The coefficients used in the DPD model were $\sigma = 3.0$ and $k_B T = 1.0$ ($\gamma = 4.5$). The interaction strength between the fluid particles was $a_f = 18.75$, and a_w, the strength of the interactions between the fluid and wall particles can be changed to mimic different wetting behaviors. The parameters for the SPH like potential and weight functions were $A = 2.0$, $r_{c1} = 0.8$, $B = 1.0$ and $r_{c2} = 1.0$. The fluid particles were injected into microchannel junction at the top of the vertical branch within 2 length units $(2r_c)$ of the upper boundary, and after equilibration, the fluid particles moved downwards in the channel. An injection rate of 100 particles per 100 steps, an interaction ratio a_w/a_f of 2, and a gravitational acceleration, g, of 0.01 (downwards in the z direction) were used in the simulation.

Figure 4.20 shows the sequential snapshots of the fluid motion in the inverted Y shape micro channel junction obtained using DPD model. In the DPD simulation, the chosen interaction ratio, a_w/a_f, led to a relatively weak wetting effect with a contact angle that was large but smaller than $\pi/2$. Again, the empirical parameter, λ, and the time step, Δt, were taken as 0.65 and 0.02 respectively. The particle distribution was roughly rather than perfectly symmetric, and only a few particles evaporated from the bulk fluid. Before the bulk fluid reached the channel intersection, a stable concave meniscus was established, with a roughly constant contact angle (**Figure 4.20**a). When the bulk fluid reached the channel junction, the curvature of the concave meniscus gradually decreased, and then became convex (**Figure 4.20**b) due to the effect of gravity acting on the accumulating mass of liquid above the meniscus. The convex meniscus moved downward gradually and then reached the apex of the channel junction (**Figure 4.20**c). Once the bulk fluid arrived at the apex, the fluid flow divided into the two inclined branches and soon formed a concave meniscus in each inclined branch with a contact angle roughly equal to that formed previously in the vertical wall (**Figure 4.20**a, d). It can be seen that the micro channel junction was then fully saturated by the invading fluid.

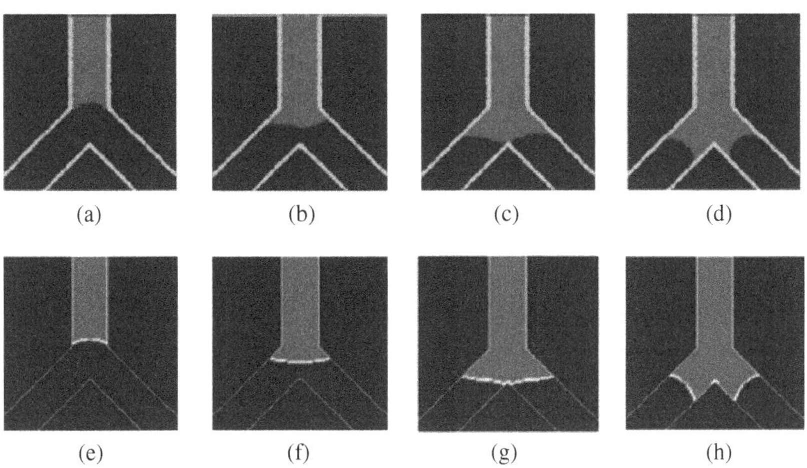

Figure 4.20 Simulations of fluid motion in an inverted Y shape micro channel junction using DPD and VOF models. The 4 figures in the upper row (a-d) show the DPD simulation snapshots and the 4 figures in the lower row (e-h) show the VOF results at 4 equivalent stages (Liu et al., 2007c).

Figure 4.20 also provides the sequential snapshots of the fluid motion obtained using VOF model. In the VOF model, the fluid properties used for simulation were $\sigma = 72.8$ dynes/cm (coefficient of surface tension) $\mu = 1.52 \times 10^{-2}$ g/(cm·s) (dynamic viscosity) $\rho = 1.0$ g/cm^3 (density) and the gravitational acceleration was $g = 980$ cm/s^2. These fluid properties are similar to those of water under normal conditions. The liquid was injected through the top of the vertical channel at a relatively high constant speed of 1.25cm/s. The advancing, receding and equilibrium contact angles in the VOF model were prescribed as 65°, 30° and 50° respectively. It is clear that the snapshots obtained using DPD and VOF models are in a very good agreement. One advantage of DPD simulations is that the solid/fluid interface and fluid/fluid interface move with the fluid particles so there is no need to track the interface explicitly. Moreover the contact angles are naturally obtained, rather than prescribed as in the VOF model.

4.3.1.3 Fluid motion in a micro channel network

A numerical study using a volume of fluid (VOF) method was presented by Huang et al. (Huang et al., 2005b) to investigate the unsaturated multiphase flow through a channel network as shown in **Figure 4.21**, together with a flow experiment based on the same channel network fabricated using polymethylmethacrylate. In their work, the overall dimensions of the apparatus are 105 mm high by 125 mm wide. The channels were fabricated with dimensions of 5 mm ± 0.05 mm wide (in the plane shown in **Figure 4.21**) by 2.5 mm ± 0.05 mm deep. Water was injected into the top entrance of the channel network using a syringe pump through a 1.6 mm inside diameter tube and drained out through one or more of the four channels located at the bottom of the apparatus. The flow rate used for the experiments was 900 ml/hr. The injected water was dyed with green food coloring for increased visibility. The apparatus was tilted 2.5° in the plane of the photograph, so gravitational forces have a component along both the horizontal and vertical channels.

In this section, the DPD model was applied to simulate the multiphase fluid motion in a complex channel network with the same pattern as the channel network in (Huang et al., 2005b). As the DPD model uses non-dimensional units, the analyses given can be extended to micro scale, depending on the proper space and time ratios chosen. In the DPD model, the size of the computational domain is $100 \times 3 \times 103$ in DPD unit in x, y and z direction. The microchannel walls were represented by 13844 frozen wall particles. In x, y and z direction, periodic boundary was applied, where on channel surface, no-slip boundary was imposed. The coefficients used in the DPD model were $\sigma = 3.0$ and $k_B T = 1.0$ ($\gamma = 4.5$). The interaction strength between the fluid particles was $a_f = 18.75$, and the interaction ratio is $a_w / a_f = 5$. The parameters for the SPH like potential and weight functions were $A = 2.0$, $r_{c1} = 0.8$, $B = 1.0$

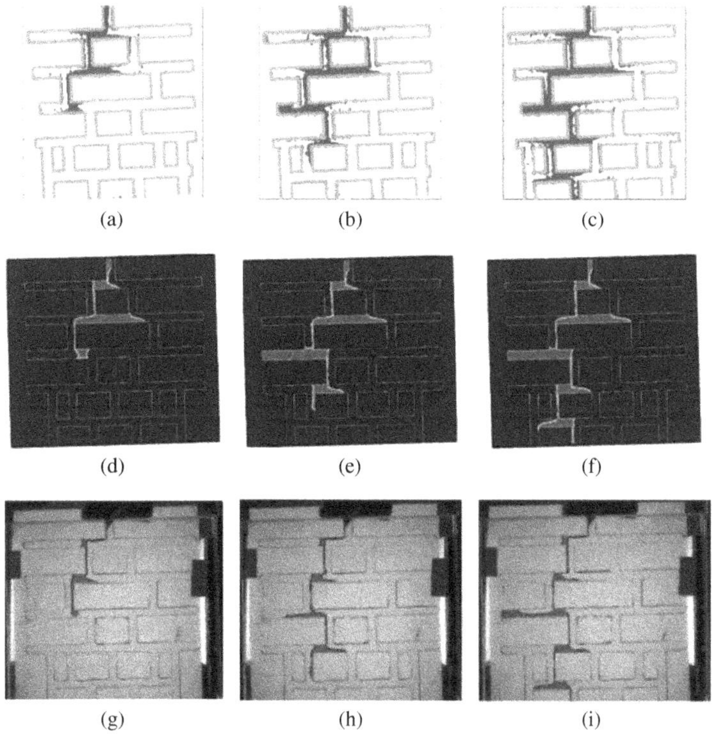

Figure 4.21 Sequential images of fluid motion into a micro channel network. The three figures in the top row (a-c) show the DPD simulation snapshots, the three figures in the middle row (d-f) show the VOF simulation results, and the three figures in the bottom row (g-i) show the experimental photographs at 3 equivalent stages (Liu et al., 2007c).

and $r_{c2} = 1.0$. The empirical parameter, λ, and the time step, Δt, were taken as 0.65 and 0.02 respectively. In the simulation, the injection rate was 50 particles per 100 steps. The gravitational force was taken 0.2 in DPD unit, and was also tilted 2.5° in the *xz* plane. This corresponds to a diagonal gravitational force with a component along both the leftward horizontal and downward vertical directions.

Figure 4.21 shows the sequential images of fluid motion into the channel network. The three Figure 4.s in the top row (**Figure 4.21**a-c) show the DPD simulation snapshots, the three Figure 4.s in the middle row (**Figure 4.21**d-f) show the VOF simulation results, and the three Figure 4.s in the bottom row (**Figure 4.21**g-i) show the experimental photographs at three equivalent stages. The DPD simulation snapshots agreed qualitatively with the experimental snapshots and VOF results at the early stage (**Figure 4.21**a, d, and g), the intermediate stage (**Figure 4.21**b, e, and h), and the late stage (**Figure 4.21**c, f, and i). The injection was conducted at two DPD cells along the left channel wall of the top entrance. This corresponds to the injection of water into the top entrance of the channel network using a syringe pump positioned next to the left top channel aperture. The flow patterns, penetration depths and formation of a quasi-steady state flow path during the late stages obtained from DPD simulation, VOF simulation and experiment are in general agreeable. There were some discrepancies if precisely examining the snapshots from the DPD simulation, VOF simulation and experiment. The discrepancies at the early stage (**Figure 4.21**a, d, g) are comparatively small. The complex spatiotemporal behavior of the multiphase flow further lead to larger discrepancy, especially in the upper part of the flow domain near the injection entrance, but the difference turns to be smaller in the lower portion of the flow domain. This is understandable since even for very simple channel geometries, the multiphase flow can exhibits very complex spatial and temporal behaviors. Small perturbations in the simulation and experiment can give rise to quite different flow modes. The VOF simulation assumed constant channel aperture with smooth channel surface, and the advancing and receding contact angles were prescribed to be constant. The channel walls in the experiment are more or less rough rather than smooth in micro-scale, which can affect the wetting behavior and flow modes of the fluid. Moreover, the experimental apparatus is not strictly in a micro scale. The DPD simulation used randomly distributed frozen wall particles, whose positions determine the roughness of the channel wall surfaces, and affect the advancing and receding contact angles.

4.3.2 Multiphase flows in porous media

4.3.2.1 Unsaturated flow through granular porous media

In this example, infiltration of liquid into an initially dry granular porous medium is simulated using DPD with a combination of short-range repulsive and long-range attractive conservative interactions (Liu et al., 2007d). **Figure 4.22**a shows the model domain and the configuration of the pore spaces and solid grains. The liquid is injected from the top of the model domain. The solid matrix of the porous medium consists of partially overlapping cylinders of different sizes. The grid cells along the left, right and bottom edges are also labelled as obstacle cells), and no-slip reflecting boundary conditions are imposed along these boundary regions. The number of fluid cells is 24618, and the number of solid cells is 26070. This corresponds to a porosity of 0.486. Similarly, 31498 wall boundary particles were used to model the solid obstacles using the approach described in Section 3.3.2, and periodic boundary conditions were used along y direction.

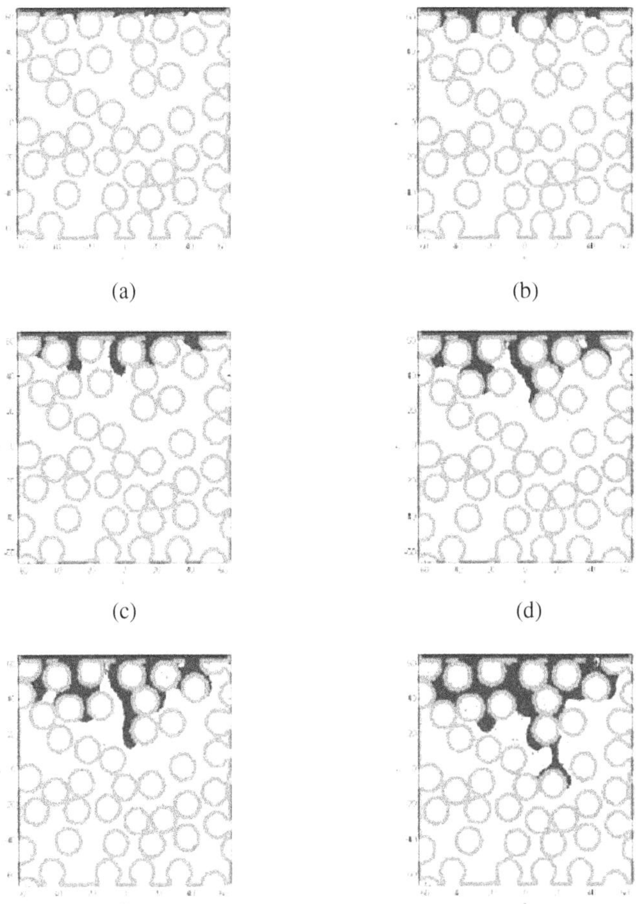

(a) (b)

(c) (d)

(e) (f)

Figure 4.22 Simulation of the infiltration process of a wetting fluid into porous media after (a) 6000, (b) 10000, (c) 14000, (d) 20000, (e) 24000, and (f) 32000 steps. The injection rate was 500 particles per 100 steps, $a_w/a_f = 5.0$ and $g = 0.05$ (Liu et al., 2007d).

The coefficients used in the DPD model were $\sigma = 3.0$ and $k_B T = 1.0$ ($\gamma = 4.5$). The interaction strength between the fluid particles was $a_f = 18.75$, and a_w, the strength of the interactions between the fluid and wall particles was changed to mimic different wetting behaviors. In the modified velocity-Verlet time integration algorithm, λ was set to 0.65, and the time step was set to $\Delta t = 0.01$. The parameters for the SPH like potential and weight functions were $A = 2.0$, $r_{c1} = 0.8$, $B = 1.0$ and $r_{c2} = 1.0$. The particle-particle interaction potentials were given by $U(r) = a_f\left(2W_1(r,0.8) - W_2(r,1.0)\right)$ for fluid-fluid particle interactions and $U(r) = a_w\left(2W_1(r,0.8) - W_2(r,1.0)\right)$ for fluid-wall particle interactions. It is found that an interaction potential of $U(r) = a_f\left(2W_1(r,0.8) - W_2(r,1.0)\right)$ is able to mimic the water/air system, though further theoretical development has to be done to further verify or calibrate the model parameters used in DPD simulations and the fluid properties.

The fluid particles were injected into the fluid cells within one DPD unit (r_c) of the upper boundary. The injected particles equilibrate with the particles that had previously entered the fluid cells and the wall particles. The injected fluid particles move downward, through the porous media, as the density of the injected particles and the concomitant pressure increased. A pressure drop through the porous media is produced due to the particle injection. The surface tension of the fluid is automatically determined by the fluid particle-particle interactions. The interaction between the wall particles and the fluid particles can be different from that between fluid particles, and these interactions can be tuned to give different wetting behaviours and capillary forces. The pressure drop, surface tension, and wetting behaviour or capillary force, together with external forces, govern the fluid flow through the porous media, which may exhibit a variety of flow regimes.

Figure 4.22 shows six different stages in a simulation of the infiltration of a wetting fluid into the porous medium. The injection rate was 500 particles per 100 time steps. The interaction strength between the wall and fluid particles was five times larger than that between fluid particles ($a_w/a_f = 5.0$). A downward gravity force of $g = 0.05$ per particle was applied. The wetting behavior is clearly shown in the figure. The fluid particles near the solid grains move rapidly along the grain surfaces due to the strong attractive interaction between the grains and the fluid particles, while the fluid particles far from the solid grains move much slower through the porous media. This leads to a small contact angle, which can be observed from the shape of the apparent liquid/gas interfaces. A small number of liquid particles evaporated from the bulk fluid.

Figure 4.23 illustrates a simulation of the infiltration of a non-wetting fluid into the same porous medium as shown in **Figure 4.22**. The injection rate was 500 particles per 100 time steps and the gravitational acceleration was $g = 0.05$, the same parameters that were used in the previous example. However, the interaction strength between the wall particles and the liquid particles is smaller than the fluid particle-particle interactions ($a_w/a_f = 0.5$), corresponding to a non-wetting liquid. Again, a few liquid particles evaporated from the bulk fluid. Unlike the previous example where the injected liquid particles rapidly wet the solid grain surface, the non-wetting behavior is obvious in this example. The fluid particles far from the solid grains advance faster through the pore spaces than those near the solid grains, and therefore it is easier to form liquid drops (see **Figure 4.23**b, c and d). The liquid drop grows with the propagation of the fluid particles, and then becomes thinner in the pore space necks between adjacent solid particles due to the interplay between the gravitational force and the interaction between fluid particles. The liquid drop eventually snaps off and breaks up (see **Figure 4.23**c and d).

The previous two examples clearly demonstrate that the ratio between the interaction strengths, a_w/a_f, has a strong influence on the wetting properties of the DPD fluid. If the interaction between the wall particles and the fluid particles

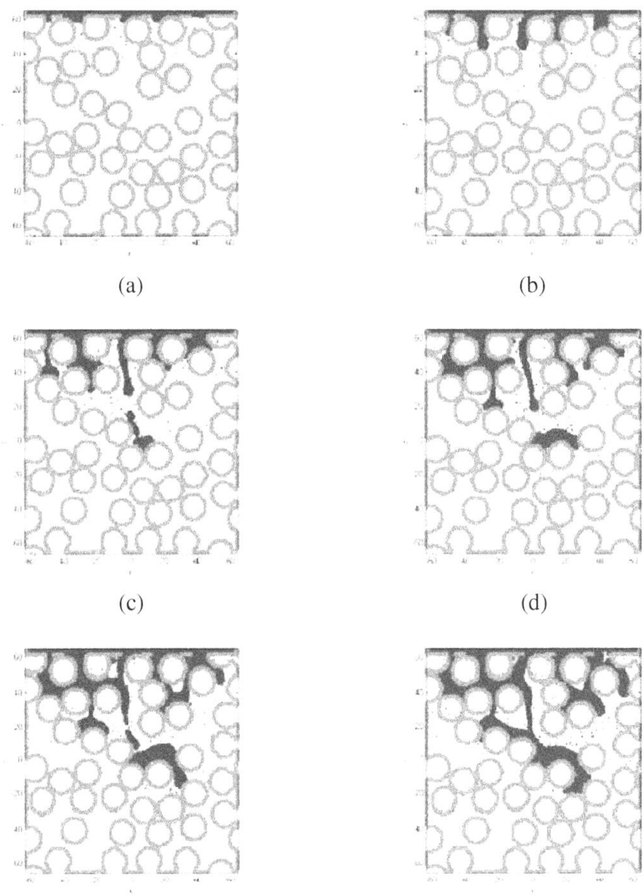

(a) (b) (c) (d) (e) (f)

Figure 4.23 Simulation of the infiltration process of a non-wetting fluid into the unsaturated porous media after (a) 6000, (b) 10000, (c) 1600, (d) 22000, (e) 26000, and (f) 32000 steps. The injection rate was 500 particles per 100 steps, $a_w/a_f = 0.5$ and $g = 0.05$ (Liu et al., 2007d).

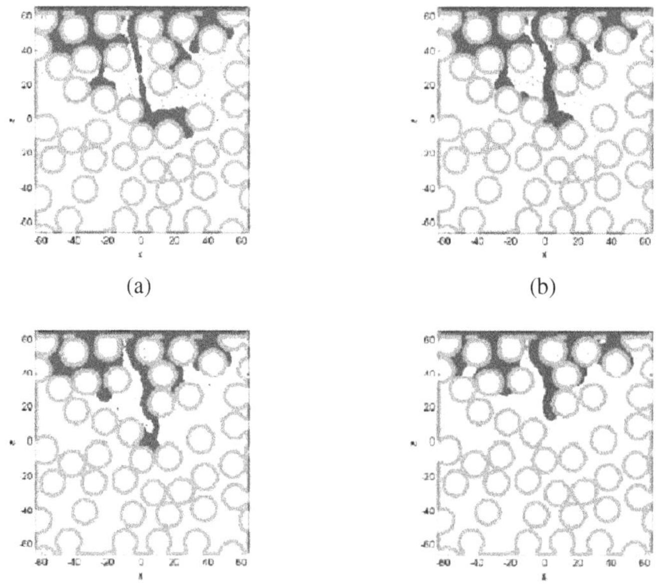

(a) (b) (c) (d)

Figure 4.24 Flow behaviors under 4 different interaction ratios, (a) $a_w/a_f = 0.5$, (b) $a_w/a_f = 1.0$, (c) $a_w/a_f = 2.5$, and (d) $a_w/a_f = 5.0$, after 24000 steps by adjusting a_w/a_f while keep other parameters unchanged, e.g. as the injection rate was 500 particles per 100 steps, and $g = 0.05$ (Liu et al., 2007d).

is equal to or smaller than that between the fluid particles ($a_w/a_f \leq 1.0$), the DPD fluid exhibits the dynamics of non-wetting fluid. A smaller a_w/a_f leads to stronger non-wetting effects. Similarly, if the interaction between the wall particles and the fluid particles is larger than that between the fluid particles ($a_w/a_f > 1.0$), the wetting behavior is obtained, and a larger interaction ratio, a_w/a_f, leads to stronger wetting effects. **Figure 4.24** shows the flow behaviors after 24000 steps for 4 different interaction ratios, (a) $a_w/a_f = 0.5$, (b) $a_w/a_f = 1.0$, (c) $a_w/a_f = 2.5$, and (d) $a_w/a_f = 5.0$, while all the other parameters were kept constant, (the injection rate was 500 particles per 100 steps, and $g = 0.05$). It is clear that when the interaction ratio is increased, the behavior changes from strongly non-wetting flow (**Figure 4.24**a), to weakly non-wetting flow (**Figure 4.24**b), then to weakly wetting flow (**Figure 4.24**c), and eventually to strongly wetting flow (**Figure 4.24**d).

4.3.2.2 Unsaturated flow through fractured porous media

In this example, the infiltration of a non-wetting fluid into a fractured granular porous medium was simulated and the results were compared with the results obtained by using the volume of fluid (VOF) method, a grid-based algorithm for tracking moving interfaces and free surfaces (Huang et al., 2005a). **Figure 4.25**a shows the model domain and configuration of the pore geometries. The solid grains are represented by a number of circular obstacles of various radii. A microfracture exists along the vertical center line within the model domain. A non-wetting fluid was injected through the top boundary. In this DPD simulation, the parameters used in the DPD model were $\sigma = 3.0$, $k_B T = 1.0$, $\gamma = 4.5$, $\lambda = 0.65$, and $\Delta t = 0.01$. The parameters for the SPH like potential and weight functions were $A = 2.0$, $r_{c1} = 0.8$, $B = 1.0$ and $r_{c2} = 1.0$. The particle-particle interaction potentials were given by $U(r) = 18.75(2W_1(r, 0.8) - W_2(r, 1.0))$ for both the fluid-fluid particle interactions and the fluid-wall particle interactions (i.e. $a_f = 18.75$ and $a_w/a_f = 1.0$). The injection rate was 500 particles per 100 steps, and the gravitational acceleration of $g = 0.01$ acting downward.

Figure 4.25 shows the DPD simulation of the infiltration process at four different stages. For all stages, the injected liquid flows preferentially through the largest pore throats and through the vertical 'microfracture' where the

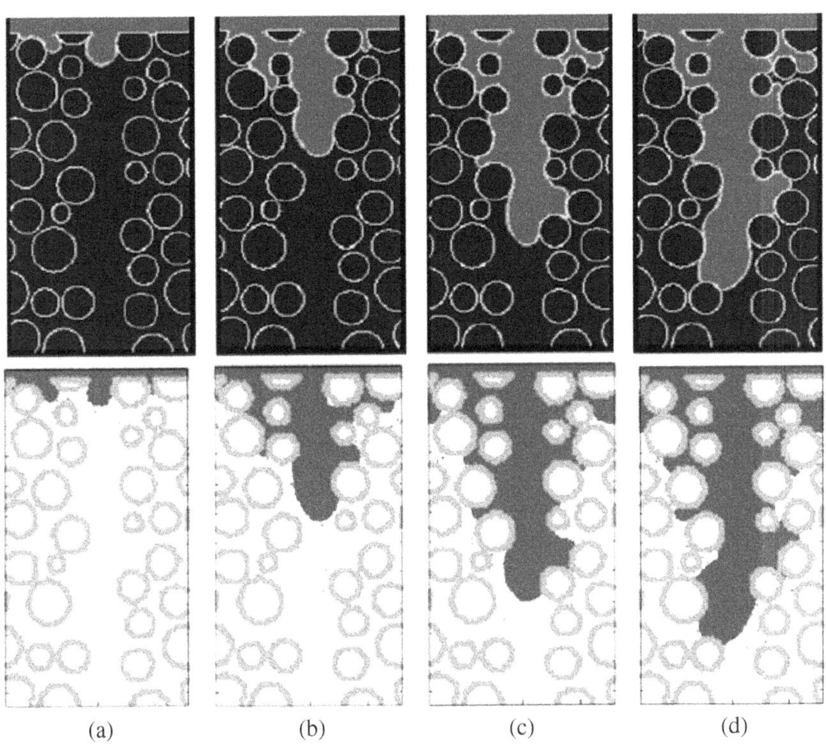

(a) (b) (c) (d)

Figure 4.25 VOF (Top) and DPD (bottom) simulation of non-wetting fluid infiltration into a fractured porous medium at 4 equivalent stages (Liu et al., 2007d).

capillary forces resisting flow are minimal. At the early stage of the infiltration (**Figure 4.25**a), the injected liquid infiltrated into the medium through the microfracture and the pore in the upper-left corner, which has a larger pore throat compared with that in the upper-right corner. As the infiltration progresses, the injected liquid moves preferentially through the microfracture (see **Figure 4.25**b-c) due to the fact that the aperture of the fracture is larger than the sizes of all the surrounding pore throats. At the same time, the injected liquid also moves downward inside the pore matrix through relatively large pore throats, but at a much slower rate. Thus the infiltration depth in the fracture is much larger than the infiltration depth in the surrounding pore matrix. At the later stages of the infiltration process, as shown in **Figure 4.25**c and **Figure 4.25**d, the preferential flow along the fracture dominates the flow inside the surrounding pores, therefore almost all the injected liquid flows along this preferential flow path, leaving a large portion of pores uninvaded by the injected liquid. During all stages of the simulation, the apparent contact angles of the liquid interfaces at the solid grain surfaces are greater than 90°, indicating that the injected liquid behaves like a non-wetting liquid.

Due to the difficulty of directly comparing the two-dimensional DPD simulation results against experiments and the difficulty of relating the particle interaction model to fluid properties, we used an alternative grid-based multiphase flow simulation approach, a grid-based Navier-Stokes equation solver coupled with the volume of fluid (VOF) interface-tracking method, to simulate the same infiltration problem to validate the DPD simulation results. The VOF method numerically solves the Navier-Stokes equations to obtain fluid velocities in discrete grid cells, coupled with an implicit interface-tracking algorithm using the volume fractions of fluid in each grid cell as a fluid phase indicator. This approach systematically incorporates the effects of inertial forces, viscosity, and gravity acting on the fluid densities, solid wall wetting, and the pressure drop across curved fluid-fluid interfaces due to surface tension. Wetting effects are modeled by imposing different contact angles depending on whether the liquid interface is advancing, receding, or essentially stationary.

The VOF model domain has a size of 1.28 cm × 2.56 cm. The average diameter of grains is 2.6 mm. The model domain is discretized into a 128 × 256 uniform grid (0.1 mm × 0.1 mm per grid cell). The input parameters to the VOF simulation include (in cgs units): a surface tension s = 32.8 dynes/cm; a dynamic viscosity of m = 1.52×10^{-1} g/(cm.s); a density of ρ = 1.0g/cm^3; and a gravitational acceleration of g=50 cm/s^2. The advancing, receding, and equilibrium contact angles are $\theta_a = 115°$, $\theta_r = 95°$ and $\theta_{eq} = 105°$. The liquid is injected across the top boundary at a rate of 5 cm/s. These input parameters give a Reynolds number (*Re*) of 9.37 and a Bond number (*Bo*) of 0.10. These two numbers indicate that the flow is still within the low Reynolds number regime and dominated mainly by the capillary forces due to surface tension and the inertial forces due to injection. In this simulation, it was assumed that the air has no impact on the dynamics of the injected liquid. This assumption is valid only when no air is trapped within the pores, but this assumption (or approximation) has a substantial computational benefit since the dynamics of only the liquid phase must be computed.

Figure 4.25 also shows the VOF simulation of the infiltration process at four different stages equivalent to those in DPD simulation. It is seen that although there are slight differences between the DPD and VOF simulation results in several of the pore spaces, the overall flow patterns within the fracture are almost the same. The slight differences in these pore spaces also originate from the different ways used to represent obstacles in two approaches. The solid obstacles are represented using "solid" occupied grid cells in the VOF model, but they are represented using frozen randomly distributed particles in the DPD model. This leads to slight differences between the flow patterns simulated using the VOF and DPD models. In general, the visual comparison between these two simulations clearly reveals that these quite different approaches give essentially the same fluid dynamics in the fractured porous medium.

4.4 Movement and suspension of macromolecules in micro channels

Understanding the dynamic behavior of macromolecules, such as DNA, is very important for fundamental research and practical applications in bio, chemical and medical engineering, especially in designing micro-devices. Recently, micro-devices enable processing, analyzing, and delivering biochemical materials in a wide range of biomedical and biological applications (Chun et al., 1999; Fan et al., 2006). For example, micro-needle can be used to efficiently and precisely deliver a small amount of drug or DNA into local tissue, skin regions, and even cells. In order to avoid pain and tissue traumas caused by traditional technologies of drug injection and delivery, a variety of micro-needles have been designed for hypodermic injection and transdermal drug delivery (Brazzle et al., 1999; Lin and Pisano, 1999). Micro-channels are the main field to deliver and control injected materials. By designing optimal structures of micro-channels or micro-channel networks, it is possible to efficiently control the injection process, either for simple fluids or complex fluids with macromolecules. It is therefore very important to understand the dynamic behavior of macromolecular when passing though micro-channel with different structures.

Recent development of experimental techniques enables us to study the dynamics and rheological properties of macromolecules such as DNA in micro-channels. For example, it is possible to use fuorescence imaging techniques to visualize the micro-structural conformations of molecules (Chu, 1991). Optical tweezers have been used to measure the extension properties of single DNA molecules (Perkins et al., 1994). By using these techniques, some experimental works have been conducted to study the mechanics of macromolecular suspension flows. Perkins et al. in 1995 measured the extension properties of tethered single DNA molecules in a uniform flow (Perkins et al., 1995). Perkins et al. in 1997 and Smith and Chu studied the dynamic behavior of single DNA molecules in an elongation flow (Perkins et al., 1997; Smith and Chu, 1998). Smith et al. observed the dynamic behavior of single DNA molecules in steady shear flows (Smith et al., 1999). The flow of molecular suspensions through a micro-channel is more complicated as it is a combination of non-uniform elongation and shear flows. Shrewsbury et al. used epi-fluorescence microscopy to characterize the flow's impact on the conformation of the molecules in microfluidic devices in which the path consists of a large, inlet reservoir connected to a long, rectangular channel followed by a large downstream reservoir (Shrewsbury et al., 2001). In the device, DNA molecules were observed to undergo elongation, non-uniform shear and compression. Near the channel wall, high shear rates results in dramatic stretching of the molecules, and may also result in chain scission of the macromolecules.

On the other hand, with the development of computational methods and computer hardware, numerical simulations of the movement and evolution of macromolecules in micro-devices have been more and more popular. Numerical simulation can provide more details on the flow field and conformations of macromolecules by tracking each molecular chain segment. The size of macromolecules is usually in the same order of magnitude as that of the channel and the equivalent Knudsen number is larger or equal to unity (Fan et al., 2006). This restricts the applicability of continuum mechanics methods to these flow problems. Molecular dynamics (MD) has been used for comparison with worm-like chain (WLC) (Koelman and Hoogerbrugge, 1993) and slip length measurements for sheared films (Cheon et al., 2002). However, but the number of beads in MD simulation is usually small and the time scale is much shorter than the time scale (of the order of second) that for gathering experimental data (Smith et al., 1999). In addition, the characteristic size of micro channels and DNA suspension can range from dozens of nanometers to several micrometers, and even to several millimeters. For meso scale problems, it is expensive for MD to directly simulate the dynamic behavior of macromolecules in micro-channels.

Compare with molecular dynamics, Monte Carlo (MC) relies on statistical mechanics and it generates states according to appropriate Boltzmann probabilities, instead of trying to reproduce the dynamics of a systems. MC can be deal with problems with larger time and space scales than MD, and it has been used to simulate DNA flow through entropic trap array where polymer is modeled by a lattice model with bond fluctuation (Tessier et al., 2002).

As the size of flow field and DNA molecules can be too large to be handled by MD simulation, various mesoscopic methods have been applied in this area. In this area, the Brownian dynamics simulation (BDS) (Northrup et al., 1984; Hur et al., 2000) is one most common approach. As a simplified version of the Langevin dynamics, Brownian dynamics corresponds to the limit where no average acceleration takes place during the simulation run. Various molecular models have been used to model the DNA molecules, such as the Kramer's bead-rod chain (Doyle and Shaqfeh, 1998), the FENE chain (Fan et al., 2003) and the worm-like chain in BDS (Hur et al., 2000). Among those molecular models in BDS, the worm-like chain is considered to be the most realistic one (Bustamante et al., 1994), comparing with experimental measurements. Larson et al. simulated a DNA molecule in an extensional flow (Larson et al., 1999), and Hur et al. in shear flow (Hur et al., 2000). Except for simulating single molecules, Brownian dynamics simulation has widely been used in simulating the rheological properties of polymer solutions. For example, the simulation of freely-draining flexible polymers in steady linear flows (Doyle et al., 1997), bead-rod chains in start-up of extensional flow (Doyle and Shaqfeh, 1998), and relaxation of dilute polymer solutions following extensional flow (Doyle et al., 1998). Brownian dynamics simulations have shown a good comparison with experiments on DNA molecules in shear flow (Hur et al., 2000). However, these models are usually only valid for simple fluid flow since the flow field has to be specified a-priori in BDS, such as the above mentioned freely-draining flexible polymers in steady linear flow, bead-rod chains in start-up of extensional flow and single DNA molecule in shear flow (Hur et al., 2000).

Although there are many other mesoscale simulation methods, such as Lattice Gas Automata (LGA), Lattice Boltzmann (LB), it is also difficult for those methods to deal with complex fluids and complex flow which may contain macromolecules. In many recent works, dissipative particle dynamics (DPD) were employed to simulate dynamic behavior of macromolecules. As a meso scale fluid simulation method, DPD is promising in simulating macromolecules suspension flow through micro channels. Macromolecules (such as DNA molecules) are generally simulated by a series of particles (beads) linked together using springs to mimic the solute molecules, while the simple DPD particles are usually used to model the solvent.

One of the first applications of DPD to modeling dilute polymer solution was presented by Schlijper et al. (Schlijper et al., 1995), who applied the DPD method to investigate the link between molecular features of polymer molecules and the rheological properties of dilute polymer solutions. Kong et al. investigated the effect of solvent quality (i.e., good solvent vs. poor solvent) on the conformation and relaxation of polymers (Kong et al., 1997), and the DPD method is applied to simulate single chains in solvents with a range of quality. Groot studied the formation of a polymer-surfactant complex in bulk solution, and DPD is used to model a system containing polymer, surfactant and water (Groot, 2000). Spenley presented scaling laws for polymer melts and dilute polymer in DPD system (Spenley, 2000). The polymer in a good solvent shows satisfactory agreement with scaling and Kirkwood theory, and the polymer melt is in excellent agreement with the predictions of Rouse theory. Symeonidis et al. demonstrated the correct static scaling laws for the radius of gyration by DPD simulations of several bead-spring representations of polymer chains in dilute solution (Symeonidis et al., 2005b). They found that the worm-like chain simulating single DNA molecules compares well with average extensions in shear flow from experiments.

Recently, the dynamics of polymer chain in different flows and micro-channels (or micro fluidic devices) were investigated. For example, Wijmans and Smit used DPD to simulate shear flow between two flat plates and to study the effects of shear flow on end-tethered polymer layers ("brushes") (Wijmans and Smit, 2002). They found that as the flow velocity changes during an oscillation cycle, the polymer chains are able to relax their configurations with respect to the shear rate. Symeonidis modeled λ-phage DNA under shear using DPD with the worm-like chain models (Symeonidis et al., 2005a). They computed the values of viscosity, diffusivity and Schmidt number and presented comparison of wormlike chain models under shear with experimental and Brownian Dynamics results. Fan et al. simulated the micro channel flows of macromolecular suspension while FENE chain is used in the DPD simulation (Fan et al., 2003). They also studied the flow of DNA suspension through periodic contraction-diffusion micro channels and obtain some typical conformations of macromolecules during evolution of DNA molecules when passing through the channel (Fan et al., 2006). Researchers from the same group also conducted other simulations using DPD. For example, Chen et al. investigated the steady-state and transient shear flow dynamics of polymer drops in a micro channel (Chen et al., 2004). Duong-Hong et al. investigated the Electroosmotic flow (EOF) of thin Debye layer and DNA molecular in nano-fluidic systems (Duong-Hong et al., 2008). Pan et al. simulated DNA separation process through entropic trapping mechanism with worm-like chain (Pan et al., 2010). The simulations show that longer DNA strands do move faster than shorter ones, as observed in experiment (Han et al., 1999; Han and Craighead, 2000). They confirmed that the delayed entrance is the cause of entropic trapping and concluded that the corner trapping is not a contributor to DNA separation.

In this section, the DPD is used to model macromolecular suspension passing through micro channels including straight channel, contracted channel, inclined channel and grooved channels with contraction and expansion structures.

4.4.1 Straight micro channel

In this example, the movement and suspension of macromolecules in a straight micro channel is simulated using DPD method with the FENE model. Different number (*nChain* = 30, 60, 90 respectively) of macromolecules with different chain length (*ChainLen* = 30, 60, 90) are modelled to investigate the dynamic behaviour and migration feature of the macromolecules. In the FENE model, $H = 6.0$, $r_{\max} = 3.0$. The conservative interaction strength (a) between simple DPD particles (a_{ss}, where s denotes simple particles), between DPD particles on macromolecules (a_{mm}, where m denotes DPD particles acting as beads on the macromolecules), between simple DPD particles and DPD particles on macromolecules (a_{sm}, or a_{ms} where $a_{sm} = a_{ms}$), and between mobile DPD particles (simple DPD particles and DPD particles on macromolecules) and fixed solid particles (a_{sf}, a_{pf} or a_{fs}, a_{fm} where f denote fixed solid particles, $a_{sf} = a_{fs}, a_{mf} = a_{fm}$) are tall taken as 18.75. The coefficient for the random force (σ) is taken as 3.0.

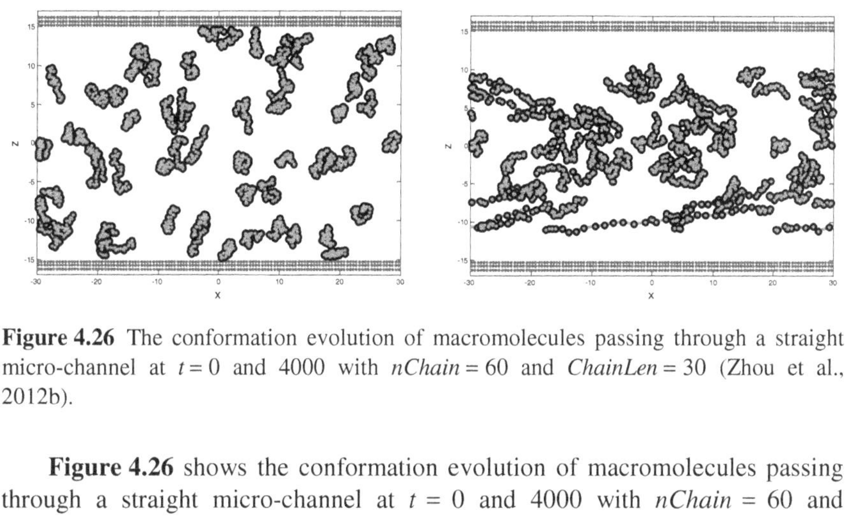

Figure 4.26 The conformation evolution of macromolecules passing through a straight micro-channel at $t = 0$ and 4000 with *nChain* = 60 and *ChainLen* = 30 (Zhou et al., 2012b).

Figure 4.26 shows the conformation evolution of macromolecules passing through a straight micro-channel at t = 0 and 4000 with *nChain* = 60 and *ChainLen* = 30. For the convenience of visualization, the solution (simple) DPD particles are ignored. At the initial stage, the macromolecules are approximately evenly distributed in a coiled state. At t = 4000, though some macromolecules located around the center of the channel are still in a coiled state, the outside macromolecules are well stretched. The macromolecules are located basically within $-10 \le z \le 10$.

Figure 4.27, **Figure 4.28** and **Figure 4.29** show the velocity, density and temperature profiles across the slit with 1) different number of macromolecular chains (*nChain* = 0, 30, 60 and 90) and the same chain length (*ChainLen* = 30) (**Figure 4.27**), 2) different chain length (*nChain* = 0, 30, 60 and 90) and the same number of macromolecules (*nChain* = 30) (**Figure 4.28**) and 3) same volume fraction of macromolecules (*ChainLen* = 30, *nChain* = 60 vs. *ChainLen* = 60, *nChain* = 30 and *ChainLen* = 30, *nChain* = 90 vs. *ChainLen* = 90, *nChain* = 30). It is clear that the number of macromolecules and the chain length do not obviously influence the temperature profile, and the system kinetic temperature is in good agreement with the pre-defined system temperature. The existence of

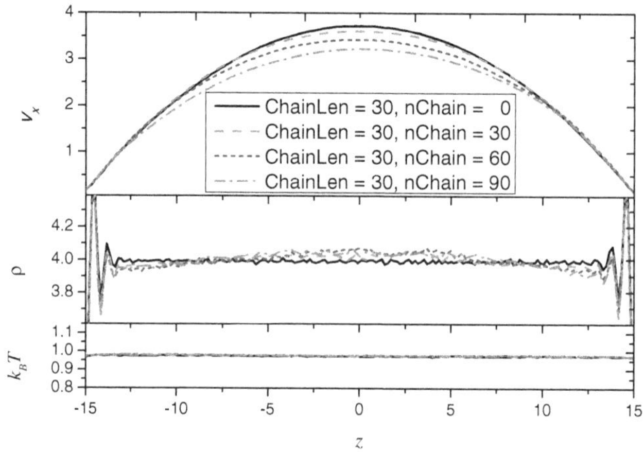

Figure 4.27 Velocity, density and temperature profiles across the slit with different number of macromolecular chains from (*nChain* = 0, 30, 60 and 90) and the same chain length (*ChainLen* = 30) (Zhou et al., 2012b).

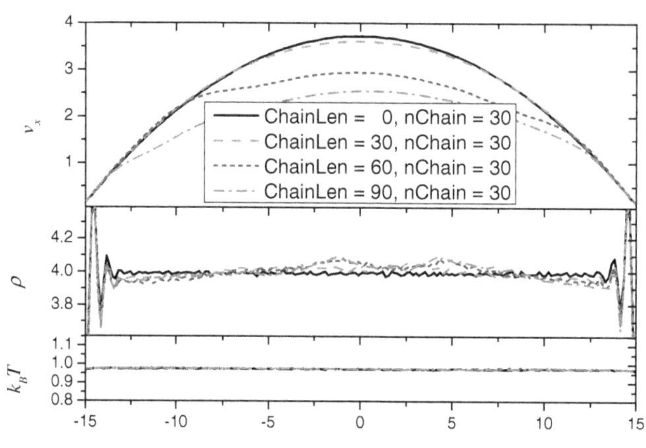

Figure 4.28 Velocity, density and temperature profiles across the slit with the same number of macromolecular chains and different chain length (*ChainLen* = 0, 30, 60, 90, *nChain* = 30) (Zhou et al., 2012b).

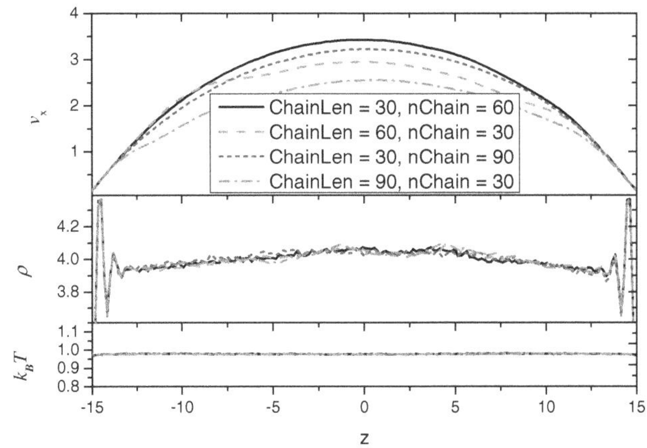

Figure 4.29 Velocity, density and temperature profiles across the slit with same volume fraction of macromolecules (*ChainLen* = 30, *nChain* = 60 vs. *ChainLen* = 60, *nChain* = 30 and *ChainLen* = 30, *nChain* = 90 vs. *ChainLen* = 90, *nChain* = 30) (Zhou et al., 2012b).

macromolecules influences slightly the density profile, and the density obtained with macromolecules oscillates around the density obtained without macromolecules, especially in the center region. This is reasonable as more macromolecules are distributed in the center region. One point is that as basically there is no macromolecule near the solid walls, the existence of macromolecules does not lead to density fluctuation in the boundary area.

Different from the temperature and density profiles, for which the existence of macromolecules have no or slight influence, the existence of macromolecules influence the velocity profile greatly. It is found that for a given chain length, increasing the number of macromolecules can lead to smaller center peak value on the velocity profile. In contrast, reducing the number of macromolecules produces bigger center peak velocity approaching the center peak value obtained without macromolecules. This is reasonable as macromolecules can influence the flow field through interaction with neighboring solute particles. Since the mass (and therefore inertia) of the macromolecules is bigger than that of solute particles, the existence of macromolecules can "drag" the movement of surrounding solute particles. This dragging effect is bigger when increasing the volume fracture of the macromolecules. In contrast, when the volume fracture of the macromolecules is small enough, macromolecules will not have big influence on the flow field, and the solution approaches that without macromolecules. Similarly, for a given number of macromolecules, increasing the chain length means bigger volume fracture of the macromolecules and larger influence on the velocity profiles. On the other hand, reducing the chain length can lead to bigger velocity profiles, which approach the velocity profile without macromolecules when chain length approaches 0 (or only one particle in a chain).

From **Figure 4.29**, it is also observed that for the same volume fraction, macromolecules with longer chain length but smaller number of chains can have bigger influence on the velocity profile. This is because longer chain length means bigger mass (and therefore bigger inertia) of a macromoleculcs, and the resultant dragging effects on surrounding particles consequently become bigger. In contrast, for a larger number of macromolecules with smaller chain length, the mass of the macromolecules is small. An extreme case is for *ChainLen* = 1, this is exactly the solution obtained without macromolecules.

4.4.2 Contracted micro channel

This example involves the movement and suspension in a square-shaped contraction expansion channel. 16800 DPD particles are randomly distributed in the flow domain (see **Figure 4.30**), among which 1800 are beads of 60 macromolecules with *ChainLen* = 30. 2868 particles are used as virtual wall particles. The temperature, density and horizontal velocity profiles are plotted along $x = \pm 25.5$, $x = \pm 16.5$ and $x = 0$ (the dashed lines on **Figure 4.30**)

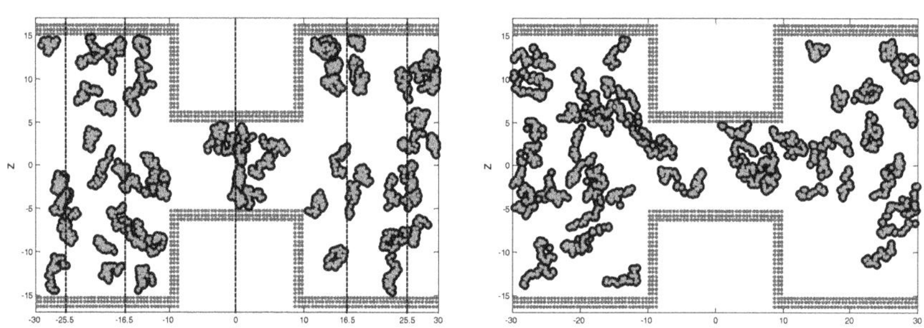

Figure 4.30 Conformation evolution of macromolecules passing through the periodic quadrate contraction micro-channel at $t = 0$ (left) and 4000 (right) with $nChain = 60$ and $ChainLen = 30$ (from (Zhou et al., 2012b)).

Figure 4.30 shows the conformation evolution of macromolecules passing through the periodic quadrate contraction micro-channel at t = 0 (left) and 4000 (right) with *nChain* = 60 and *ChainLen* = 30. Again, at the initial stage, the macromolecules are distributed evenly in a coiled state. At t = 4000, the flow is fully developed. Compared to the distribution of particle and macromolecule in a straight channel, the square-shaped contrition channel influences the extension of the macromolecules. Hence few macromolecules can be fully stretched.

Figure 4.31 shows the velocity, density and temperature profiles at different horizontal positions (x) across the periodic quadrate contraction micro channel at t = 4000 (left: without macromoeclue, i.e., *nChain* = 0, *ChainLen* = 0; right:

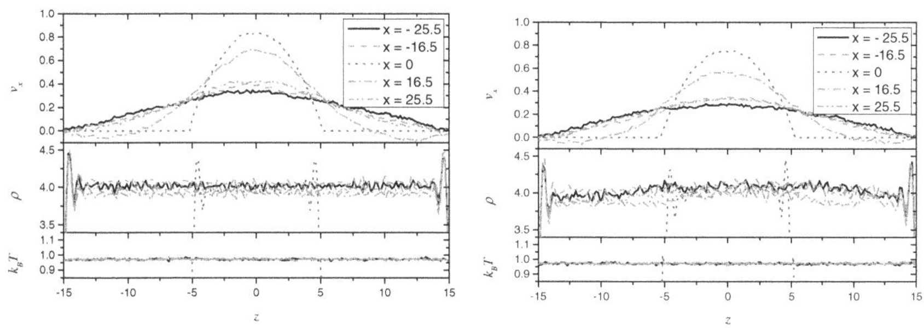

(a) (b)

Figure 4.31 Velocity, density and temperature profiles at different horizontal positions (x) across the periodic quadrate contraction micro channel at $t = 4000$ (left: *nChain* = 0, *ChainLen* = 0; right: *nChain* = 60, *ChainLen* = 30) (Zhou et al., 2012b).

with macromolecules while *nChain* = 60, *ChainLen* = 30). Comparing **Figure 4.31**a and **Figure 4.31**b, it can conclude that

- For temperature profiles: Whether there are macromolecules or not, the system kinetic temperatures are all in good agreement with the pre-defined system temperature with very small oscillations. In other words, the existence of macromolecules does not influence the system temperature.
- For density profiles: Due to the contraction effects of the channel, DPD particles flowing into the contracted channel are constrained, and this leads to the slightly higher density distribution before the contraction section than that after the contraction section. The existence of macromolecules exaggerates the density differences before and after the contracted section due to the dragging effects by the macromolecules on simple solute particles.
- For velocity profiles: As DPD particles are restrained when flowing into the contracted section of the micro channel, the velocity profiles on the left-hand-side of the contracted section is smaller than those at the same locations on the right-hand-side of the contracted section. At $x = 16.5$, recirculatory flows can be found at the top and bottom corners. The existence of macromolecules can alleviate the recirculatory effects due to the dragging phenomenon by the macromolecules on simple solute particles.

4.4.3 Inclined micro channel

This example involves the movement and suspension of macromolecules in an inclined micro channel. Similar to the previous example, 16800 DPD particles are randomly distributed in the flow domain (see **Figure 4.32**), among which 1800 are beads of 60 macromolecules with *ChainLen* = 30. 2868 particles are used as virtual wall particles. The temperature, density and horizontal velocity profiles are plotted along $x = \pm 25.5$, $x = \pm 16.5$ and $x = 0$ (the dashed lines on **Figure 4.32**).

Figure 4.32 shows the conformation evolution of macromolecules passing through a periodic sloping micro channel at t = 0 (left) and 4000 (right). Again, the macromolecules are initialized in a coiled state. After the flow is fully developed, some macromolecules are in a fully stretched state, while some are still in a coiled state. The extension of the macromolecules is dependent on the slope of the inclined channel. If the slope is very small, the inclined channel will behave somewhat similar to the straight channel, and macromolecules can be fully stretched. If the slope is very big, the inclined channel will behave somewhat similar to square-shaped contraction channel, and it is not easy for the macromolecules to be fully stretched.

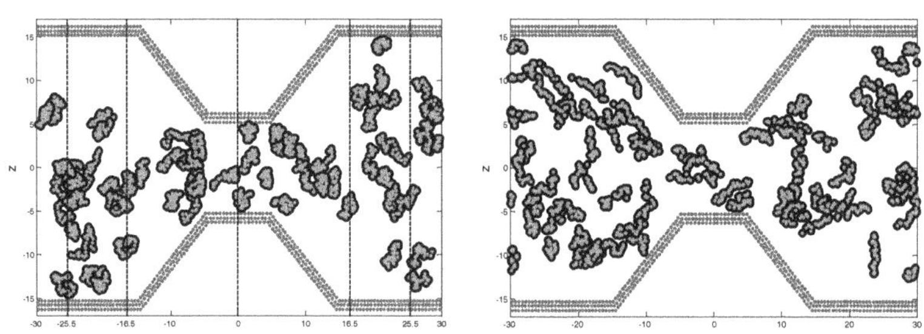

Figure 4.32 Conformation evolution of macromolecules passing through a periodic sloping micro channel at t = 0 (left) and 4000 (right) with *nChain* = 60 and *ChainLen* = 30 (Zhou et al., 2012b).

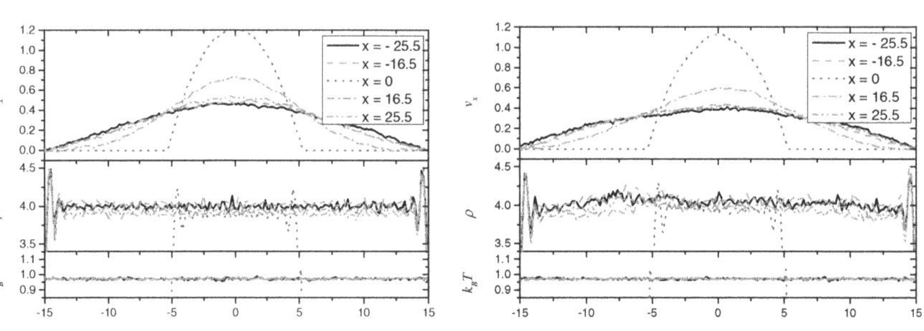

(a) (b)

Figure 4.33 Velocity, density and temperature profiles at different horizontal positions (x) across the periodic sloping at t = 4000 (left: *nChain* = 0, *ChainLen* = 0; right: *nChain* = 60, *ChainLen* = 30) (Zhou et al., 2012b).

Figure 4.33 shows velocity, density and temperature profiles at different horizontal positions (x) across the periodic sloping at t = 4000. Again, the existence of macromolecules has no and slight influence on the temperature and density profile, and influences the velocity profile and recirculatory flow pattern. The influence is dependent on the slop of the inclined channel. The performance of the inclined channel is between the straight channel and square-shaped contraction channel.

4.4.4 Grooved micro-channel

Grooved channels with regular contraction and expansion sections are frequently used in micro-devices and the movement and evolution of macro-molecules are not yet not fully understood. Except for the experimental work by Shrewsbury, there are only a few works available on this subject so far. Kasiteropoulou et al. used DPD method to model simple fluid flow in periodically grooved micro-channels and to seek further insight on the effects of wall roughness on fluid flow (Kasiteropoulou et al., 2012). Fan et al. simulated the suspension flows of macro-molecules with FENE chain and worm-like chain through simple a micro slit, and found varieties of conformations of macro-molecular in micro-channels, including periodic contraction-diffusion micro-channels (Fan et al., 2003). Duong-Hong et al. (2008) used DPD to simulate electro-osmotic flow and DNA molecular in simple and complex two-dimensional and 3D geometries frequently used in nano-fluidic devices, and the results are in very good agreement with theoretical results (Duong-Hong et al., 2008). Pan et al. used DPD with worm-like chain to model the process of DNA separation in a micro-device, and found that longer DNA strands move faster than shorter ones (Pan et al., 2010).

In this example, the movement and suspension of macromolecules in a grooved micro-channel with contraction and expansion structures are modeled by the DPD method with FENE chain model. Before modeling macromolecular suspension, the simple flow (DPD particles are not chained to form complex structures) are simulated.

4.4.4.1 DPD simulation of simple flow in a grooved channel

As shown in **Figure 4.34**, the grooved micro-channel is modeling by two stationary parallel solid plates with contraction and expansion sections, characterized by periodically placed rectangular protruding bottleneck. The computational domain is $-100 \leq x \leq 100$, $-1.5 \leq y \leq 1.5$ and $-10 \leq z \leq 10$ while $-5 \leq z \leq 5$ at rectangular protruding bottleneck. A total number of 48996 simple DPD particles are used, including 40800 fluid particles placed in the planar slit and 8196 wall particles located in three layers parallel to the x-y plane

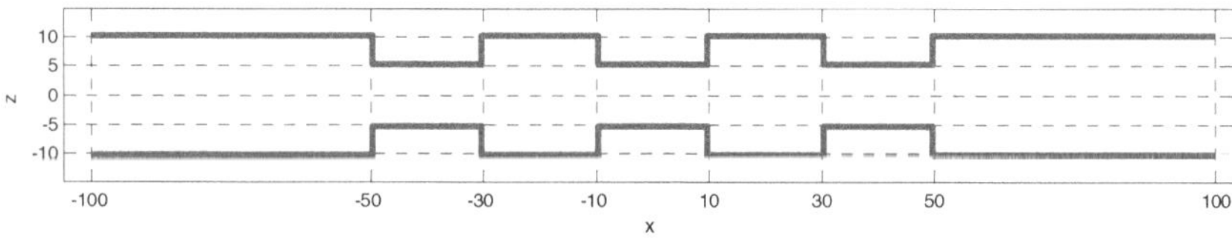

Figure 4.34 Schematic view of the grooved micro-channel simulated. The computational domain is divided into 400 bins in the x-direction, one bin in y-direction, and 94 bins in z-direction (Zhou et al., 2013).

in each side. The periodic boundary conditions are applied to fluid boundaries in the x and y directions. On the surface of solid walls, we applied Maxwellian reflection boundary conditions to yield the no slip boundary condition.

As discussed in Chapter 3, the physical properties of a DPD fluid are characterized by the parameters in the DPD formulation. Some parameters can be determined by fitting the relevant data of the real fluid, some are chosen to maintain the numerical accuracy in simulating simple cases, such as the ideal gas and equilibrium states. For complex systems, there is no solid physical basis to determine the parameters characterized interaction strengths between different components. According to Groot and Warren's work (Groot and Warren, 1997), to satisfy the compressibility of water, the coefficients of the conservation and random forces should be $a_{ij} \approx 75k_BT/\rho$ and $\sigma = 3.0$, with $\lambda = 0.65$ in the verlet-type algorithm. This conclusion can also be extended to the dilute polymer solution (Fan et al., 2003; Fan et al., 2006). In this simulation, the unit of energy (k_BT) is 1.0 and the density (ρ) is 4.0. Thus the corresponding $a_{ij} = a_{fluid} =$ 18.75, if i and j both denote fluid particles or beads in the molecular chains. There are no physical bases on how the solid particles interact with solvent particles or beads in the molecular chains. For simple, we assumed $a_{wall} = 5.0$ and $a_{ij} = \sqrt{a_{fluid}a_{wall}}$ when calculating the interaction between fluid and wall particles. To generate the flow, a field force, $g = 0.01$, are applied to each fluid particle and beads in the x direction. To solve the equation of motion, a modified velocity-verlet algorithm with time step at $\Delta t = 0.01$ is used. The computational domain is divided into 400×94 bins in the xz plane and local data are collected in each bin. By averaging the sampled data in each bin over 10000 time steps, we can obtain all local flow properties, such as temperature, stress tensor and pressure.

By averaging the velocity vectors of all particles in each bin, we can get the velocity field. **Figure 4.35** shows the velocity field in x and y direction respectively. It is easy to get the streamline of the flow field as we have already got the velocity of all bins. **Figure 4.36** shows the streamline of the flow field. It is observed that the horizontal velocity is smaller in the region near the wall. As the width of the channel reduces, the horizontal velocity increases. In areas with a sudden contraction, for example, flow contracts from cross-section $x = -60$ to

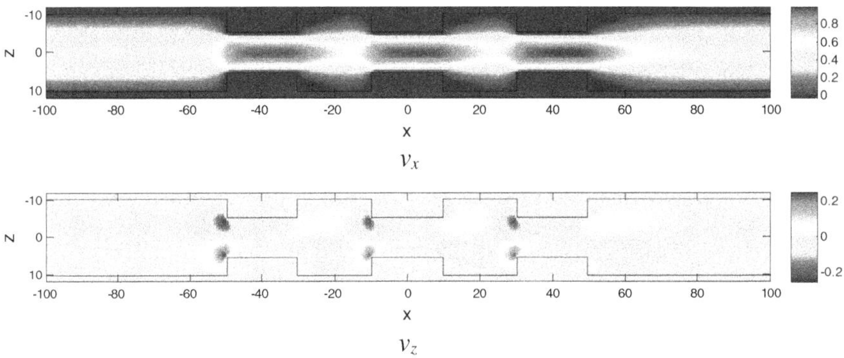

Figure 4.35 Velocity field (top: horizontal velocity v_x; bottom: vertical velocity v_z) (Zhou et al., 2013).

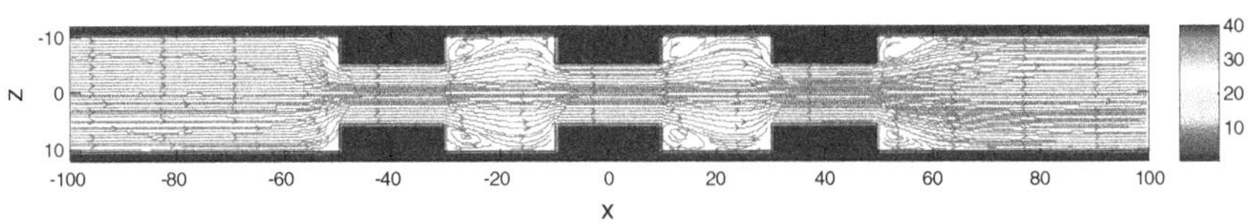

Figure 4.36 The streamline of the simple flow field in the grooved channel (Zhou et al., 2013).

cross-section $x = -40$, forming a vena contraction. This contraction rate is about 50% (half of the original channel width). In a sudden expansion section, for example, fluid flow from cross-section $x = -40$ to cross-section $x = -20$, the horizontal velocity gradually reduces. It is observed that vortices have been generated at the top and bottom corners behind the contraction structure (e.g. $-30 \le x \le -20$).

In order to better show the difference of horizontal velocity at different position, the profiles of horizontal velocity at $x = -90$, -55, -40 and -25 are plotted in **Figure 4.37**. The horizontal velocity profile at x is calculated by average the local horizontal velocity of bins between $x-5$ and $x+5$ along the x direction. For example, the profile of horizontal velocity at $x = -90$ is actually mean value of horizontal velocity between $x = -95$ and $x = -85$ along the x direction.

It is found in **Figure 4.37** that the structure of the grooved channel with sudden contraction and expansion strongly affects the velocity distribution. There are obvious differences in the velocity distribution of different cross-sections. The horizontal velocity profile at $x = -90$, far away from sudden contraction and sudden enlargement, is close to quadratic. This indicates that the flow at $x = -90$ is close to Poiseuille flow. The horizontal velocity profile at

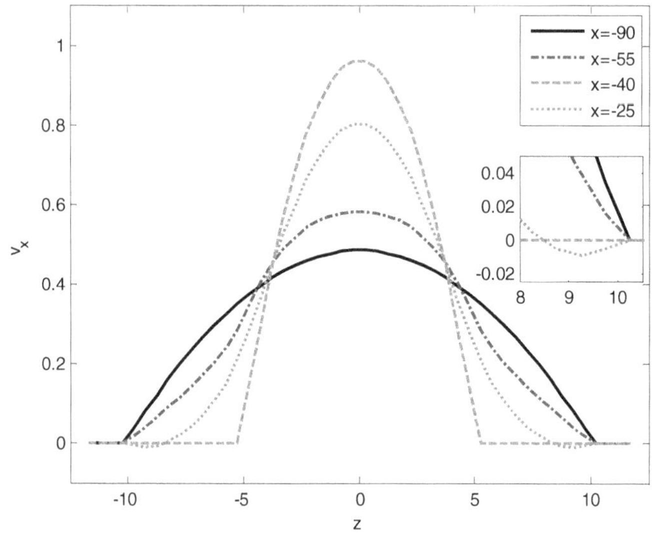

Figure 4.37 The horizontal velocity profiles at x = −90, −55, −40 and −25 (Zhou et al., 2013).

$x = -55$, facing a sudden contraction, is higher in center but lower in marginal regions. The horizontal velocity profile at $x = -40$, in the center region of the contraction section, is close to quadratic too but is much sharper than the profile at $x = -90$. The horizontal velocity profile at $x = -25$, just behind a sudden enlargement, is negative in marginal region, which correspond to the vortices in **Figure 4.36**. It is a coincidence that velocity profiles at $x = -55$, −40 and −25 intersect at two points.

The profiles of density and temperature at $x = -90$, −55, −40 and −25 are also plotted to show the difference of local properties at different cross-sections (see **Figure 4.38**). Form **Figure 4.38**, we can see that temperature is almost uniform across the channel at each cross-section except for regions near the wall. The temperature drops near the wall due to low wall temperature. Density is uniform in most regions except for marginal regions near the wall, where density fluctuations exist. The peak density value near the wall is higher than that in the center of the channel due to mass conservation. These fluctuations in density are not only due to the interaction of fluid particles and wall particles, but also due to bounce-back conditions. The structure of quadrate contraction has slight influences on the density distribution at different cross-sections. We can easily find that the density of particles in the region just facing the contraction construct is slight higher than other regions. It may be due to the quadrate contraction blocking the flow from the left to the right.

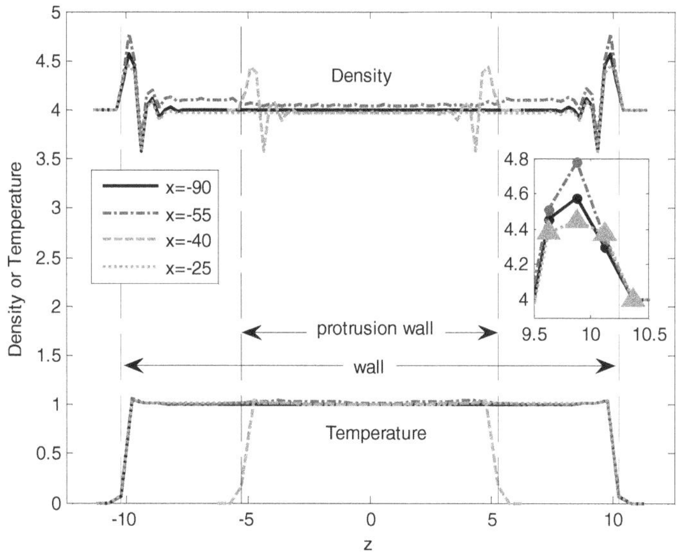

Figure 4.38 Density and temperature profiles at different x across the slit (Zhou et al., 2013).

The density profile at cavity ($x = -25$) is nearly constant and almost coincide with the profile at $x = -90$ far away from protrusion and cavity. But it should be noted that a small difference exists between the two profiles for regions very close the wall. This phenomenon may be caused by particle trapping inside the cavities of the vortices. To confirm this assumption, we investigate the particle trajectories.

The trajectory of particles in DPD simulation does not strictly flow along the streamline due to the random force. **Figure 4.39** shows four typical trajectories of four concerned particles at different instants:

- The first trajectory shows a particle passing through the region with protrusions and cavities very smoothly without any block. These trajectories usually exist in the center region of z direction.
- The second trajectory shows a particle passing through the region with protrusions and cavities smoothly except in the cavity ($-30 \le x \le -10$). The particle hesitates in the cavity for a significant amount of time. These trajectories usually exist in the region which is close to the protrusion wall.
- The third trajectory shows that a particle is trapped by the vortices inside the cavity after through two protrusions and one cavity. These trajectories usually exist in the region very close to the protrusion wall.

- The fourth trajectory shows a particle escaping from the vortices inside the cavity after being trapped. This trajectory is the extension from the third one. A detailed trajectory analysis shows that there are no particles that stay trapped inside the cavities for the whole simulation time. This phenomenon indicates that this trapping effect does not depend on the initial particles configuration. Particles that are initially located inside the cavities do not stay there during the entire simulation. Meanwhile, particles that are initially distributed outside the cavities may enter the cavity and are trapped during the simulation.

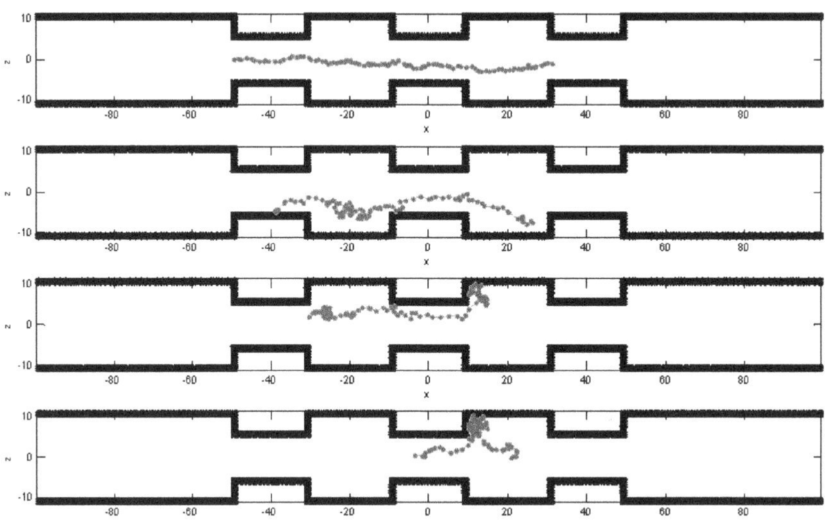

Figure 4.39 Four typical trajectories of particles (Zhou et al., 2013).

4.4.4.2 DPD simulation of macromolecular suspension in a grooved channel

In this section, the DPD particles and FENE chains are used to to model the suspension of macromolecules by just adding FENE chains to the flow field of grooved micro-channel flow. 40500 fluid particles and 1800 FENE chain beads (30 FENE chains each with 60 beads) are used in the simulations to model the dynamic behavior of macromolecular suspension in grooved micro-channel. The parameters in the spring force law of FENE are taken to be $H = 6.0$, $r_m = 1.5$. Other parameters remain the same as those mentioned in the previous simulation.

Figure 4.40a shows the initial configurations of some selected macromolecules in the grooved micro-channel. In the initial moment ($t = 0$), all chains randomly and relatively uniform distributed in the entire flow field, and most of them are coiled. As flow develops, the FENE chains move, rotate and stretch. When fully developed ($t = 8000$), as shown in **Figure 4.40**b, most chains are distributed in regions between $z = -5$ and $z = 5$, Macromolecules located nearer to walls are stretched better than those in the center region, where macromolecules are usually coiled. The coiled chains are found in the central part of the slit, where shear rate is low, while the longest chains are found in the region close to wall, where shear rate is high.

Figure 4.41 shows the velocity field and **Figure 4.42** shows the streamline of the flow field. From **Figure 4.41** and **Figure 4.42**, the movement and evolution of macromolecules with FENE chain do not show significant difference in velocity fields and streamline, compared with those from the simple flow.

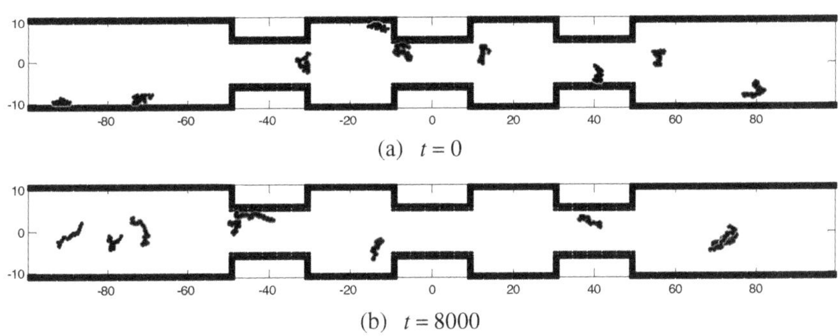

Figure 4.40 The initial configuration and final conformations of macromolecules in the grooved micro-channel (Zhou et al., 2013).

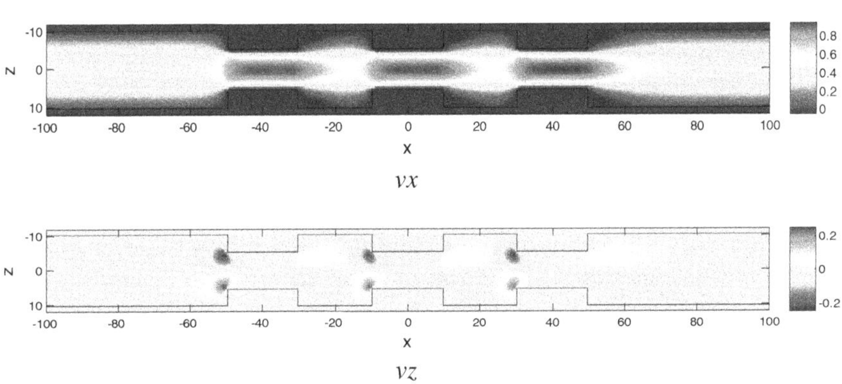

Figure 4.41 Velocity field (top: horizontal velocity v_x; bottom: vertical velocity v_z) (Zhou et al., 2013).

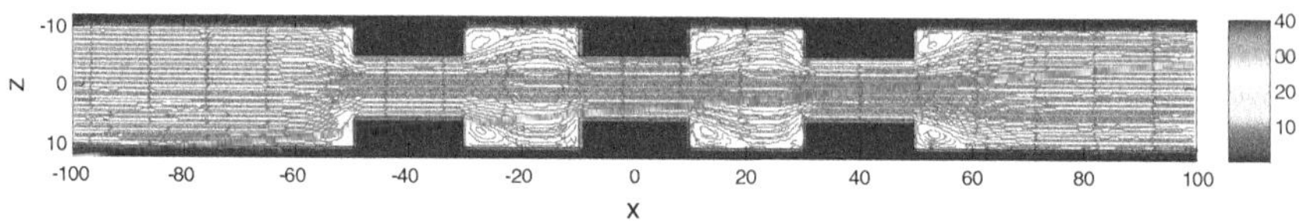

Figure 4.42 The streamline of the flow field (Zhou et al., 2013).

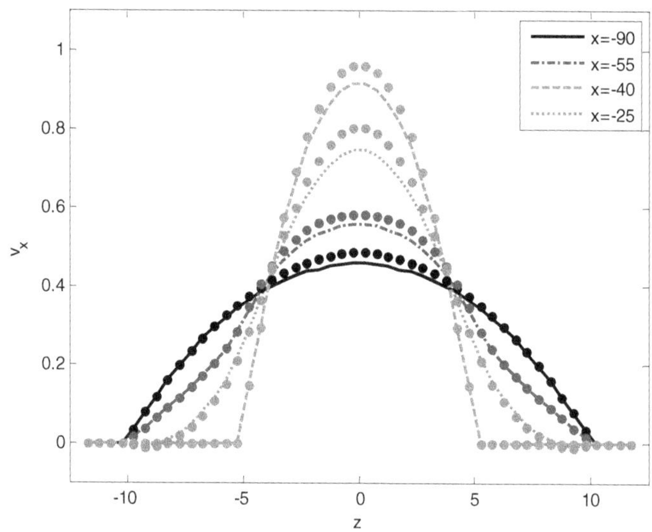

Figure 4.43 Velocity profiles at different horizontal positions (x). (Lines indicate the horizontal velocity profiles from macromolecules suspension, while dots indicate those from simple fluid flow) (Zhou et al., 2013).

It is not easy to identify different properties between flow of macromolecule suspension and flow of simple fluid by directly observing the flow fields. Hence, we draw the profiles of horizontal velocity at $x = -90$, -55, -40 and -25 (see Figure 4.43). In Figure 4.43, lines indicate the horizontal velocity profiles from macromolecules suspension, while dots indicate those from simple fluid flow. It can be observed that macromolecular chains have remarkable influences on the velocity distribution across the micro-channel. The existence of macromolecular chains will drop velocity in center region of the channel. It is observed that the impact of the macromolecular chains on the velocity distribution is more significant in the region $-5 \le z \le 5$, and this result is consistent with the phenomenon that most chains are distributed in the region $-5 \le z \le 5$ as shown in **Figure 4.40**.

Figure 4.44 shows the profiles of density and temperature at $x = -90$, -55, -40 and -25. It is seen that the existence of macromolecule chains does not

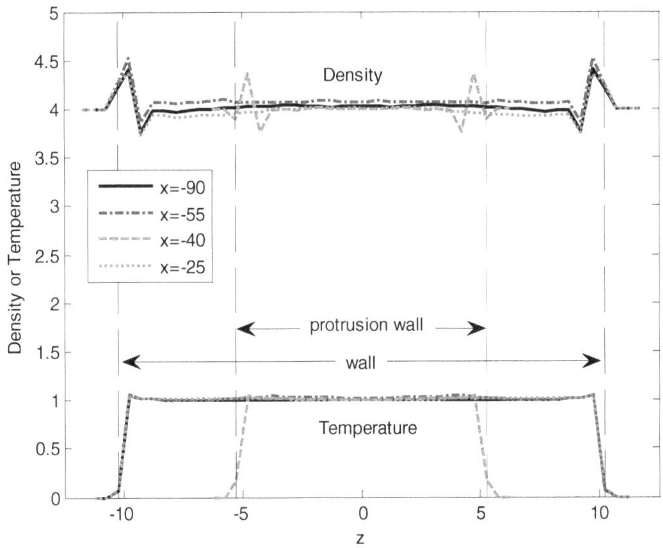

Figure 4.44 Density and temperature profiles at different x across the slit for macromolecular suspension in a grooved channel (Zhou et al., 2013).

significantly influence the temperature distribution: the temperature is almost uniform across the channel and is very close to the given temperature, but only drops near the wall due to low wall temperature. Meanwhile, macromolecular chains have slight influence on the density distribution across the micro-channel: the density of central region is slightly higher instead of uniform across the channel. Region where macromolecular chains are relatively dense has a slight higher particles density. Whether there exists macromolecules or not, a fluctuation in density exists in the regions near the wall.

In order to observe the dynamic behavior of macromolecules passing through the channel, serial frames of cconformations of marked macromolecules in the channel are plotted in **Figure 4.45**. It is observed that macromolecules kept changing their own conformation due to the random force in DPD formula, which corresponds to the thermal fluctuations in mesoscopic scale. From the conformations of a single macromolecule in the serial frame, we can find that macromolecule trend to stretch longer when passing the region close to wall than middle region of the channel. The structure of the grooved channel with sudden contraction and enlargement affect the conformation evolution of the macromolecule significantly. To better illustrate this effect, we labeled each macromolecule in **Figure 4.45**. When facing a sudden contraction, macromolecule tends to coil itself, or adjust the direction of extension to parallel the protruding wall, to pass the channel smoothly. For example, Macromolecule # 1 changes from a stretching state to become curled before and close to the contraction section. Macromolecule # 2 changes from a vertically stretching state to become a horizontal stretching state, and become curled before, in and after the contraction section. When facing a sudden enlargement, macromolecule can stretch and rotate, just as shown by Macromolecule # 3. A detailed conformation and trajectory observation of macromolecule show that there is no macromolecule trapped by the vortices inside the cavity for the whole simulation time, while simple DPD particle can be trapped by the vortices. Therefore, it is reasonable to draw such a conclusion: a macromolecule tends to adjust its own conformation and angle to suit the flow and pass the channel quickly.

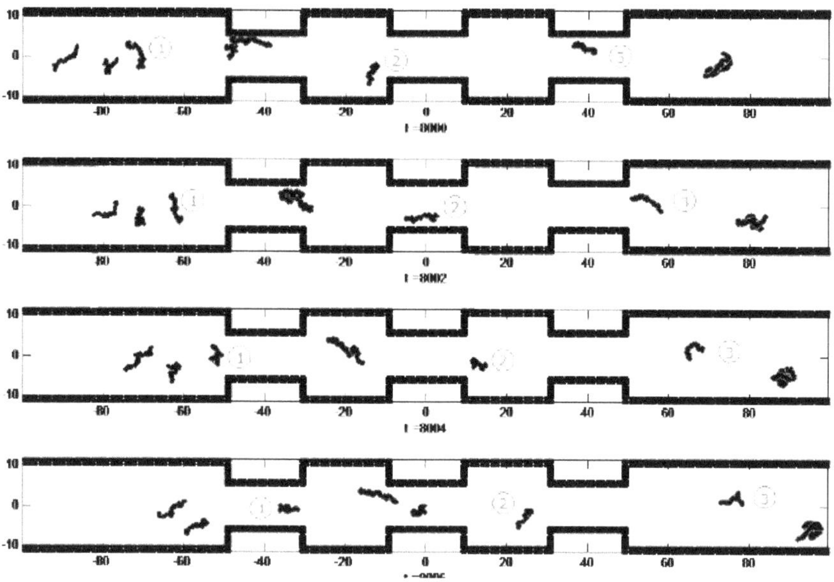

Figure 4.45 Conformation evolution of marked macromolecules passing through the channel (Zhou et al., 2013).

4.5 Movement and deformation of single cells

Dynamical behaviors of migration and deformation variations of cells in blood vessels are caused by pathological changes in mechanical properties of cells, which may be closely related to severe cell diseases. In cancer, the variations are due to internal factors such as genetic mutation. While in malaria, the variations are due to external factors such as parasites and bioactive lipids. Through biochemical reactions, both internal and external factors can change the internal structure and mechanical behavior of living cells. These variations are often facilitated by the altering in the mechanical behavior of living cells such as large changes of elastic modulus. For example, Healthy red blood cells (RBCs) can squeeze through capillaries with diameters smaller than itself to deliver oxygen to various parts of the body due to its high deformability. While the elastic modulus of RBCs infected by the protozoan plasmodium may increase by more than a factor of 10 due to intracellular structural changes (Suresh, 2006). The pathological RBCs are too stiff to deform sufficiently to traverse narrow capillaries. Instead they may disrupt the blood flow and block the capillaries, possibly leading to anemia and can even cause death. Modern physiology and medicine have established the relationship of mechanical variations between healthy human cells and pathological cells. For instance, compared to healthy cells, diseased cells such as cancer cells are known to have different stiffness and elasticity (Lee and Lim, 2007). Such differences could be used to distinguish between normal and diseased cells (Bathe et al., 2002; Hou et al., 2009). Recently, increased micro-fluidic devices were designed to diagnose and treat cells disease such as cancer as difference cells can have different mechanical properties (Suresh, 2007). It is therefore an important step to understand how cells with different mechanical properties respond to physical loads.

Continuum cell models are the main approach to model the mechanic dynamics of cell. The continuum models treat the cell as comprising materials with certain continuum material properties. Appropriate constitutive material models and the associated parameters can be derived from experimental observations (Lim et al., 2006). Generally, continuum models can be classified into two main categories, namely solid models and liquid drop models. The solid models usually assumed the whole cell to be homogeneous without considering the distinct cortical layer. By assuming homogeneity, the mechanical parameters can be reduced. This greatly simplifies the experimental data analysis. The corresponding material models can be incompressible elastic solid (linear elastic solid model) or the viscoelastic solid (linear viscoelastic solid model). The homogeneous viscoelastic solid model was first proposed in 1981 (Schmid-Schönbein et al., 1981), which was used to study the human leukocytes undergoing micropipette aspiration with small-strain deformation. For large cell deformations, this model may not work. The elastic model is a simplified version of the viscoelastic model, and it neglects the time factor (Lim et al., 2006). This model was used to model endothelial cells in micropipette aspiration (Theret et al., 1988). It was found when the pipette radius is very small compared to the local radius of the cell surface. The cell can be approximated as an incompressible elastic half-space. Based on this model, Mijailovich et al. constructed a finite element model to compute cell deformation during magnetic bead twisting experiment (Mijailovich et al., 2002). Numerical simulations show that adhesive forces are sufficient to keep the bead firmly attached to the cell surface throughout the range of working torques. Practice shows the solid models can usually achieved equilibrium after certain amount of loading. For instance, even when the suction pressure greatly exceeded the critical suction pressure, endothelial cells and chondrocytes were unable to flow into the pipette (Jones et al., 1999).

On the other hand, liquid drop models treat the cell as a liquid drop, which can model large cell deformations. The Newtonian liquid drop model was developed by Yeung and Evans in 1989 to simulate the flow of cells into the micropipette (Yeung and Evans, 1989). The model describes the cell as a homogeneous Newtonian viscous liquid drop enclosed by a cortex with constant, isotropic tension but without bending resistance (Evans and Yeung, 1989). Newtonian liquid drop model can simulate large cell deformations well when the progress of cells in micro-pipette aspiration is slow. However, for a fast period less than 5s, Newtonian liquid drop model could not explain why the cells would exhibit as a fast elastic recoil, analogous to the initial rapid entry in the aspiration experiment (Evans and Yeung, 1989). In order to consider the effects of the nucleus on cell deformation, the compound drop model was developed, which assumed the nucleus as an encapsulated liquid drop (Hochmuth et al., 1993). Compare with above mentioned Newtonian liquid drop, the compound drop model can effectively explain the rapid initial response in micro-pipette aspiration and fast recoil on recovery (Tran-Son-Tay et al., 1998). The compound drop model was also used to model cell under shear flow (Agresar et al., 1998) and extensional flows (Kan et al., 1999), cell adhesion, and migration (N'dri et al., 2003) as well as shear thinning and membrane elasticity (Marella and Udaykumar, 2004). Recently, Leong et al. presented a modified compound drop model, which can consider stiffness, elasticity, and viscosity of both the cortex and the nucleus to model breast cancer cell entry into a constricted micro-channel. The modeled cell entry behavior agrees with experimental observations (Leong et al., 2011).

The continuum cell models are easy to implement and straightforward to use in computing the mechanical properties of the cells if the biomechanical response at the cell level is needed. It provides less insight into the detailed molecular mechanical events. For this reason, more accurate micro- and nano-structural models were developed. However, almost all micro- and nano-structural models were only applicable to red human blood cell (RBC). The RBC membrane is composed of a lipid bilayer and an attached cytoskeleton. The cytoskeleton consists primarily of spectrin proteins, which form the network by linking short actin filaments. Discher et al. (Discher et al., 1998) and Li et al. (Li et al., 2005) developed the spectrin-level RBC model. The spectrin-level model corresponds to an effective spectrin network where each spring represents a single spectrin tetramer. The RBC is represented by a network of springs in combination with bending rigidity and constraints for surface-area and volume conservation. The spectrin-level RBC model was successfully validated against experimental data of the mechanical response of an individual cell. However it involves limited degrees of freedom and application of the model in flow simulations requires prohibitively expensive computations. For this reason, Pivkin and Karniadakis developed a coarse-grained model based on the spectrin-level RBC model using mean-field theory and then applied it to DPD simulations in capillaries of 10 μm in diameter while the blood velocity is typically about 1 mm/s (Pivkin and Karniadakis, 2008). This RBC, as a collection of DPD particles, is immersed in DPD fluid. The RBC particles interact with the fluid particles through DPD potentials and the temperature of the system is controlled through the DPD thermostat. Initially, the fluid is at rest and the RBC is placed in the middle of the capillary. A body force is applied in the axial direction to drive the flow in the tube. The RBC was found to deform under the flow conditions and, after some transition period, assumed the parachute-type shape, which is commonly observed in experiments (Tomaiuolo et al., 2007). After the body force is turned off, the DPD fluid slows down and eventually returns to rest, while the RBC recovers its equilibrium biconcave shape. A more systematic and rigorous procedure to derive coarse-grained RBC models was present by Fedosov et al. (Fedosov et al., 2010). The RBC is modeled by DPD and captures the elastic response at small and large deformations, which agrees very well with experiments of RBC stretching by optical tweezers. In addition, they also develop a stress-free model which avoids a number of pitfalls of existing RBC models, such as non-smooth or poorly controlled equilibrium shape and dependence of the mechanical properties on the initial triangulation quality. Fedosov et al. also extended this model to model adhesive dynamics of RBCs in Malaria (Fedosov et al., 2011).

Zhou et al. developed a similar approach for modeling other cells in an attempt to simulate the breast cancer cells through a constricted micro-channel (Zhou et al., 2012a). Based on the finite extensible non-linear elastic (FENE) bead spring chain model, they successfully used DPD particles to construct a phenomenological breast cancer cell **Figure 4.46**. The model was then used to investigate the transportation and deformation of cell through a constricted micro-channel **Figure 4.47**. It is found that before entering the contracted micro-channel, the cell gradually deforms with reducing velocity. As the cell gradually enters the micro-channel, it elongates until its major axis reaches maximum. As the front of the cell gradually leaves the micro-channel, it restores its original structure with increasing velocity. Those results are agreeable with experimental observations (Hou et al., 2009) and the results obtained from compound drop model (Leong et al., 2011). The obtained patterns of cell deformation, contraction and expansion as well the recovery of its original shape are similar to experimental observations. Size effects of the contraction micro-channel and influences of cell membrane properties are also studied.

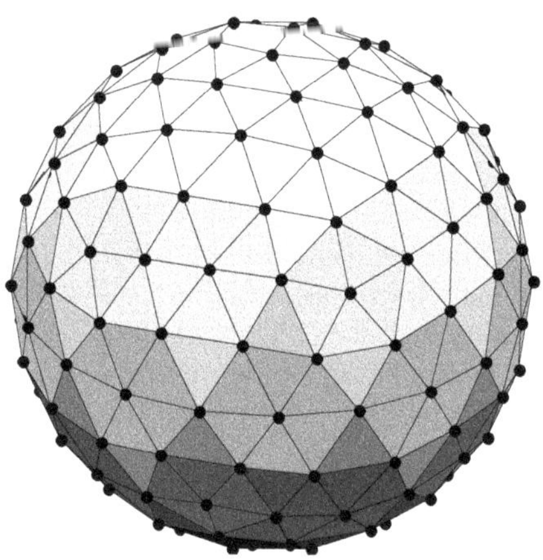

Figure 4.46 Particle model for cell membrane represented by a network of springs linked DPD particles (Zhou et al., 2012a).

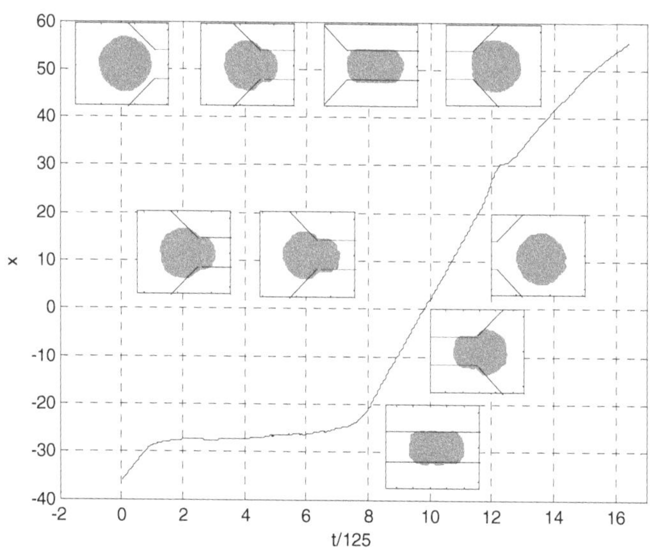

Figure 4.47 Snapshots of a single cell passing through a micro channel, and the distance-to-origin profile. The origin is at the point of entry of cell into the micro channel (Zhou et al., 2012a).

4.6 Concluding remarks

This chapter presents the applications of DPD in modeling the dynamic flow process of mesoscopic complex fluids including liquid drop dynamics (drop formation, oscillation, coalescence, collision, impacting, and spreading) and the saturated or unsaturated flows in mesoscopic structures (micro channels, fractures and porous media). It is seen that except for its success in modeling the equilibration process of complex fluids, the DPD method is also effective in simulating the dynamic flows of mesoscopic fluids, either simple or complex, in complex flow geometries. The DPD method is also able to reproduce continuum hydrodynamic when the time step in the time integration scheme approaches zero and the system size is large enough for hydrodynamic (continuum) concepts and models to be valid.

This chapter also shows the difference of the conventional purely repulsive pairwise particle-particle interaction force and the newly developed particle-particle interaction force with short-range repulsion and long-distance attraction. It is demonstrated that the purely repulsive pairwise particle-particle interaction force is able to model the behaviour of a homogeneous fluid, and can be used to simulate saturated single-phase flow. In contrast, particle-particle interaction force with short-range repulsion and long-distance attraction is able to model processes with co-existing liquid and gas phases, such as multiphase drop dynamics and multiphase flow through partially saturated geometries. In general, it can be expected that DPD simulations with this type of interaction potential can be used to simulate gases, liquids, solids and multiphase systems, depending on the average particle density, the temperature and the details of the particle-particle interactions, just as shown in Section 4.2 and 4.3. It is noted that the ratio between the strengths of the fluid-wall particle-particle interactions and the fluid-fluid particle-particle interactions has a significant influence on the dynamics of fluid flow in partially saturated porous media, and different flow modes including wetting and non-wetting fluid flows can be modeled appropriately by adjusting the interaction strength ratio. DPD simulations of multiphase flow do not require contact angle models since contact angles can be naturally inferred from the shape of the moving particle distributions, and can vary spatially and temporally, depending on the dynamic balance of viscous, capillary and gravitational forces.

For the modeling of movement and suspension of macromolecules or movement and deformation of single cell, the spring-bead chain models are necessary for simulating the connectivity of neighboring DPD particles on macromolecules or cell. No matter what kind of spring-bead chain model is used, one key point is to match the modeling parameters (such as spring constants) with the physical properties of real materials/fluids. For example, based on an analytical theory, Fedosov et al. developed a DPD model for simulating red blood cell (RBC) to predict RBC mechanics, rheology and dynamics, while the modeled membrane properties can be uniquely related to experimentally determined RBC macroscopic properties (Fedosov et al., 2010).

The computational effort of DPD simulation can be greatly reduced by using proper algorithm and parallel computing techniques. Similar to molecular dynamics simulation, the main computational effort in DPD simulations arises from the calculation of interaction forces between particles. A special neighbor-list algorithm can be used in the DPD code that allows the resultant neighbor-list to remain valid for a number of time steps, typically 10-20 time steps (Rapaport, 2004). This neighbor-list algorithm can greatly reduce the computational time in calculating interaction forces between particles. In addition, the computational performance of the DPD models can be greatly enhanced using parallel computing techniques such as Message Passing Interface (MPI) and Graphic Processing Unit (GPU). The parallelization procedure of a DPD code is very similar to existing parallelization techniques developed for MD simulations and SPH simulations (Liu and Liu, 2003).

References

Agresar, G., Linderman, J., Tryggvason, G. and Powell, K. (1998) An adaptive, cartesian, front-tracking method for the motion, deformation and adhesion of circulating cells, *Journal of Computational Physics* **143**(2): 346–380.

Apfel, R. E., Tian, Y., Jankovsky, J., Shi, T., Chen, X., Holt, R. G., Trinh, E., Croonquist, A., Thornton, K. C. and Sacco Jr, A. (1997) Free oscillations and surfactant studies of superdeformed drops in microgravity, *Physical Review Letters* **78**(10): 1912.

Bathe, M., Shirai, A., Doerschuk, C. M. and Kamm, R. D. (2002) Neutrophil transit times through pulmonary capillaries: The effects of capillary geometry and fmlp-stimulation, *Biophysical Journal* **83**(4): 1917–1933.

Boistelle, R. and Astier, J. (1988) Crystallization mechanisms in solution, *Journal of Crystal Growth* **90**(1): 14–30.

Bracke, M., De Voeght, F. and Joos, P. (1989). The kinetics of wetting: The dynamic contact angle, Trends in colloid and interface science iii, Springer: 142–149.

Brazzle, J. D., Mohanty, S. K. and Frazier, A. B. (1999). Hollow metallic micromachined needles with multiple output ports, *Symposium on Micromachining and Microfabrication*, 257–266, International Society for Optics and Photonics.

Bustamante, C., Marko, J. F., Siggia, E. D. and Smith, S. (1994) Entropic elasticity of lambda-phage DNA, *Science* **265**(5178): 1599–1600.

Chen, S., Phan-Thien, N., Fan, X. J. and Khoo, B. C. (2004) Dissipative particle dynamics simulation of polymer drops in a periodic shear flow, *Journal Of Non-Newtonian Fluid Mechanics* **118**(1): 65–81.

Cheon, M., Chang, I., Koplik, J. and Banavar, J. (2002) Chain molecule deformation in a uniform flow—a computer experiment, *Europhysics Letters* **58**(2): 215.

Chu, S. (1991) Laser manipulation of atoms and particles, *Science* **253**(5022): 861–866.

Chun, K., Hashiguchi, G. and Fujita, H. (1999) Fabrication of array of hollow microcapillaries used for injection of genetic materials into animal/plant cells, *Japanese Journal of Applied Physics* **38**: 279.

Clark, A. T., Lal, M., Ruddock, J. N. and Warren, P. B. (2000) Mesoscopic simulation of drops in gravitational and shear fields, *Langmuir* **16**(15): 6342–6350.

Cox, R. (1998) Inertial and viscous effects on dynamic contact angles, *Journal of Fluid Mechanics* **357**: 249–278.

Cupelli, C., Henrich, B., Glatzel, T., Zengerle, R., Moseler, M. and Santer, M. (2008) Dynamic capillary wetting studied with dissipative particle dynamics, *New Journal of Physics* **10**(4): 043009.

De Gennes, P.-G. (1985) Wetting: Statics and dynamics, *Reviews of Modern Physics* **57**(3): 827.

Discher, D. E., Boal, D. H. and Boey, S. K. (1998) Simulations of the erythrocyte cytoskeleton at large deformation. II. Micropipette aspiration, *Biophysical Journal* **75**(3): 1584–1597.

Doyle, P. S. and Shaqfeh, E. S. (1998) Dynamic simulation of freely-draining, flexible bead-rod chains: Start-up of extensional and shear flow, *Journal Of Non-Newtonian Fluid Mechanics* **76**(1): 43–78.

Doyle, P. S., Shaqfeh, E. S. and Gast, A. P. (1997) Dynamic simulation of freely draining flexible polymers in steady linear flows, *Journal of Fluid Mechanics* **334**: 251–291.

Doyle, P. S., Shaqfeh, E. S., McKinley, G. H. and Spiegelberg, S. H. (1998) Relaxation of dilute polymer solutions following extensional flow, *Journal of Non-Newtonian Fluid Mechanics* **76**(1): 79–110.

Dragila, M. I. and Weisbrod, N. (2004) Fluid motion through an unsaturated fracture junction, *Water Resources Research* **40**(2): W02403.

Duong-Hong, D., Wang, J. S., Liu, G. R., Chen, Y. Z., Han, J. Y. and Hadjiconstantinou, N. G. (2008) Dissipative particle dynamics simulations of electroosmotic flow in nano-fluidic devices, *Microfluidics and Nanofluidics* **4**(3): 219–225.

Español, P. (1997) Fluid particle dynamics: A synthesis of dissipative particle dynamics and smoothed particle dynamics, *Europhysics Letters* **39**(6): 605.

Espanol, P. and Revenga, M. (2003) Smoothed dissipative particle dynamics, *Physical Review E* **67**(2): 26705.

Evans, E. and Yeung, A. (1989) Apparent viscosity and cortical tension of blood granulocytes determined by micropipet aspiration, *Biophysical Journal* **56**(1): 151–160.

Fan, X. J., Phan-Thien, N., Chen, S., Wu, X. H. and Ng, T. Y. (2006) Simulating flow of DNA suspension using dissipative particle dynamics, *Physics of Fluids* **18**(6).

Fan, X. J., Phan-Thien, N., Yong, N. T., Wu, X. and Xu, D. (2003) Microchannel flow of a macromolecular suspension, *Physics of Fluids* **15**(1).

Fedosov, D. A., Caswell, B. and Karniadakis, G. E. (2010) Systematic coarse-graining of spectrin-level red blood cell models, *Computer Methods in Applied Mechanics and Engineering* **199**(29): 1937–1948.

Fedosov, D. A., Caswell, B. and Karniadakis, G. E. (2011) Wall shear stress-based model for adhesive dynamics of red blood cells in malaria, *Biophysical Journal* **100**(9): 2084–2093.

Groot, R. D. (2000) Mesoscopic simulation of polymer-surfactant aggregation, *Langmuir* **16**(19): 7493–7502.

Groot, R. D. and Rabone, K. L. (2001) Mesoscopic simulation of cell membrane damage, morphology change and rupture by nonionic surfactants, *Biophysical Journal* **81**(2): 725–736.

Groot, R. D. and Warren, P. B. (1997) Dissipative particle dynamics: Bridging the gap between atomistic and mesoscopic simulation, *Journal of Chemical Physics* **107**(11): 4423.

Han, J. and Craighead, H. (2000) Separation of long DNA molecules in a microfabricated entropic trap array, *Science* **288**(5468): 1026–1029.

Han, J., Turner, S. and Craighead, H. (1999) Entropic trapping and escape of long DNA molecules at submicron size constriction, *Physical Review Letters* **83**(8): 1688.

Harlow, F. H. (1964) The particle-in-cell computing method for fluid dynamics, *Methods in Computational Physics* **3**: 319–343.

Hirt, C. W. and Nichols, B. D. (1981) Volume of fluid (vof) method for the dynamics of free boundaries, *Journal of Computational Physics* **39**(1): 201–225.

Hochmuth, R., Ting-Beall, H., Beaty, B., Needham, D. and Tran-Son-Tay, R. (1993) Viscosity of passive human neutrophils undergoing small deformations, *Biophysical Journal* **64**(5): 1596–1601.

Hoffman, R. L. (1975) A study of the advancing interface. I. Interface shape in liquid—gas systems, *Journal of Colloid And Interface Science* **50**(2): 228–241.

Hoogerbrugge, P. J. and Koelman, J. (1992) Simulating microscopic hydrodynamic phenomena with dissipative particle dynamics, *Europhysics Letters* **19**: 155.

Hou, H., Li, Q., Lee, G., Kumar, A., Ong, C. and Lim, C. (2009) Deformability study of breast cancer cells using microfluidics, *Biomedical Microdevices* **11**(3): 557–564.

Huang, H., Meakin, P. and Liu, M. (2005a) Computer simulation of two-phase immiscible fluid motion in unsaturated complex fractures using a volume of fluid method, *Water Resources Research* **41**(12): W12413.

Huang, H., Meakin, P., Liu, M. and McCreery, G. E. (2005b) Modeling of multiphase fluid motion in fracture intersections and fracture networks, *Geophysical Research Letters* **32**(19).

Hur, J. S., Shaqfeh, E. S. and Larson, R. G. (2000) Brownian dynamics simulations of single DNA molecules in shear flow, *Journal of Rheology* **44**: 713.

Jones, W. R., Ping Ting-Beall, H., Lee, G. M., Kelley, S. S., Hochmuth, R. M. and Guilak, F. (1999) Alterations in the young's modulus and volumetric properties of chondrocytes isolated from normal and osteoarthritic human cartilage, *Journal of biomechanics* **32**(2): 119–127.

Kan, H.-C., Shyy, W., Udaykumar, H., Vigneron, P. and Tran-Son-Tay, R. (1999) Effects of nucleus on leukocyte recovery, *Annals of Biomedical Engineering* **27**(5): 648–655.

Karniadakis, G. and Beskok, A. (2002) Micro flows: Fundamentals and simulation. Springer, New York.

Karniadakis, G. E., Beskok, A. and Aluru, A. (2005) Microflows and nanoflows: Fundamentals and simulation. Springer, Berlin.

Kasiteropoulou, D., Karakasidis, T. and Liakopoulos, A. (2012) A dissipative particle dynamics study of flow in periodically grooved nanochannels, *International Journal For Numerical Methods In Fluids* **68**(9): 1156–1172.

Koelman, M. V. A. and Hoogerbrugge, P. J. (1993) Dynamics simulation of hard-sphere suspensions under steady shear, *Europhysics Letter* **21**(3): 363–368.

Kong, Y., Manke, C., Madden, W. and Schlijper, A. (1997) Effect of solvent quality on the conformation and relaxation of polymers via dissipative particle dynamics, *The Journal of Chemical Physics* **107**: 592.

Kwicklis, E. M. and Healy, R. W. (1993) Numerical investigation of steady liquid water flow in a variably saturated fracture network, *Water Resources Research* **29**(12): 4091–4102.

Larson, R., Perkins, T., Smith, D. and Chu, S. (1999). The hydrodynamics of a DNA molecule in a flow field, Flexible polymer chains in elongational flow, Springer: 259–282.

Lee, G. Y. and Lim, C. T. (2007) Biomechanics approaches to studying human diseases, *Trends in Biotechnology* **25**(3): 111–118.

Leong, F. Y., Li, Q., Lim, C. T. and Chiam, K.-H. (2011) Modeling cell entry into a micro-channel, *Biomechanics and Modeling in Mechanobiology* **10**(5): 755–766.

Li, J., Dao, M., Lim, C. and Suresh, S. (2005) Spectrin-level modeling of the cytoskeleton and optical tweezers stretching of the erythrocyte, *Biophysical Journal* **88**(5): 3707–3719.

Li, Z., Hu, G. H., Wang, Z. L., Ma, Y. B. and Zhou, Z. W. (2013) Three dimensional flow structures in a moving droplet on substrate: A dissipative particle dynamics study, *Physics of Fluids* **25**: 072103.

Lim, C., Zhou, E. and Quek, S. (2006) Mechanical models for living cells—a review, *Journal of biomechanics* **39**(2): 195–216.

Lin, L. and Pisano, A. P. (1999) Silicon-processed microneedles, *Journal of Microelectromechanical Systems* **8**(1): 78–84.

Liu, G. R. and Liu, M. B. (2003) Smoothed particle hydrodynamics: A meshfree particle method. World Scientific, Singapore.

Liu, M. B., Chang, J. Z. and Li, H. Q. (2007a). Numerical modeling of injection flow of drug agents for controlled drug delivery, Engineering in Medicine and Biology Society, 2007 (EMBS 2007). 29th Annual International Conference of the IEEE, 1152–1155.

Liu, M. B., Meakin, P. and Huang, H. (2006) Dissipative particle dynamics with attractive and repulsive particle-particle interactions, *Physics of Fluids* **18**(1): 017101.

Liu, M. B., Meakin, P. and Huang, H. (2007b) Dissipative particle dynamics simulation of fluid motion through an unsaturated fracture and fracture junction, *Journal of Computational Physics* **222**(1): 110–130.

Liu, M. B., Meakin, P. and Huang, H. (2007c) Dissipative particle dynamics simulation of multiphase fluid flow in microchannels and microchannel networks, *Physics of Fluids* **19**(3).

Liu, M. B., Meakin, P. and Huang, H. (2007d) Dissipative particle dynamics simulation of pore-scale flow, *Water Resources Research* **43**.

Marella, S. V. and Udaykumar, H. (2004) Computational analysis of the deformability of leukocytes modeled with viscous and elastic structural components, *Physics of Fluids* **16**: 244.

Merabia, S. and Pagonabarraga, I. (2006) A mesoscopic model for (de) wetting, *European Physical Journal E* **20**(2): 209–214.

Mijailovich, S. M., Kojic, M., Zivkovic, M., Fabry, B. and Fredberg, J. J. (2002) A finite element model of cell deformation during magnetic bead twisting, *Journal of Applied Physiology* **93**(4): 1429–1436.

Moeendarbary, E., Ng, T. Y. and Zangeneh, M. (2009) Dissipative particle dynamics: Introduction, methodology and complex fluid applications—a review, *International Journal of Applied Mechanics* **1**(04): 737–763.

N'dri, N., Shyy, W. and Tran-Son-Tay, R. (2003) Computational modeling of cell adhesion and movement using a continuum-kinetics approach, *Biophysical Journal* **85**(4): 2273–2286.

Nativ, R., Adar, E., Dahan, O. and Geyh, M. (1995) Water recharge and solute transport through the vadose zone of fractured chalk under desert conditions, *Water Resources Research* **31**(2): 253–261.

Northrup, S. H., Allison, S. A. and McCammon, J. A. (1984) Brownian dynamics simulation of diffusion - influenced bimolecular reactions, *The Journal of Chemical Physics* **80**: 1517.

Orive, G., Hernandez, R. M., Gascón, A. R. g., Domínguez-Gil, A. and Pedraz, J. L. (2003) Drug delivery in biotechnology: Present and future, *Current opinion in biotechnology* **14**(6): 659–664.

Pan, H., Ng, T., Li, H. and Moeendarbary, E. (2010) Dissipative particle dynamics simulation of entropic trapping for DNA separation, *Sensors and Actuators A-Physical* **157**(2): 328–335.

Perkins, T. T., Quake, S. R., Smith, D. E. and Chu, S. (1994) Relaxation of a single DNA molecule observed by optical microscopy, *Science-AAAS-Weekly Paper Edition-including Guide to Scientific Information* **264**(5160): 822–825.

Perkins, T. T., Smith, D. E. and Chu, S. (1997) Single polymer dynamics in an elongational flow, *Science* **276**(5321): 2016–2021.

Perkins, T. T., Smith, D. E., Larson, R. G. and Chu, S. (1995) Stretching of a single tethered polymer in a uniform flow, *Science* **268**: 83–83.

Persoff, P. and Pruess, K. (1995) Two - phase flow visualization and relative permeability measurement in natural rough - walled rock fractures, *Water Resources Research* **31**(5): 1175–1186.

Pivkin, I. V. and Karniadakis, G. E. (2008) Accurate coarse-grained modeling of red blood cells, *Physical Review Letters* **101**(11): 118105.

Rapaport, D. C. (2004) The art of molecular dynamics simulation. Cambridge University Press, Cambridge.

Scanlon, B. R., Tyler, S. W. and Wierenga, P. J. (1997) Hydrologic issues in arid, unsaturated systems and implications for contaminant transport, *Reviews of Geophysics* **35**(4): 461–490.

Schlijper, A. G., Hoogerbrugge, P. J. and Manke, C. W. (1995) Computer simulation of dilute polymer solutions with the dissipative particle dynamics method, *Journal Of Rheology* **39**: 567.

Schmid-Schönbein, G., Sung, K., Tözeren, H., Skalak, R. and Chien, S. (1981) Passive mechanical properties of human leukocytes, *Biophysical Journal* **36**(1): 243–256.

Shrewsbury, P. J., Muller, S. J. and Liepmann, D. (2001) Effect of flow on complex biological macromolecules in microfluidic devices, *Biomedical Microdevices* **3**(3): 225–238.

Smith, D. E., Babcock, H. P. and Chu, S. (1999) Single-polymer dynamics in steady shear flow, *Science* **283**(5408): 1724–1727.

Smith, D. E. and Chu, S. (1998) Response of flexible polymers to a sudden elongational flow, *Science* **281**(5381): 1335–1340.

Spenley, N. (2000) Scaling laws for polymers in dissipative particle dynamics, *Europhysics Letters* **49**(4): 534.

Suresh, S. (2006) Mechanical response of human red blood cells in health and disease: Some structure-property-function relationships, *Journal Of Materials Research* **21**(08): 1871–1877.

Suresh, S. (2007) Biomechanics and biophysics of cancer cells, *Acta Materialia* **55**(12): 3989–4014.

Sussman, M., Smereka, P. and Osher, S. (1994) A level set approach for computing solutions to incompressible two-phase flow, *Journal Of Computational Physics* **114**(1): 146–159.

Symeonidis, V., Karniadakis, G. and Caswell, B. (2005a) Simulation of λ-phage DNA in microchannels using dissipative particle dynamics, *Technical Sciences* **53**(4).

Symeonidis, V., Karniadakis, G. E. and Caswell, B. (2005b) Dissipative particle dynamics simulations of polymer chains: Scaling laws and shearing response compared to DNA experiments, *Physical Review Letters* **95**(7): 076001.

Tessier, F., Labrie, J. and Slater, G. W. (2002) Electrophoretic separation of long polyelectrolytes in submolecular-size constrictions: A monte carlo study, *Macromolecules* **35**(12): 4791–4800.

Theret, D. P., Levesque, M., Sato, M., Nerem, R. and Wheeler, L. (1988) The application of a homogeneous half-space model in the analysis of endothelial cell micropipette measurements, *Journal of Biomechanical Engineering* **110**(3): 190–199.

Tiwari, A. and Abraham, J. (2006) Dissipative-particle-dynamics model for two-phase flows, *Physical Review E* **74**(5): 056701.

Tomaiuolo, G., Preziosi, V., Simeone, M., Guido, S., Ciancia, R., Martineelli, V., Rinaldi, C. and Rotoli, B. (2007) A methodology to study the deformability of red blood cells flowing in microcapillaries in vitro, *Annali Dell'Istituto Superiore Di sanità* **43**(2): 186–192.

Tran-Son-Tay, R., Kan, H.-C., Udaykumar, H., Damay, E. and Shyy, W. (1998) Rheological modelling of leukocytes, *Medical and Biological Engineering and Computing* **36**(2): 246–250.

Unverdi, S. O. and Tryggvason, G. (1992) A front-tracking method for viscous, incompressible, multi-fluid flows, *Journal of Computational Physics* **100**(1): 25–37.

Warren, P. (2003) Vapour-liquid coexistence in many-body dissipative particle dynamics, *arXiv preprint cond-mat/0306027.*

Wijmans, C. and Smit, B. (2002) Simulating tethered polymer layers in shear flow with the dissipative particle dynamics technique, *Macromolecules* **35**(18): 7138–7148.

Yeung, A. and Evans, E. (1989) Cortical shell-liquid core model for passive flow of liquid-like spherical cells into micropipets, *Biophysical Journal* **56**(1): 139–149.

Zhang, M. K., Chen, S. and Shang, Z. (2012) Numerical simulation of a droplet motion in a grooved microchannel, *Acta Physica Sinica* **61**(3).

Zhou, L. V., Liu, M. B. and Chang, J. Z. (2012a) Dissipative particle dynamics simulations of cell micropipetting, *Imech Internal Report*.

Zhou, L. V., Liu, M. B. and Chang, J. Z. (2012b) Dissipative particle dynamics simulations of macromolecules in micro-channels, *Acta Polymerica Sinica* **7**: 720–727.

Zhou, L. W., Liu, M. B. and Chang, J. Z. (2013) Movement and evolution of macromolecules in a grooved micro-channel, *Interaction and Multiscale Mechanics* **6**(2): 157–172.

Chapter 5

Smoothed Particle Hydrodynamics — Methodology

In this chapter, the smoothed particle hydrodynamic (SPH) method is introduced. Different from the dissipative particle dynamics (DPD), which is a meso-scale particle method, SPH is originally a macro-scale particle method, and has been extended to problems with different scales. It uses particles to represent the problem domain and to act as the computational frame for field variable approximations. As a meshfree, Lagrangian, particle method, SPH has some particular features. It has special advantages over the traditional grid-based numerical methods, and the most significant one is the adaptation nature of the SPH method: particle approximation is performed using particles in the local smoothing domain that may change with time. Because of this adaptation nature of the SPH approximation, the formulation of SPH is robust against the arbitrariness of the particle distribution. Therefore, it can naturally handle problems with extremely large deformation: the most attractive feature of the SPH method.

The meshfree nature of the SPH method is also due to the above-mentioned adaptive formulation using only particles for field variable approximations. The SPH approximation does not require a pre-defined mesh of connectivity for the particles in the process of computation, and it works often well even without any particle refinement operation. This meshfree nature is very attractive for problems where the traditional FEM or FDM encounters difficulties.

Besides the meshfree and adaptive nature, another attractive feature of the SPH method is the harmonic combination of the Lagrangian formulation and particle approximation. Unlike the nodes in many other meshfree methods, which are only used as interpolation points, the SPH particles also carry material properties, and are allowed to move driven by the internal interactions and external forces, functioning as both approximation points and material components.

The pith and marrow of the method are fully embodied in the three terms of *SMOOTHED PARTICLE HYDRODYNAMICS*. The first term SMOOTHED represents the nature of *smoothed* approximation by using a special weighted average over the neighboring particles; while the third term HYDRODYNAMICS is the right niche of the method in the application to hydrodynamics problems. It is this harmonic combination of the adaptive, Lagrangian and particle nature in the SPH method that leads to various practical applications in different areas in engineering and science, as will be demonstrated in Chapter 6.

This chapter is outlined as follows.

- In Section 5.1, the history and development of SPH are briefly introduced..
- In Section 5.2, the basic ideas of the numerical approximations of the SPH are discussed.
- In Section 5.3, commonly used SPH smoothing functions are reviewed and the general constructing conditions of SPH smoothing functions are described.
- In Section 5.4, the consistency concepts including kernel consistency and particle consistency are described. A number of consistency restoring numerical approaches for the conventional SPH method are introduced.
- In Section 5.5, some numerical aspects of the SPH method are discussed, which include artificial viscosity, artificial heat, smoothing length, symmetrization of particle interaction and tensile instability.
- In Section 5.6, some remarks and conclusion are given.

5.1 History and development

Smoothed particle hydrodynamics (SPH) is a "truly" meshfree, Lagrangian particle method originally used for continuum scale applications, and may be regarded as the oldest modern meshfree particle method. It was first invented to solve astrophysical problems in three-dimensional open space (Gingold and Monaghan, 1977; Lucy, 1977), since the collective movement of those particles is similar to the movement of a liquid or gas flow, and it can be modeled by the governing equations of the classical Newtonian hydrodynamics. Typical applications include the simulations of binary stars and stellar collisions (Benz, 1988; Monaghan, 1992; Frederic and James, 1999), supernova (Hultman and Pharayn, 1999; Thacker and Couchman, 2001), collapse as well as the formation of galaxies (Monaghan and Lattanzio, 1991; Berczik, 2000), coalescence of black holes with neutron stars (Lee and Kluzniak, 1999; Lee, 2000), single and multiple detonation of white dwarfs (Senz et al., 1999), and even the evolution of the universe (Monaghan, 1990). It also has been extended to a vast range of problems in both fluid and solid mechanics because of the strong ability to incorporate complex physics into the SPH formulations (Liu and Liu, 2003; Cleary et al., 2007; Liu and Liu, 2010; Monaghan, 2012).

In SPH, the state of a system is represented by a set of particles, which possess material properties and interact with each other within the range controlled by a weight function or smoothing function (Fulk and Quinn, 1996; Liu and Liu, 2003; Liu et al., 2003a). The discretization of the governing equations is based on these discrete particles, and a variety of particle-based formulations have been used to calculate the local density, velocity and acceleration of the fluid. The fluid pressure is calculated from the density using an equation of state, the particle acceleration is then calculated from the pressure gradient and the density. For viscous flows, the effects of physical viscosity on the particle accelerations can also be included. As a Lagrangian particle method, SPH conserves mass exactly. In SPH, there is no explicit interface tracking for multiphase flows – the motion of the fluid is represented by the motion of the particles, and fluid surfaces or fluid-fluid interfaces move with particles representing their phase defined at the initial stage.

SPH has some special advantages over the traditional grid based numerical methods, summarized below.

1. SPH is a particle method of Lagrangian nature, and the algorithm is Galilean invariant. It can obtain the time history of the material particles. The advection and transport of the system can thus be calculated.
2. By properly deploying particles at specific positions at the initial stage before the analysis, the free surfaces, material interfaces, and moving boundaries can all be traced naturally in the process of simulation regardless the complicity of the movement of the particles, which have been very challenging to many Eulerian methods. Therefore, SPH is an ideal choice for modeling free surface and interfacial flow problems.
3. SPH is a particle method without using a grid/mesh. This distinct meshfree feature of the SPH method allows a straightforward handling of very large deformations, since the connectivity between particles are generated as part of the computation and can change with time. Typical examples include the SPH applications in high energy phenomena such as explosion, underwater explosion, high velocity impact, and penetrations.
4. In SPH method, a particle represents a finite volume in continuum scale. This is quite similar to the classic molecular dynamics (MD)

method (Allen and Tildesley, 1987; Frenkel and Smit, 2002) that uses a particle to represent an atom or a molecule in nano-scale, and the dissipative particle dynamics (DPD) method (Hoogerbrugge and Koelman, 1992; Liu et al., 2014) that uses a particle to represent a small cluster of molecules in meso-cale. Thus, it is natural to generalize or extend SPH to smaller scales, or to couple SPH with molecular dynamics and dissipative particle dynamics for multiple scale applications, especially in biophysics, and biochemistry.

5. SPH is suitable for problems where the object under consideration is not a continuum. This is especially true in bio- and nano-engineering at micro and nano scale, and astrophysics at astronomic scale. For such problems, SPH can be a natural choice for numerical simulations.
6. SPH is comparatively easier in numerical implementation, and it is more natural to develop three-dimensional numerical models than grid based methods.

The early SPH algorithms were derived from the probability theory, and statistical mechanics are extensively used for numerical estimation (Gingold and Monaghan, 1977; Lucy, 1977). These algorithms did not conserve linear and angular momentum. However, they can give reasonably good results for many astrophysical phenomena. For the simulations of fluid and solid mechanics problems, there are challenges to reproduce faithfully the partial differential equations governing the corresponding fluid and solid dynamics. These challenges involve accuracy and stability of the numerical schemes in implementing the SPH methods.

With the development of the SPH method, and the extensive applications to a wide range of problems, more attractive features have been showcased while some inherent drawbacks have also been identified. Different variants or modifications have been proposed to improve the original SPH method. For example, Gingold and Monaghan found the non-conservation of linear and angular momentum of the original SPH algorithm, and then introduced an SPH algorithm that conserves both linear and angular momentum (Gingold and Monaghan, 1982). Hu and Adams also invented an angular-momentum conservative SPH algorithm for incompressible viscous flows (Hu and Adams, 2006).

Many researchers have conducted investigations on the SPH method on the numerical aspects in accuracy, stability, convergence and efficiency. Swegle et al. identified the tensile instability problem that can be important for materials with strength (Swegle et al., 1995). Morris noted the particle inconsistency problem that can lead to poor accuracy in the SPH solution (Morris, 1996a). Over the past years, different modifications or corrections have been tried to restore the consistency and to improve the accuracy of the SPH method. Monaghan proposed symmetrization formulations that were reported to have better effects (Monaghan, 1982; 1985; 1992). Johnson and his co-workers gave an axis-symmetry normalization formulation so that, for velocity fields that yield constant values of normal velocity strains, the normal velocity strains can be exactly reproduced (Johnson, 1996; Johnson and Beissel, 1996). Randles and Libersky derived a normalization formulation for the density approximation and a normalization for the divergence of the stress tensor (Randles and Libersky, 1996). Chen et al. proposed a corrective smoothed particle method (CSPM) which improves the simulation accuracy both inside the problem domain and around the boundary area (Chen et al., 1999a; Chen and Beraun, 2000). The CSPM has been improved by Liu et al. in resolving problems with discontinuity such as shock waves in a discontinuous SPH (DSPH) (Liu et al., 2003b). Liu et al. also proposed a finite particle method (FPM), which uses a set of basis function to approximate field variables at a set of arbitrarily distributed particles (Liu et al., 2005; Liu and Liu, 2006). FPM can be regarded as an improved version of SPH and CSPM with better performance in particle consistency. Batra et al. concurrently developed a similar idea to FPM, and it is named as modified SPH (MSPH) (Batra and Zhang, 2004) with applications mainly in solid mechanics. Fang et al. further improved this idea for simulating free surface flows (Fang et al., 2006), and they later developed a regularized Lagrangian finite point method for the simulation of incompressible viscous flows (Fang and Parriaux, 2008; Fang et al., 2009). Ouyang and her research team recently presented a number of corrected SPH methods (Jiang et al., 2011a; Jiang et al., 2011b) and applied the corrected SPH methods mainly for modeling non-Newtonian free surface flows (Ren et al., 2011; Xu et al., 2013).

A stress point method was invented to improve the tensile instability and zero energy mode problems (Dyka and Ingel, 1995; Dyka et al., 1997; Randles and Libersky, 2000; Vignjevic et al., 2000). Other notable modifications or corrections of the SPH method include the moving least square particle hydrodynamics (MLSPH) (Dilts, 1999, 2000), the integration kernel correction (Bonet and Kulasegaram, 2000), the reproducing kernel particle method (RKPM) (Chen et al., 1996; Liu et al., 1996), the correction for stable particle method (Belytschko et al., 1998; Rabczuk et al., 2004), and several other particle consistency restoring approaches (Liu and Liu, 2003; Zhang and Batra, 2004; Liu and Liu, 2006; Jiang et al., 2011a; Jiang et al., 2011b). Belytschko and his co-workers have conducted a series of stability and convergence analyses on meshfree particle methods, and some of the numerical techniques and analyses can also be applicable to SPH (Belytschko et al., 1996; Belytschko et al., 1998; Belytschko et al., 2000).

5.2 Basic concepts of SPH approximation

The kernel approximation in the SPH method involves an integral representation of a function and its derivatives using a smoothing function. The smoothing function is also called kernel, smoothing kernel, smoothing kernel function, or sometimes even weight function in the SPH literature (Monaghan, 1992; Fulk, 1994; Morris, 1996a; Liu et al., 2003a). A detailed discussion on smoothing function, basic requirements and constructing conditions will be given in Section 5.3.

5.2.1 Kernel approximation of a function

The conventional SPH method was originally developed for hydrodynamics problems in which the governing equations are in strong form of partial differential equations of field variables such as density, velocity, energy, and etc. There are basically two steps in obtaining an SPH formulation. The first step is to represent a function and/or its derivatives in continuous form as integral representation, and this step is also termed as kernel approximation. In this kernel approximation step, the approximation of a function and its derivatives are based on the evaluation of the smoothing kernel function and its derivatives. The second step is usually referred to as particle approximation. In this step, the computational domain is first discretized by representing the domain with a set of initial distribution of particles representing the initial settings of the problem. After discretization, field variables on a particle are approximated by a summation of the values over the nearest neighbor particles.

The kernel approximation of a function $f(\mathbf{x})$ used in the SPH method starts from the following identity

$$f(\mathbf{x}) = \int_{\Omega} f(\mathbf{x}')\delta(\mathbf{x}-\mathbf{x}')d\mathbf{x}', \tag{5.1}$$

where f is a function of the position vector $\mathbf{x}$, and $\delta(\mathbf{x}-\mathbf{x}')$ is the Dirac delta function given by

$$\delta(\mathbf{x}-\mathbf{x}') = \begin{cases} \infty & \mathbf{x}=\mathbf{x}' \\ 0 & \mathbf{x}\neq\mathbf{x}' \end{cases} \tag{5.2}$$

and

$$\int_{\Omega} \delta(\mathbf{x}-\mathbf{x}')d\mathbf{x}' = 1. \qquad (5.3)'$$

In equation (5.1), Ω is the volume of the integral that contains $\boldsymbol{x}$. Equation (5.1) implies that a function can be represented in an integral form. Since the Dirac delta function is used, the integral representation in equation (5.1) is exact and rigorous, as long as $f(\mathbf{x})$ is defined and continuous at $\mathbf{x}$ when $\Omega \to 0$.

The Delta function $\delta(\mathbf{x}-\mathbf{x}')$ with only a "point" support, and hence equation (5.1) cannot be used for establishing discrete numerical models. We thus replace the Delta function $\delta(\mathbf{x}-\mathbf{x}')$ by a smoothing (kernel) function $W(\mathbf{x}-\mathbf{x}',h)$ with a support of finite spatial dimension h, the kernel approximation of $f(\mathbf{x})$, becomes

$$f(\mathbf{x}) \approx \int_{\Omega} f(\mathbf{x}')W(\mathbf{x}-\mathbf{x}',h)d\mathbf{x}', \tag{5.4}$$

where h is the smoothing length defining the support or influence area of the smoothing function W. Note that as long as W is not the Dirac delta function, the integral representation shown in equation (5.4) can only be an approximation, except for very special simple cases. Therefore equation (5.4) is written as

$$< f(\mathbf{x}) > = \int_{\Omega} f(\mathbf{x}')W(\mathbf{x}-\mathbf{x}',h)d\mathbf{x}'. \tag{5.5}$$

with the brackets representing the kernel "approximation".

A smoothing function W is *usually* chosen to be an *even* function for reasons given in Section 5.3. It should also satisfy a number of conditions. The first one is the *normalization condition* that states

$$\int_{\Omega} W(\mathbf{x}-\mathbf{x}',h)d\mathbf{x}' = 1. \tag{5.6}$$

This condition is also termed as *unity condition* since the integration of the smoothing function produces the unity.

The second condition is the *Delta function property* that is observed when the smoothing length approaches zero

$$\lim_{h\to 0} W(\mathbf{x}-\mathbf{x}',h) = \delta(\mathbf{x}-\mathbf{x}'). \tag{5.7}$$

The third condition is the *compact condition*

$$W(\mathbf{x}-\mathbf{x}',h) = 0 \quad \text{when} \quad |\mathbf{x}-\mathbf{x}'| > \kappa h. \tag{5.8}$$

where κ is a constant related to the smoothing function for a particle at **x**, and κh defines the effective (non-zero) area of the smoothing function. This effective area is usually called as the *support domain* of the smoothing function for a point at **x** (or the support domain of that point). Using this compact condition, integration over the entire problem domain is localized over the support domain of the smoothing function.

In some SPH literatures, the kernel approximation is often said to have h^2 accuracy or second order accuracy (Monaghan, 1982, 1992; Hernquist, 1993; Fulk, 1994; Morris, 1996a; Monaghan, 2005). The is true as long as the function is first order differentiable, and can be observed easily using Taylor series expansion on equation (5.5).

Note from equation (5.8) that the support domain of the smoothing function is $|\mathbf{x}-\mathbf{x}'| \le \kappa h$, the errors in the SPH integral representation can be estimated using the Taylor series expansion of $f(\mathbf{x}')$ around $\mathbf{x}$. If $f(\mathbf{x})$ is differentiable, we have

$$\begin{aligned} < f(\mathbf{x}) > &= \int_{\Omega} [f(\mathbf{x}) + f'(\mathbf{x})(\mathbf{x}'-\mathbf{x}) + r((\mathbf{x}'-\mathbf{x})^2)] W(\mathbf{x}-\mathbf{x}',h) d\mathbf{x}' \\ &= f(\mathbf{x}) \int_{\Omega} W(\mathbf{x}-\mathbf{x}',h) d\mathbf{x}' + f'(\mathbf{x}) \int_{\Omega} (\mathbf{x}'-\mathbf{x}) W(\mathbf{x}-\mathbf{x}',h) d\mathbf{x}' + r(h^2) , \end{aligned} \quad (5.9)$$

where r stands for the residual. Note that W is an even function with respect to **x**, and $(\mathbf{x}'-\mathbf{x})W(\mathbf{x}-\mathbf{x}',h)$ should be an odd function. Hence we should have

$$\int_{\Omega} (\mathbf{x}'-\mathbf{x}) W(\mathbf{x}-\mathbf{x}',h) d\mathbf{x}' = 0 . \quad (5.10)$$

Using equations (5.6) and (5.10), equation (5.9) becomes

$$< f(\mathbf{x}) > = f(\mathbf{x}) + r(h^2) . \quad (5.11)$$

It is now clear that SPH kernel approximation of an arbitrary field function is said to have second order accuracy, as long as an even smooth function is used.

5.2.2 Kernel approximation of derivatives

The approximation for the spatial derivative $\nabla \cdot f(\mathbf{x})$ is obtained simply by substituting $f(\mathbf{x})$ with $\nabla \cdot f(\mathbf{x})$ in equation (5.5), which gives

$$< \nabla \cdot f(\mathbf{x}) > = \int_{\Omega} [\nabla \cdot f(\mathbf{x}')] W(\mathbf{x}-\mathbf{x}',h) d\mathbf{x}' , \quad (5.12)$$

where the divergence in the integral is operated with respect to the primed coordinate. Considering

$$[\nabla\cdot f(\mathbf{x}')]W(\mathbf{x}-\mathbf{x}',h)=\nabla\cdot[f(\mathbf{x}')W(\mathbf{x}-\mathbf{x}',h)] \\ -f(\mathbf{x}')\cdot\nabla[W(\mathbf{x}-\mathbf{x}',h)], \tag{5.13}$$

gives

$$<\nabla\cdot f(\mathbf{x})>=\int_{\Omega}\nabla\cdot[f(\mathbf{x}')W(\mathbf{x}-\mathbf{x}',h)]d\mathbf{x}'-\int_{\Omega}f(\mathbf{x}')\cdot\nabla[W(\mathbf{x}-\mathbf{x}',h)]d\mathbf{x}'. \tag{5.14}$$

Using the *divergence theorem*, the first integral on the right hand side (RHS) of equation (5.14) can be converted into an integral over the surface S of the domain of the integration Ω:

$$<\nabla\cdot f(\mathbf{x})>=\int_{S}f(\mathbf{x}')W(\mathbf{x}-\mathbf{x}',h)\cdot\vec{\mathbf{n}}dS-\int_{\Omega}f(\mathbf{x}')\cdot\nabla W(\mathbf{x}-\mathbf{x}',h)d\mathbf{x}', \tag{5.15}$$

where $\vec{\mathbf{n}}$ is the unit vector normal to the surface S. Since the smoothing function W is usually defined to have compact support (see equation (5.8)), and the value of W on the surface of the integral in equation (5.15) is zero in SPH (In the gradient smoothing method (GSM) (Liu and Xu, 2008; Xu et al., 2010; Li et al., 2011; Wang et al., 2013), we can do the alternative by Heaviside smoothing functions so that the 2nd term vanishes, leaving only the first term). Therefore, the surface integral on the right hand side of equation (5.15) is also zero. Hence, the kernel approximation of the derivatives can be written from equation (5.15) as

$$<\nabla\cdot f(\mathbf{x})>=-\int_{\Omega}f(\mathbf{x}')\cdot\nabla W(\mathbf{x}-\mathbf{x}',h)d\mathbf{x}'. \tag{5.16}$$

It is clear that the differential operation on a function is transformed into a differential operation on the smoothing function. In other words, the SPH kernel approximation of the derivative of a field function allows the spatial gradient to be determined from the values of the function and the derivatives of the smoothing function W, rather than from the derivatives of the function itself.

Kernel approximation of higher order derivatives can be obtained in a similar way by substituting $f(\mathbf{x})$ with the corresponding derivatives in equation (5.5), using integration by parts, divergence theorem and some trivial transformations. Another approach is to repeatedly use equation (5.16) to obtain the

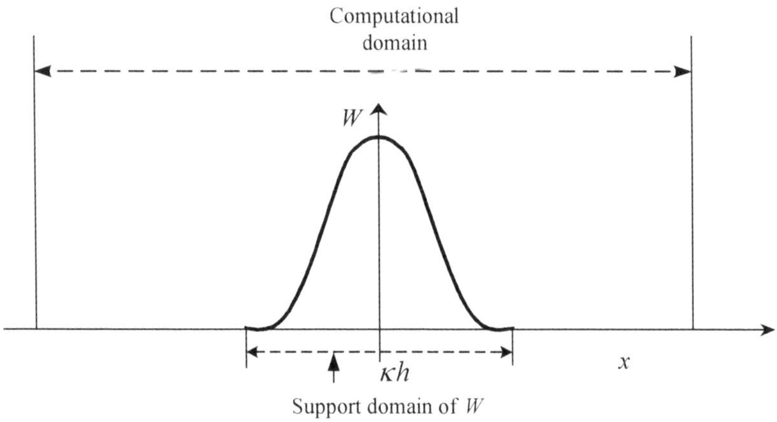

Figure 5.1 Schematic illustration of the scenarios in which the support domain is located within the problem domain. For such scenarios, the SPH kernel approximations are of second order accuracy.

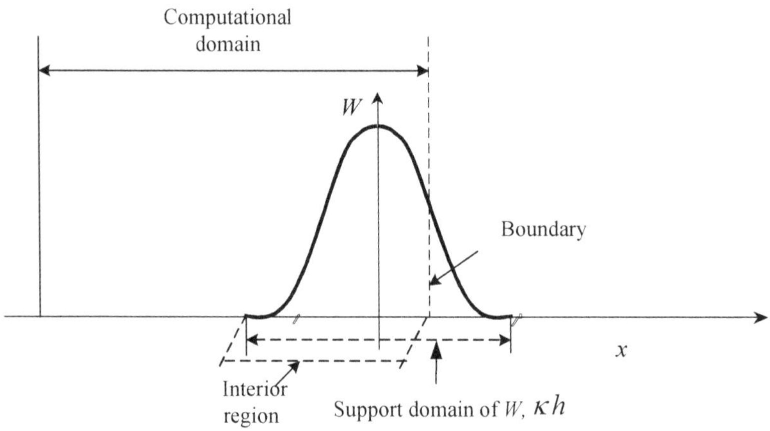

Figure 5.2 Schematic illustration of the scenarios in which the support domain intersects with the problem domain. For such scenarios, the SPH kernel approximations are not exactly of second order accuracy.

kernel approximation of the higher order derivatives, since any higher order derivative can always be regarded as the first order derivative of its next lower order derivative.

Following similar analyses based on Taylor series expansion, it is easy to show that the kernel approximation of the derivative is also of second order accuracy. Since the SPH kernel approximations for a field function and its derivatives are of second order accuracy, that is why the SPH method has usually been referred as a method of second order accuracy. However, equation (5.11) is not always true because equations (5.6) and (5.10) are sometimes not satisfied. For example, in a 1D problem space, if the support domain is within the problem domain under consideration, the integration of the smoothing function is unity (equation (5.6)), and the integration of the first moment of the smoothing function (see equation (5.10)) is zero. Also the surface integral in equation (5.15) is zero. Hence the SPH kernel approximations are of second order accuracy, and this is shown in **Figure 5.1**.

However, there are scenarios in which the support domain intersects with the problem domain boundary, as shown in **Figure 5.2**. Therefore, the smoothing function W is truncated by the boundary, and the integration of the smoothing function is no longer unity. The integration of the first moment term of the smoothing function and the surface integral in equation (5.15) are also no longer zero. At such scenarios, the SPH kernel approximations are not of second order accuracy.

5.2.3 Particle approximation

The second step of SPH method is the particle approximation, which involves representing the problem domain using a set of particles, and then estimating field variables on this set of particles. Considering a problem domain Ω filled with a set of particles (usually arbitrarily distributed, see **Figure 5.3** for illustration in a two-dimensional domain). These particles can either be centered particles initially generated using existing mesh generation tools or concentrated particles initially generated using some kind of space discretization model such

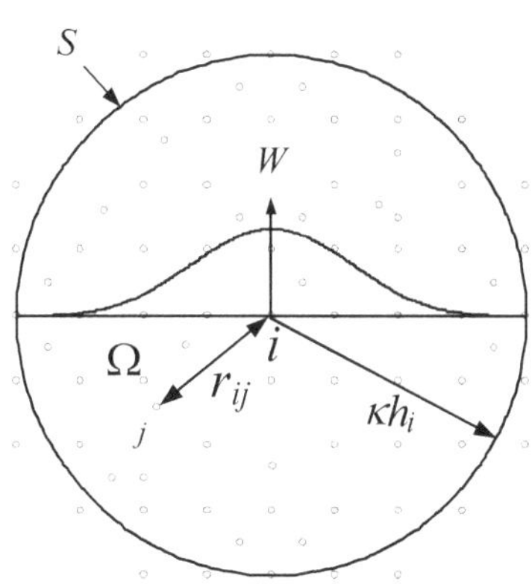

Figure 5.3 SPH particle approximations in a two-dimensional problem domain Ω with a surface S. W is the smoothing function that is used to approximate the field variables at particle i using averaged summations over particles j within the support domain with a cut off distance of κh_i.

as the particle-fill model in AUTODYN (Century Dynamics Incorporated, 1997). The state of the system is represented by these particles, each associated with field properties. These particles can be used not only for integration, interpolation or differencing, but also for representing the material. The volume of a subsection is lumped on the corresponding particle. Therefore one particle i is associated with a fixed lumped volume ΔV_i without fixed shape. If the particle mass and density are concerned, the lumped volume can also be replaced by the corresponding mass to density ratio m_i/ρ_i. These particles can be fixed in an Eulerian frame or move in a Lagrangian frame.

After representing the computational domain with a finite number of particles, the continuous form of kernel approximation expressed in equations (5.5) can be written in discretized form of a summation of the neighboring particles as follows

$$< f(\mathbf{x}) > = \sum_{j=1}^{N} \frac{m_j}{\rho_j} f(\mathbf{x}_j) W(\mathbf{x}-\mathbf{x}_j, h), \tag{5.17}$$

where N is the total number of particles within the influence area of the particle at $\mathbf{x}$. It is the total number of particles that are within the support domain which has a cutoff distance, characterized by the smoothing length, h, multiplied by a scalar constant κ. This procedure of summation over the neighboring particles is referred to as particle approximation, which states that the value of a function at a particle can be approximated by using the average of the values of the function at all the particles in the support domain weighted by the smoothing function. Following the same procedure, the particle approximation of a derivative can be obtained as

$$< \nabla \cdot f(\mathbf{x}) > = -\sum_{j=1}^{N} \frac{m_j}{\rho_j} f(\mathbf{x}_j) \cdot \nabla W(\mathbf{x}-\mathbf{x}_j, h), \tag{5.18}$$

where the gradient ∇W in the above equation is evaluated at particle j. Equation (5.18) states that the value of the gradient of a function at a particle located at $\mathbf{x}$ can be approximated by using the summation of those values of the function at all the particles in the support domain weighted by the gradient of the smoothing function. The particle approximation in equations (5.17) and (5.18) converts the continuous form of kernel approximation of a field function and its derivatives to the discrete summations over a set of particles. The use of particle summations to approximate the integral is, in fact, a key approximation that makes the SPH method simple without using a background mesh for numerical integration, and it is also the key factor influencing the solution accuracy of the SPH method.

One important aspect is that the particle approximation in the SPH method introduces the *mass* and *density* of the particle into the equations. This can be conveniently applied to hydrodynamic problems in which the density is a key field variable. This is probably one of the major reasons for the SPH method being particularly popular for dynamic fluid flow problems. If the SPH particle approximation is applied to solid mechanics problems, special treatments are required. One of the ways is to use the SPH approximation to create shape functions, and to establish the discrete system equations (Liu, 2002).

The particle approximation is, however, related to some numerical problems inherent in the SPH method, such as the particle inconsistency and the tensile instability, as will be addressed in the following sections. One basic reason is that the discrete summation is only taken over the particles themselves (collocation). In general, in meshfree methods, to achieve stability and accuracy, the number of sampling points for integration should be more than the field nodes (particles). This is especially true for meshfree methods based on weak forms for solid mechanics problems (Liu, 2002). Otherwise, it may (not always) lead to some kind of instability problems.

5.2.4 Techniques for deriving SPH formulations

By using the above-described procedure of kernel approximation and particle approximation, SPH formulations for partial differential equations can always be derived. There are in fact a number of ways to derive SPH formulation of PDEs. Benz used one approach to derive the SPH equations for PDEs that is to multiply each term in the PDEs with the smoothing function, and integrate over the volume with the use of integration by parts and Taylor expansions (Benz, 1990). Monaghan employed a straightforward approach of directly using equations (5.17) and (5.18) (Monaghan, 1992). In that approach, the following two identities are employed to place the density inside the gradient operator

$$\nabla \cdot f(\mathbf{x}) = \frac{1}{\rho}[\nabla \cdot (\rho f(\mathbf{x})) - f(\mathbf{x}) \cdot \nabla \rho], \tag{5.19}$$

$$\nabla \cdot f(\mathbf{x}) = \rho \left[\nabla \cdot \left(\frac{f(\mathbf{x})}{\rho} \right) + \frac{f(\mathbf{x})}{\rho^2} \cdot \nabla \rho \right]. \tag{5.20}$$

The above two identities may be substituted into the integral in equation (5.12). The same procedure of the particle approximation to obtain equation (5.18) is applied to each gradient term on the right hand side of equations (5.19) and (5.20). Note that each expression at the outside of every gradient term is evaluated at the particle itself, the results from equations (5.19) and (5.20) for the divergence of $f(\mathbf{x})$ at particle i are obtained as

$$<\nabla \cdot f(\mathbf{x}_i)> = \frac{1}{\rho_i}\left[\sum_{j=1}^{N} m_j \left[f(\mathbf{x}_j) - f(\mathbf{x}_i)\right] \cdot \nabla_i W_{ij}\right], \tag{5.21}$$

and

$$<\nabla \cdot f(\mathbf{x}_i)> = \rho_i\left[\sum_{j=1}^{N} m_j \left[\left(\frac{f(\mathbf{x}_j)}{\rho_j^2}\right) + \left(\frac{f(\mathbf{x}_i)}{\rho_i^2}\right)\right] \cdot \nabla_i W_{ij}\right]. \tag{5.22}$$

One of the good features for the above two equations is that the field function $f(\mathbf{x})$ appears pairwisely and involves asymmetric and symmetric SPH formulations. These asymmetric and symmetric formulations can help to improve the numerical accuracy in SPH simulations (Monaghan, 1992; Liu and Liu, 2003; Liu and Liu, 2006).

Besides the above-mentioned two identities, some other rules of operation can be convenient in deriving the SPH formulations for complex system equations (Liu and Liu, 2003). For example, for two arbitrary functions of field variables f_1 and f_2, the following rules exist

$$\langle f_1 \pm f_2 \rangle = \langle f_1 \rangle \pm \langle f_2 \rangle, \tag{5.23}$$

$$\langle f_1 f_2 \rangle = \langle f_1 \rangle \langle f_2 \rangle. \tag{5.24}$$

Hence, an SPH approximation of the sum of functions equals to the sum of the SPH approximations of the individual function, and an SPH approximation of a product of functions equals to the product of the SPH approximations of the individual functions.

If f_1 is a constant denoted by c, we should have

$$\langle c f_2 \rangle = c\langle f_2 \rangle. \tag{5.25}$$

It is clear that the SPH approximation operator is a linear operator. It is also easy to show that the SPH approximation operator is commutative, i.e.,

$$\langle f_1 + f_2 \rangle = \langle f_2 + f_1 \rangle, \tag{5.26}$$

and

$$\langle f_1 f_2 \rangle = \langle f_2 f_1 \rangle . \tag{5.27}$$

For convenience, the SPH approximation operator “<>” is omitted in later sections.

5.2.5 SPH formulations for the Navier-Stokes (N-S) equations

Using the afore-mentioned kernel and particle approximation techniques with necessary numerical tricks, it is possible to derive SPH formulations for partial differential equations governing the physics of fluid flows. For example, for Navier-Stokes (N-S) equations controlling the general fluid dynamic problems, we have

$$\begin{cases} \dfrac{D\rho}{Dt} = -\rho \dfrac{\partial v^\beta}{\partial x^\beta} \\ \dfrac{Dv^\alpha}{Dt} = \dfrac{1}{\rho}\dfrac{\partial \sigma^{\alpha\beta}}{\partial x^\beta} + F^\alpha \\ \dfrac{De}{Dt} = \dfrac{\sigma^{\alpha\beta}}{\rho}\dfrac{\partial v^\alpha}{\partial x^\beta} \end{cases} , \tag{5.28}$$

where the Greek superscripts α and β are used to denote the coordinate directions, the summation in the equations is taken over repeated indices, and the total time derivatives are taken in the moving Lagrangian frame. The scalar density ρ, and internal energy e, the velocity component v^α (component of velocity vector $\mathbf{v}$ in α direction), and the total stress tensor $\sigma^{\alpha\beta}$ are the dependent variables. F^α (component of external force $\mathbf{F}$ in α direction) is the external forces such as gravity. The spatial coordinates x^α and time t are the independent variables. The total stress tensor $\sigma^{\alpha\beta}$ is made up of two parts, one part of isotropic pressure p and the other part of viscous stress τ, i.e., $\sigma^{\alpha\beta} = -p\delta^{\alpha\beta} + \tau^{\alpha\beta}$. For Newtonian fluids, the viscous shear stress should be proportional to the shear strain rate denoted by ε through the dynamic viscosity μ, i.e., $\tau^{\alpha\beta} = \mu\varepsilon^{\alpha\beta}$, where $\varepsilon^{\alpha\beta} = \dfrac{\partial v^\beta}{\partial x^\alpha} + \dfrac{\partial v^\alpha}{\partial x^\beta} - \dfrac{2}{3}(\nabla \cdot \mathbf{v})\delta^{\alpha\beta}$.

Substituting the SPH approximations for a function and its derivative (as shown in equations (5.17) and (5.18)) to the N-S equations, the SPH equations of motion for the N-S equations can be written as

$$\begin{cases} \dfrac{D\rho_i}{Dt} = \sum_{j=1}^{N} m_j v_{ij}^{\beta} \dfrac{\partial W_{ij}}{\partial x_i^{\beta}} \\ \dfrac{D\mathbf{v}_i^{\alpha}}{Dt} = -\sum_{j=1}^{N} m_j \left(\dfrac{\sigma_i^{\alpha\beta}}{\rho_i^2} + \dfrac{\sigma_j^{\alpha\beta}}{\rho_j^2} \right) \dfrac{\partial W_{ij}}{\partial x_i^{\beta}} + F_i^{\alpha} \\ \dfrac{De_i}{Dt} = \dfrac{1}{2} \sum_{j=1}^{N} m_j \left(\dfrac{p_i}{\rho_i^2} + \dfrac{p_j}{\rho_j^2} \right) v_{ij}^{\beta} \dfrac{\partial W_{ij}}{\partial x_i^{\beta}} + \dfrac{\mu_i}{2\rho_i} \varepsilon_i^{\alpha\beta} \varepsilon_i^{\alpha\beta} \end{cases} , \tag{5.29}$$

where $\mathbf{v}_{ij} = \mathbf{v}_i - \mathbf{v}_j$. Equation (5.29) is a set of commonly used SPH equations for the N-S equations. It should be noted that by using different numerical tricks, it is possible to get other different forms of SPH equations for the same partial differential equations. The obtained SPH formulations may have special features and advantages suitable for different applications (Liu and Liu, 2003). One typical example is the approximation of density. If the field function is the density, equation (5.17) can be re-written as

$$\rho_i = \sum_{j=1}^{N} m_j W_{ij} . \tag{5.30}$$

This is another approach to obtain density directly from the SPH summation of the mass of all particles in the support domain of a given particle, rather than from the continuum equation. Compared to the SPH formulations on density change in (5.29), this summation density approach conserves mass exactly, but suffers from serious boundary deficiency due to the particle inconsistency. A frequently used way to remediate the boundary deficiency is the following normalization form by the summation of the smoothing function itself (Randles and Libersky, 1996; Chen et al., 1999a)

$$\rho_i = \frac{\sum_{j=1}^{N} m_j W_{ij}}{\sum_{j=1}^{N} \left(\dfrac{m_j}{\rho_j} \right) W_{ij}} . \tag{5.31}$$

5.3 SPH smoothing function

5.3.1 Review on commonly used smoothing functions

One of the central issues for meshfree methods is how to effectively perform function approximation based on a set of nodes scattered in an arbitrary manner without using a predefined mesh or grid that provides the connectivity of the nodes. In the SPH method, the smoothing function is used for kernel and particle approximations. It is of utmost importance in the SPH method as it determines the pattern to interpolate, and defines the cutoff distance of the influencing area of a particle.

Many researchers have investigated the smoothing kernel, hoping to improve the performance of the SPH method, and/or to generalize the requirements for constructing the smoothing kernel function. Fulk numerically investigated a number of smoothing kernel functions in one-dimensional space, and the obtained results are basically valid for regularly distributed particles (Fulk, 1994; Fulk and Quinn, 1996). Swegle et al. revealed the tensile instability, which is closely related to the smoothing kernel function (Swegle et al., 1995). Morris studied the performances of several different smoothing functions, and found that by properly selecting the smoothing function, the accuracy and stability property of the SPH simulation can be improved (Morris, 1996b, a). Omang provided investigations on alternative kernel functions for SPH in cylindrical symmetry (Omang et al., 2005). Jin and Ding investigated the criterions for smoothed particle hydrodynamics kernels in stable field (Jin and Ding, 2005). Capuzzo-Dolcetta gave a criterion for the choice of the interpolation kernel in SPH (Capuzzo-Dolcetta and Di Lisio, 2000). Cabezon and his co-workers proposed a one-parameter family of interpolating kernels for SPH studies (Cabezon et al., 2008).

Different smoothing functions have been used in the SPH method as shown in published literatures. Various requirements or properties for the smoothing functions have been discussed. Major properties or requirements are now summarized and described in the following discussion.

1. The smoothing function must be normalized over its support domain (*Unity*)

$$\int_{\Omega} W(\mathbf{x}-\mathbf{x}', h)d\mathbf{x}' = 1. \tag{5.32}$$

   This normalization property ensures that the integral of the smoothing function over the support domain to be unity. It can be shown in the next section that it also ensures the zero-th order consistency (C^0) of the integral representation of a continuum function.

2. The smoothing function should be compactly supported (*Compact support*), i.e.,

$$W(\mathbf{x}-\mathbf{x}')=0\ ,\ \text{for}\ \left|\mathbf{x}-\mathbf{x}'\right|>\kappa h\ . \tag{5.33}$$

The dimension of the compact support is defined by the smoothing length h and a scaling factor κ, where h is the smoothing length, and κ determines the spread of the specified smoothing function. $\left|\mathbf{x}-\mathbf{x}'\right|\leq \kappa h$ defines the support domain of the particle at point $\mathbf{x}$. This compact supportness property transforms an SPH approximation from a global operation to a local operation. This will later lead to a set of sparse discretized system matrices, and therefore is very important as far as the computational efforts are concerned.

3. $W(\mathbf{x}-\mathbf{x}')\geq 0$ for any point at $\mathbf{x}'$ within the support domain of the particle at point $\mathbf{x}$ (*Positivity*). This property states that the smoothing function should be non-negative in the support domain. It is not mathematically necessary as a convergent condition, but it is important to ensure a physically meaningful (or stable) representation of some physical phenomena. A few smoothing functions used in some literatures are negative in parts of the support domain. However in hydrodynamic simulations, negative value of the smoothing function can have serious consequences that may result in some unphysical parameters such as negative density and energy.

4. The smoothing function value for a particle should be monotonically decreasing with the increase of the distance away from the particle (*Decay*). This property is based on the physical consideration in that a nearer particle should have a bigger influence on the particle under consideration. In other words, with the increase of the distance of two interacting particles, the interaction force decreases.

5. The smoothing function should satisfy the Dirac delta function condition as the smoothing length approaches to zero (*Delta function property*)

$$\lim_{h\to 0} W(\mathbf{x}-\mathbf{x}',h)=\delta(\mathbf{x}-\mathbf{x}')\ . \tag{5.34}$$

This property makes sure that as the smoothing length tends to be zero, the approximation value approaches the function value, i.e. $\langle f(\mathbf{x})\rangle = f(\mathbf{x})$.

6. The smoothing function should be an even function (*Symmetric property*). This means that particles from same distance but different positions should have equal effect on a given particle. This is not a very

rigid condition, and it is sometimes violated in some meshfree particle methods that provide higher consistency.

7. The smoothing function should be sufficiently smooth (*Smoothness*). This property aims to obtain better approximation accuracy. For the approximations of a function and its derivatives, a smoothing function needs to be sufficiently continuous to obtain good results. A smoothing function with smoother value of the function and derivatives would usually yield better results and better performance in numerical stability. This is because the smoothing function will not be sensitive to particle disorder, and the errors in approximating the integral interpolants are small, provided that the particle disorder is not extreme (Monaghan, 1992; Fulk, 1994; Liu and Liu, 2003).

Any function having the above properties may be employed as SPH smoothing functions, and many kinds of smoothing functions have been used. Lucy in the original SPH paper (Lucy, 1977) used a bell-shaped function

$$W(\mathbf{x}-\mathbf{x}',h)=W(R,h)=\alpha_d\begin{cases}(1+3R)(1-R)^3 & R\le 1\\ 0 & R>1\end{cases}, \tag{5.35}$$

where α_d is $5/4h$, $5/\pi h^2$ and $105/16\pi h^3$ in one-, two- and three-dimensional space, respectively, so that the condition of unity can be satisfied for all the three dimensions. R is the relative distance between two points (particles) at points $\boldsymbol{x}$ and $\mathbf{x}'$, $R=\dfrac{r}{h}=\dfrac{|\mathbf{x}-\mathbf{x}'|}{h}$, where r is the distance between the two points.

Gingold and Monaghan in their original paper (Gingold and Monaghan, 1977) selected the following Gaussian kernel to simulate the non-spherical stars

$$W(R,h)=\alpha_d e^{-R^2}, \tag{5.36}$$

where α_d is $1/\pi^{1/2}h$, $1/\pi h^2$ and $1/\pi^{3/2}h^3$, respectively, in one-, two- and three-dimensional space, for the unity requirement. The Gaussian kernel is sufficiently smooth even for high orders of derivatives, and is regarded as a "golden" selection since it is very stable and accurate especially for disordered particles. It is, however, not really compact, as it never goes to zero theoretically, unless R approaches to infinity. Because it approaches zero numerically very fast, it is practically compact. Note that it is computationally more expensive since it can take a longer distance for the kernel to approach zero. This can result in a large support domain with more particles for particle approximations.

The most frequently used smoothing function may be the cubic B-spline function, which was originally used by Monaghan and Lattanzio (Monaghan and Lattanzio, 1985).

$$W(R,h)=\alpha_d\times\begin{cases}2/3-R^2+R^3/2 & 0\le R<1\\ (2-R)^3/6 & 1\le R<2\,.\\ 0 & R\ge 2\end{cases} \tag{5.37}$$

In one-, two- and three-dimensional space, $\alpha_d=1/h,\ 15/7\pi h^2$ and $3/2\pi h^3$, respectively. The cubic spline function has been the most widely used smoothing function in the emerged SPH literatures since it closely resembles a Gaussian function while having a narrower compact support. However, the second derivative of the cubic spline is a piecewise linear function, and accordingly, the stability properties can be inferior to those of smoother kernels.

Morris have introduced higher order (quartic and quintic) splines that are more closely approximating the Gaussian and more stable (Morris, 1996b, a). The quartic spline is

$$W(R,h)=\alpha_d\times\begin{cases}(R+2.5)^4-5(R+1.5)^4+10(R+0.5)^4 & 0\le R<0.5\\ (2.5-R)^4-5(1.5-R)^4 & 0.5\le R<1.5\\ (2.5-R)^4 & 1.5\le R<2.5\\ 0 & R>2.5\end{cases}, \tag{5.38}$$

where α_d is $1/24h$ in one-dimensional space. The quintic spline is

$$W(R,h)=\alpha_d\times\begin{cases}(3-R)^5-6(2-R)^5+15(1-R)^5 & 0\le R<1\\ (3-R)^5-6(2-R)^5 & 1\le R<2\\ (3-R)^5 & 2\le R<3\\ 0 & R>3\end{cases}, \tag{5.39}$$

where α_d is $120/h$, $7/478\pi h^2$ and $3/359\pi h^3$ in one-, two- and three-dimensional space, respectively.

Johnson et al. used the following quadratic smoothing function to simulate high velocity impact problems (Johnson et al., 1996)

$$W(R,h)=\alpha_d\left(\frac{3}{16}R^2-\frac{3}{4}R+\frac{3}{4}\right)\qquad 0\le R\le 2, \tag{5.40}$$

where in one-, two- and three-dimensional space, $\alpha_d = 1/h,\ 2/\pi h^2$ and $5/4\pi h^3$, respectively. Unlike other smoothing functions, the derivative of this quadratic smoothing function always increases as the particles move closer, and always decreases as they move apart. This was regarded by the authors as an important improvement over the cubic spline function, and it was reported to relieve the problem of compressive instability.

Recently, Yang et al. presented a new kernel function, which consists of two cosine functions (and therefore it is referred to as double cosine kernel function) (Yang et al., 2014b) as

$$W(R,h)=\alpha_d\begin{cases}4\cos\left(\dfrac{\pi}{\kappa}R\right)+\cos\left(\dfrac{2\pi}{\kappa}R\right)+3, & 0\le R\le\kappa\\ 0, & \kappa<R\end{cases}, \tag{5.41}$$

where α_d is $1/(6\kappa h)$, $\pi/[(3\pi^2-16)(\kappa h)^2]$ and $\pi/[(4\pi^2-30)(\kappa h)^3]$ in one, two and three dimensional space, respectively. The double cosine kernel function is sufficiently smooth, and is associated with an adjustable support domain. It also has smaller second order momentum, and therefore it can have better accuracy in terms of kernel approximation.

Some higher order smoothing functions that are devised from lower order forms have been constructed, such as the super-Gaussian kernel (Monaghan and Lattanzio, 1985)

$$W(R,h)=\alpha_d\left(\frac{3}{2}-R^2\right)e^{-R^2} \qquad 0\le R\le 2, \tag{5.42}$$

where α_d is $1/\sqrt{\pi}$ in one-dimensional space. One disadvantage of the high order smoothing function is that the kernel is negative in some region of its support domain. This may lead to unphysical results for hydrodynamic problems (Fulk, 1994).

The smoothing function has been studied mathematically in detail by Liu and his co-workers. They proposed a systematical way to construct a smoothing function that may meet different needs (Liu et al., 2003a). A new quartic smoothing function has been constructed to demonstrate the effectiveness of the approach for constructing a smoothing function as follows.

$$W(R,h)=\alpha_d\begin{cases}\left(\dfrac{2}{3}-\dfrac{9}{8}R^2+\dfrac{19}{24}R^3-\dfrac{5}{32}R^4\right) & 0\le R\le 2\\ 0 & R>2\end{cases}. \tag{5.43}$$

where α_d is $1/h$, $15/7\pi h^2$ and $315/208\pi h^3$ in one-, two- and three-dimensional space, respectively. Note that the centre peak value of this quartic smoothing function is defined as $2/3$. The quartic smoothing function behaves very much like the widely used cubic B-spline function given in Equation (5.37), but has only one piece, and hence is much more convenient and efficient to use. More discussions on this quartic smoothing function will be given in the next section.

5.3.2 Generalizing constructing conditions

Major requirements of an SPH smoothing function have been addressed in Section 3.1. Some of these requirements can be derived by conducting Taylor series analysis. This analysis is carried out at the stage of the SPH kernel approximation for a function and its derivatives. It shows that, to exactly approximate a function and its derivatives, certain conditions need to be satisfied. These conditions can then be used to construct the smoothing functions.

Considering the SPH kernel approximation for a field function $f(\mathbf{x})$ as shown in equation (5.5), if $f(\mathbf{x})$ is sufficiently smooth, applying Taylor series expansion of $f(\mathbf{x}')$ in the vicinity of $\mathbf{x}$ yields

$$\begin{aligned} f(\mathbf{x}') &= f(\mathbf{x}) + f'(\mathbf{x})(\mathbf{x}'-\mathbf{x}) + \frac{1}{2} f''(\mathbf{x})(\mathbf{x}'-\mathbf{x})^2 + \ldots \\ &= \sum_{k=0}^{n} \frac{(-1)^k h^k f^{(k)}(\mathbf{x})}{k!} \left(\frac{\mathbf{x}-\mathbf{x}'}{h} \right)^k + r_n \left(\frac{\mathbf{x}-\mathbf{x}'}{h} \right), \end{aligned} \tag{5.44}$$

where r_n is the remainder of the Taylor series expansion. Substituting equation (5.44) into equation (5.5) leads to

$$f(\mathbf{x}) = \sum_{k=0}^{n} A_k f^{(k)}(\mathbf{x}) + r_n \left(\frac{\mathbf{x}-\mathbf{x}'}{h} \right), \tag{5.45}$$

where

$$A_k = \frac{(-1)^k h^k}{k!} \int_{\Omega} \left(\frac{\mathbf{x}-\mathbf{x}'}{h} \right)^k W(\mathbf{x}-\mathbf{x}', h) d\mathbf{x}'. \tag{5.46}$$

Comparing the LHS with the RHS of equation (5.45), in order for $f(\mathbf{x})$ to be approximated to n-th order, the coefficients A_k must equal to the counterparts for $f^{(k)}(\mathbf{x})$ at the LHS of equation (5.45). Therefore, after trivial transformation, the following conditions for the smoothing function W can be obtained as follows

$$\left.\begin{aligned} M_0 &= \int_\Omega W(\mathbf{x}-\mathbf{x}',h)d\mathbf{x}' = 1 \\ M_1 &= \int_\Omega (\mathbf{x}-\mathbf{x}')W(\mathbf{x}-\mathbf{x}',h)d\mathbf{x}' = 0 \\ M_2 &= \int_\Omega (\mathbf{x}-\mathbf{x}')^2 W(\mathbf{x}-\mathbf{x}',h)d\mathbf{x}' = 0 \\ &\vdots \\ M_n &= \int_\Omega (\mathbf{x}-\mathbf{x}')^n W(\mathbf{x}-\mathbf{x}',h)d\mathbf{x}' = 0 \end{aligned}\right\}, \tag{5.47}$$

where M_k is the the k-th moments of the smoothing function. Note that the first equation in (5.47) is, in fact, the unity condition expressed in equation (5.32), and the second equation in (5.47) stands for the symmetric property. Satisfaction of these two conditions ensures the first order consistency for the SPH kernel approximation for a function.

Also performing Taylor series analysis for the SPH kernel approximation of the derivatives of a field function $f(\mathbf{x})$, using the concept of integration by parts, and divergence theorem with some trivial transformation, the following equations

$$W(\mathbf{x}-\mathbf{x}',h)\,|_S = 0 \tag{5.48}$$

and

$$\left.\begin{aligned} M_0' &= \int_\Omega W'(\mathbf{x}-\mathbf{x}',h)d\mathbf{x}' = 0 \\ M_1' &= \int_\Omega (\mathbf{x}-\mathbf{x}')W'(\mathbf{x}-\mathbf{x}',h)d\mathbf{x}' = 1 \\ M_2' &= \int_\Omega (\mathbf{x}-\mathbf{x}')^2 W'(\mathbf{x}-\mathbf{x}',h)d\mathbf{x}' = 0 \\ &\vdots \\ M_n' &= \int_\Omega (\mathbf{x}-\mathbf{x}')^n W'(\mathbf{x}-\mathbf{x}',h)d\mathbf{x}' = 0 \end{aligned}\right\} \tag{5.49}$$

can be obtained. Equation (5.48) actually specifies that the smoothing function vanishes on the surface of the support domain. This is compatible to the compactness condition of the smoothing function. Equation (5.49) defines the conditions with which the derivatives of the smoothing function should be satisfied. Note that equations (5.47) and (5.49) are actually compatible considering integration by parts, divergence theorem and the boundary value vanishing effects (equation (5.48)) of the smoothing function.

Performing Taylor series analysis on the SPH kernel approximation for the second derivatives, similar equations can be obtained. Except for the requirements on the second derivatives of the momentums, the first derivative of the smoothing function also needs to vanish on the surface, which is

$$W'(\mathbf{x}-\mathbf{x}',h)\,|_S=0\,. \tag{5.50}$$

Equations (5.47)-(5.50) can be used to construct smoothing functions. It can be seen that the conditions of smoothing functions can be classified into two groups. The first group shows the ability of a smoothing function to reproduce polynomials. Satisfying the first group, the function can be approximated to n-th order accuracy. The second group defines the surface values of a smoothing function as well as its first derivatives, and is the requirements of the property of compact support for the smoothing function and its first derivative. Satisfying these conditions, the first two derivatives of the function can be exactly approximated to the n-th order.

5.3.3 Constructing SPH smoothing functions

By using above-mentioned conditions, it is possible to have a systematic way to construct the SPH smoothing functions. If the smoothing function is assumed to be a polynomial dependent only on the relative distance of the concerned points, it can be assumed to have the following form in the support domain with an influence width of κh.

$$W(\mathbf{x}-\mathbf{x}',h)=W(R)=a_0+a_1R+a_2R^2+...+a_nR^n\,. \tag{5.51}$$

It is clear that a smoothing function in the above-mentioned form is a distance function since it depends on the relative distance. It is easy to show that for the second derivative of the smoothing function to exist, a_1 should vanish. Substituting this polynomial form smoothing function into the conditions (equations (5.47)-(5.50)), the parameters a_0, a_2, …, a_n can be calculated from the resultant linear equations, and then the smoothing function can be determined.

There are several issues that need further consideration. Firstly, a smoothing function derived from this set of conditions (see equations (5.47)) will not necessarily be positive in the entire support domain, especially when high order reproducibility is required. Such a negative smoothing function may result in unphysical solutions, for example, negative density (mass) and negative energy. For this reason, smoothing functions used in SPH literatures are generally non-negative for CFD problems. On the other hand, for the even moments ($k = 2, 4, 6...$) to be zero, a smoothing function has to be negative in some parts of the region. This implies that one cannot have both non-negativity and high-order reproducibility at the same time.

Secondly, in constructing a smoothing function, the center peak value is a factor that needs to be considered. The center peak value of a smoothing function is very important since it determines how much the particle itself will contribute to the approximation. Revisiting equations (5.47), if a positive smoothing function is used, the highest order of accuracy for the function approximation is second order. Therefore, the second momentum ($M_2 = \int_\Omega (\mathbf{x}-\mathbf{x}')^2 W(\mathbf{x}-\mathbf{x}',h)d\mathbf{x}'$) can be used as a rough indicator to measure the accuracy of the kernel approximation. The smaller the second moment M_2 is, the more accurate the kernel approximation is. The center peak value of a smoothing function is closely related to M_2. A positive smoothing function with a large center peak value will have a smaller second moment M_2. This implies that a smoothing function is closer to the Delta function, and therefore is more accurate in terms of kernel approximations.

Thirdly, in some circumstances, a piecewise smoothing function is preferable since the shape of the piecewise smoothing function is easier to be controlled by changing the number of the pieces and the locations of the connection points. For example, consider the general form of a smoothing function with two pieces,

$$W(R) = \begin{cases} W_1(R) & 0 \le R < R_1 \\ W_2(R) & R_1 \le R < R_2 \\ 0 & R_2 \le R \end{cases} . \tag{5.52}$$

The function itself and the first two derivatives at the connection points should be continuous, i.e., $W_1(R_1) = W_2(R_1)$, $W_1'(R_1) = W_2'(R_1)$ and $W_1''(R_1) = W_2''(R_1)$. Considering the requirements at these points as well as the compact support property, one possible form of the smoothing function is

$$W(R) = \alpha_d \begin{cases} b_1(R_1 - R)^n + b_2(R_2 - R)^n & 0 \le R < R_1 \\ b_2(R_2 - R)^n & R_1 \le R < R_2 \\ 0 & R_2 \le R \end{cases} . \tag{5.53}$$

It is also feasible to construct smoothing function with more pieces using similar expressions.

To show the effectiveness of this approach to constructing general SPH smoothing functions Liu et al. (Liu et al., 2003a) derived a new quartic smoothing function using the following conditions

- the unity condition,
- compact support of the smoothing function,
- compact support of the first derivative of the smoothing function,
- centre peak value.

By using these constructing conditions, the constructed smoothing function is given as $W(R,h)=\alpha_d\left(\frac{2}{3}-\frac{9}{8}R^2+\frac{19}{24}R^3-\frac{5}{32}R^4\right)$, for $0\le R\le 2$, where α_d is $1/h$, $15/7\pi h^2$ and $315/208\pi h^3$ in one-, two- and three-dimensional space, respectively. Note that the centre peak value of this quartic smoothing function is defined as $2/3$.

As defined, this quartic function satisfies the normalization condition, while the function itself and its first derivative have compact support. It is very close to the most commonly used cubic spline (equation (5.37)) with the same center peak

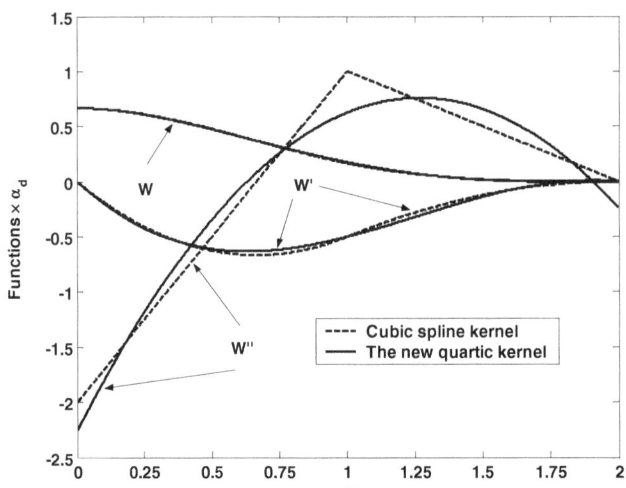

Figure 5.4 The quartic smoothing function constructed by Liu et al. by using the smoothing function constructing conditions (Liu et al., 2003a). The shapes of the quartic function and its first derivative are very close to the shapes of the cubic spline function and its first derivative. However, this one piece smoothing function is expected to produce better accuracy as it has smaller second moment. It is also expected to be more stable since it has a continuous second derivative.

value of $2/3$, and monotonically decreases with the increase of the distance as show in **Figure 5.4**. However, this quartic function produces a smaller second momentum than the cubic spline function, and therefore can produce better accuracy for kernel approximation. Also this quartic smoothing function has a smoother second derivative than the cubic spline smoothing function, thus the stability properties should be superior to those of the cubic spline function, as reported by many researchers that a smoother second derivative can lead to less instability in SPH simulation (Swegle et al., 1994; Morris, 1996b).

5.4 Numerical aspects of SPH

5.4.1 Artificial viscosity

In order to simulate problems of hydrodynamics, special treatments or methods are required to allow the algorithms to be capable of modeling shock waves, or else the simulation will develop unphysical oscillations in the numerical results around the shocked region. A shock wave is not a true physical discontinuity, but a very narrow transition zone whose thickness is usually in the order of a few molecular mean free paths. Application of the conservation of mass, momentum, and energy conditions across a shock wave front requires the simulation of transformation of kinetic energy into heat energy. Physically, this energy transformation can be represented as a form of viscous dissipation. This idea leads to the development of the von Neumann-Richtmyer artificial viscosity (von Neumann and Richtmyer, 1950) that is given by

$$\Pi_1 = \begin{cases} a_1 \Delta x^2 \rho (\nabla \cdot \mathbf{v})^2 & \nabla \cdot \mathbf{v} < 0 \\ 0 & \nabla \cdot \mathbf{v} \geq 0 \end{cases}, \tag{5.54}$$

where Π_1 is the von Neumann-Richtmyer artificial viscosity, and needs only to be present during material compression. a_1 is an adjustable non-dimensional constant. Note that this von Neumann-Richtmyer artificial viscosity is, in fact, a quadratic expression of velocity divergence.

It is found that adding the following linear artificial viscosity term Π_2 has the advantage of further smoothing the oscillations that are not totally dampened by the quadratic artificial viscosity term

$$\Pi_2 = \begin{cases} a_2 \Delta x c \rho \nabla \cdot \mathbf{v} & \nabla \cdot \mathbf{v} < 0 \\ 0 & \nabla \cdot \mathbf{v} \geq 0 \end{cases}, \tag{5.55}$$

where c is the speed of sound, and a_2 is an adjustable non-dimensional constant.

The quadratic von Neumann-Richtmyer artificial viscosity Π_1 and the linear artificial viscosity Π_2 are widely used today for removing numerical oscillations in hydrodynamic simulations using FDM, FVM, FEM, and etc. The introduced artificial viscosity terms spread the shock wave over several mesh cells and regularize the numerical instability caused by the sharp spatial variation (discontinuity). The artificial viscosity terms are usually added to the physical pressure term, and help to diffuse sharp variations in the flow and to dissipate the energy of high frequency term. Formulation and magnitude needed for an artificial viscosity has undergone many refinements over the last few decades.

The SPH method was first applied to treat problems with low or no dissipation. Later an artificial viscosity was developed (Monaghan and Gingold, 1983; Monaghan and Poinracic, 1985; Monaghan, 1987) to allow shocks to be simulated. This Monaghan type artificial viscosity Π_{ij} is the most widely used artificial viscosity so far in the SPH literatures. It not only provides the necessary dissipation to convert kinetic energy into heat at the shock front, but also prevent unphysical penetration for particles approaching each other (Lattanzio et al., 1986; Monaghan, 1989). The detailed formulation is as follows

$$\Pi_{ij} = \begin{cases} \dfrac{-\alpha_{\Pi}\bar{c}_{ij}\phi_{ij} + \beta_{\Pi}\phi_{ij}^2}{\bar{\rho}_{ij}} & \mathbf{v}_{ij}\cdot\mathbf{x}_{ij} < 0 \\ 0 & \mathbf{v}_{ij}\cdot\mathbf{x}_{ij} \geq 0 \end{cases}, \tag{5.56}$$

where

$$\varphi_{ij} = \frac{h_{ij}\mathbf{v}_{ij}\cdot\mathbf{x}_{ij}}{\left|\mathbf{x}_{ij}\right|^2 + \phi^2}, \tag{5.57}$$

$$\bar{c}_{ij} = \frac{1}{2}\left(c_i + c_j\right), \tag{5.58}$$

$$\bar{\rho}_{ij} = \frac{1}{2}\left(\rho_i + \rho_j\right), \tag{5.59}$$

$$h_{ij} = \frac{1}{2}\left(h_i + h_j\right), \tag{5.60}$$

$$\mathbf{v}_{ij} = \mathbf{v}_i - \mathbf{v}_j, \mathbf{x}_{ij} = \mathbf{x}_i - \mathbf{x}_j. \tag{5.61}$$

In the above equations, $\alpha_{\Pi}, \beta_{\Pi}$ are constants that are all typically set around 1.0 (Evrard, 1988; Monaghan, 1988). The factor $\phi = 0.1h_{ij}$ is inserted to prevent numerical divergences when two particles are approaching each other. c and $\mathbf{v}$

represent the speed of sound and the particle velocity vector, respectively. The viscosity associated with α_{Π} produces a bulk viscosity, while the second term associated with β_{Π}, which is intended to suppress particle interpenetration at high Mach number, is similar to the von Neumann-Richtmyer artificial viscosity. The artificial viscosity given by equation (5.56) is added into the pressure terms in the SPH equations.

Since the Monaghan type artificial viscosity introduces a shear viscosity into the flows especially in regions away from the shock, an artificial viscosity depending on the divergence of the velocity field was employed by Herquist and Katz (Hernquist and Katz, 1989)

$$\Pi_{ij} = \frac{q_i}{\rho_i^2} + \frac{q_j}{\rho_j^2}, \tag{5.62}$$

where

$$q_i = \begin{cases} \alpha_{\Pi} h_i \rho_i c_i \left|\nabla \cdot \mathbf{v}_i\right| + \beta_{\Pi} h_i^2 \rho_i \left|\nabla \cdot \mathbf{v}_i\right|^2 & \nabla \cdot \mathbf{v} < 0 \\ 0 & \nabla \cdot \mathbf{v} \geq 0 \end{cases}. \tag{5.63}$$

Other modifications for the Monaghan type artificial viscosity were also proposed (Morris et al., 1997), and are still under investigation.

5.4.2 Artificial heat

The Monaghan type artificial viscosity described above often provides good results when modeling shocks. However, excessive heating can occur under some severe circumstances such as the wall heating from the classic example of a stream of gas being brought to rest against a rigid wall. Noh fixed this problem by adding an artificial heat conduction term to the energy equation (Noh, 1987). An SPH form of artificial heat term was derived by Monaghan (Monaghan, 1995) as follows and is added to the energy equation if necessary (Fulk, 1994)

$$H_i = 2\sum_{j=1}^{N} \frac{\bar{q}_{ij}}{\bar{\rho}_{ij}} \frac{e_i - e_j}{\left|\mathbf{x}_{ij}\right|^2 + \varphi^2} \mathbf{x}_{ij} \cdot \nabla_i W_{ij}, \tag{5.64}$$

where

$$q_i = \alpha_{\Pi} h_i \rho_i c_i \left|\nabla \cdot \mathbf{v}_i\right| + \beta_{\Pi} h_i^2 \rho_i \left|\nabla \cdot \mathbf{v}_i\right|^2, \tag{5.65}$$

$$q_j = \alpha_{\Pi} h_j \rho_j c_j \left|\nabla \cdot \mathbf{v}_j\right| + \beta_{\Pi} h_j^2 \rho_j \left|\nabla \cdot \mathbf{v}_j\right|^2, \tag{5.66}$$

$$\bar{q}_{ij} = q_i + q_j \,. \tag{5.67}$$

5.4.3 Smoothing length

The smoothing length h is very important in the SPH method, which has direct influence on the efficiency of the computation and the accuracy of the solution. If h is too small, there may be not enough particles in the support domain of dimension κh to exert forces on a given particle, which results in low accuracy. If the smoothing length is too large, all details of the particle or local properties may be smoothed out, and the accuracy suffers, too. The particle approximations used by the SPH method depend on having a sufficient and necessary number of particles within the support domain of κh. The computational effort or speed also depends on this particle number. In one, two and three dimensions, the number of neighboring particles (including the particle itself) should be about 5, 21, 57 respectively if the particles are placed in a lattice with a smoothing length of 1.2 times the particle spacing, and $\kappa = 2$.

In early implementation of SPH, the global particle smoothing length was used which depended on the initial average density of the system. Later, the smoothing length was improved to solve problems where the fluid expands or contracts locally so as to maintain consistent accuracy throughout the space by assigning each particle an individual smoothing length according to the variation of the local number density of each particle (Monaghan, 1982, 1992)

For problems that are not isotropic such as shock problems, the smoothing length needs to be adapted both in space and time (Hernquist and Katz, 1989; Steinmetz and Mueller, 1993; Nelson and Papaloizou, 1994). Using a tensor smoothing length that is based on an ellipsoidal kernel rather than the traditional spherical kernel, Shapiro, Owen and their co-workers developed an adaptive SPH (ASPH) (Shapiro et al., 1996; Owen et al., 1998). Liu et al. also developed an ASPH model for simulating flows in micro channel (Liu and Liu, 2005) and impact and penetration problems (Liu et al., 2006).

There are many ways to dynamically evolve h so that the number of the neighboring particles remains relatively constant. The simplest approach is to update the smoothing length according to the averaged density

$$h = h_0 \left(\frac{\rho_0}{\rho} \right)^{1/d} , \tag{5.68}$$

where h_0 and ρ_0 are the initial smoothing length and the initial density respectively. d is the number of dimensions.

Benz suggested another method (Benz, 1989) to evolve the smoothing length, which takes the time derivative of the smoothing function in terms of the continuity equation

$$\frac{dh}{dt} = -\frac{1}{d}\frac{h}{\rho}\frac{d\rho}{dt} . \tag{5.69}$$

Equation (5.69) can be discretized using the SPH approximations and calculated with the other differential equations in parallel.

5.4.4 Symmetrization of particle interaction

If the smoothing length is set to vary both in time and space, each particle has its own smoothing length. If h_i is not equal to h_j, the influencing domain of particle i may cover particle j but not necessarily vice versa. Therefore, it is possible for particle i to exert a force on particle j without j exerting the same corresponding reaction on i. This is a violation of the Newton's Third Law. In order to overcome this problem, some measures must be taken to preserve the symmetry of particle interactions.

One approach in preserving the symmetry of particle interaction is to modify the smoothing length. There are different ways to perform the modification to produce a symmetric smoothing length. One way to obtain the symmetric smoothing length is to take the arithmetic mean or the average of the smoothing lengths of the pair of interacting particles (Benz, 1989)

$$h_{ij} = \frac{h_i + h_j}{2} . \tag{5.70}$$

Other ways can also be used to get the symmetric smoothing length using the geometric mean of the smoothing lengths of the pair of the interacting particles

$$h_{ij} = \frac{2h_i h_j}{h_i + h_j} , \tag{5.71}$$

or the maximal value of the smoothing lengths

$$h_{ij} = \min(h_i, h_j) , \tag{5.72}$$

or the minimal value of the smoothing lengths

$$h_{ij} = \max(h_i, h_j) . \tag{5.73}$$

The smoothing function can then be obtained using the symmetric smoothing length

$$W_{ij} = W(r_{ij}, h_{ij}) . \quad (5.74)$$

There are advantages and disadvantages in these different ways to determine the symmetric smoothing length h_{ij}. Taking the arithmetic mean or the maximal value of the smoothing lengths tends to use more neighboring particles and sometimes may overly smooth out the interactions among surrounding particles. Taking the geometric mean or the minimal value of the smoothing length tends to possess less neighboring particles.

Another approach to preserve symmetry of particle interaction is to use directly the average of the smoothing function values (Hernquist and Katz, 1989) without using a symmetric smoothing length, i.e.

$$W_{ij} = \frac{1}{2}(W(h_i) + W(h_j)) . \quad (5.75)$$

These two approaches in preserving the symmetry of the particle interactions are both widely used in the implementation of SPH. No detailed comparison study on these two approaches has been reported so far.

5.4.5 Tensile instability

When using the SPH method for hydrodynamics with material strength, one numerical problem called tensile instability (Balsara, 1995; Dyka and Ingel, 1995; Swegle et al., 1995; Dyka et al., 1997; Belytschko et al., 2000; Monaghan, 2000; Bonet and Kulasegaram, 2001; Belytschko and Xiao, 2002; Randles et al., 2003; Lanson and Vila, 2007; Sigalotti and Lopez, 2008) may arise. The tensile instability is the situation that when particles are under tensile stress state, the motion of the particle becomes unstable. It could result in particle clumping or even complete blowup in the computation.

According to Swegle, the tensile instability depends neither on the artificial viscosity, nor on the time integration scheme (Swegle et al., 1995). It is closely related to the selection of smoothing kernel function. In a one-dimensional von Neumann stability analysis, Swegle et al. gave a criterion for being stable or instable in terms of the stress state and the second derivative of the smoothing function, i.e., a sufficient condition for the unstable growth is

$$W_{\alpha\alpha}\sigma^{\alpha\alpha} > 0 , \quad (5.76)$$

where $W_{\alpha\alpha}$ is the second derivative of the smoothing function.

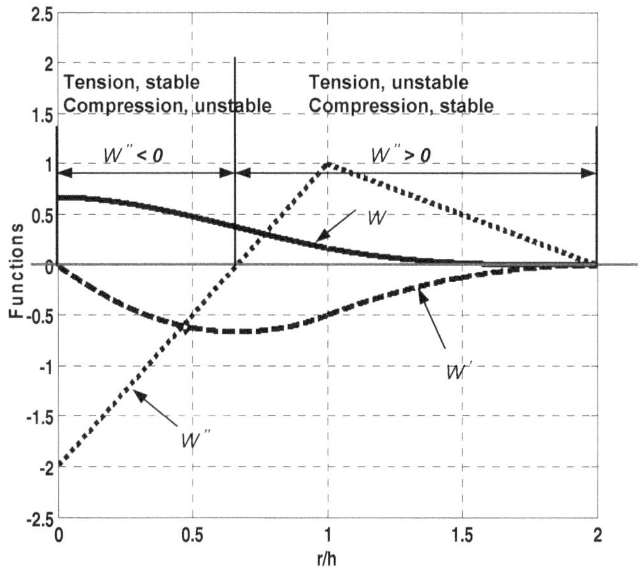

Figure 5.5 Schematic illustration of tensile instability with the cubic spline function and its first and second derivatives.

In the SPH method, the cubic spline smoothing function (illustrated in **Figure 5.5**) is mostly commonly used. The initial smoothing length is usually set to be equal to the particle spacing. Under such circumstances, the first nearest neighbor particles are located at $r/h = 1$; and the next nearest neighbor particles are at $r/h = 2$. As can be seen from , the second derivatives of the cubic spline function from $r/h = 1$ to $r/h = 2$ are always positive. Therefore it is expected that, according to equation (5.76), the SPH method with the cubic spline function would be stable in a compressed state but could be unstable in a tensile state in this region.

Several remedies have been proposed to improve or avoid such tensile instability. Morris suggested using special smoothing functions since the tensile instability is closely related to the second order derivative of the smoothing function (Morris, 1996a). Though successful in some cases, they do not always yield satisfactory results generally. Chen and his co-workers proposed the corrective smoothed particle method (CSPM), which was reported to improve the tensile instability (Chen et al., 1999b). Monaghan and his colleagues proposed an artificial force to stabilize the computation (Monaghan, 2000; Gray et al., 2001).

The basic reason of tensile instability is that the SPH method is essentially a collocation method, in which the particle approximations are conducted ONLY over the particles that represent the entire system. This leads to insufficient "sampling" points for establishing equations, and can result in numerical instability problem (Liu, 2002). The situation is very much similar to the so-called "node integration" in the implementation of the element free Galerkin method (EFG) (Beissel and Belytschko, 1996). In the EFG method, the instability is restored by adding stabilization terms in the Galerkin weak form. Based on this analysis, we need somehow bring in more information from other points, in addition to these particles.

One of such a method is to make use of the information at additional points in the support domain, rather than use only these collocation particles. Dyka et al. first introduced additional stress points other than the normal particles in a one-dimensional algorithm aimed at removing the tensile instability in SPH (Dyka and Ingel, 1995; Dyka et al., 1997). The stress points were also shown to be stable in tension and contributed considerably to the accuracy in wave propagation problems. Later, this approach has been further extended to multi-dimensional space by staggering the SPH particles using stress points so that there are essentially an equal number in each set of points (Randles and Libersky, 2000; Randles et al., 2003; Randles and Libersky, 2005). Basically in this approach, two sets of particles are used. One set of SPH particles carry velocity, and are referred to as "velocity particles". The other companion set of particles carry all required field variables except for the velocity, and are referred to as "stress particles".

Randles and Libersky pointed out that, the tensile instability for problems involving material strength generally is latent. The growth rate of damages in solid continuum models is often much faster than the grow rate of the tensile instability (Randles and Libersky, 2000).

Except for problems with material strength which can experience tensile instability, fluid mechanics problems sometimes can also meet tensile instability. Melean et al. showed the tensile instability in a formation of viscous drop (Melean et al., 2004), and the instability can be removed by using the artificial stress proposed by Monaghan (Monaghan, 2000; Gray et al., 2001). Yang et al. presents an improved SPH method for modeling viscous liquid drop without tensile instability using a hyperbolic shaped kernel function which possesses non-negative second derivatives (Yang et al., 2014a).

5.5 Consistency of the SPH method

5.5.1 Consistency in kernel approximation (kernel consistency)

For a constant (0th order polynomial) function $f(\mathbf{x}) = c$ (where c is a constant) to be exactly reproduced by the SPH kernel approximation, following equation (5.4), we require

$$f(\mathbf{x}) = \int cW(\mathbf{x}-\mathbf{x}',h)d\mathbf{x}' = c\,, \tag{5.77}$$

or

$$\int W(\mathbf{x}-\mathbf{x}',h)d\mathbf{x}' = 1\,. \tag{5.78}$$

Equation (5.78) is exactly the normalization condition described previously.

Further, for a linear function $f(\mathbf{x}) = c_0 + c_1\mathbf{x}$ (where c_0 and c_1 are constants) to be exactly reproduced, we must have

$$f(\mathbf{x}) = \int (c_0 + c_1\mathbf{x}')W(\mathbf{x}-\mathbf{x}',h)d\mathbf{x}' = c_0 + c_1\mathbf{x}\,. \tag{5.79}$$

Using equation (5.77), equation (5.79) can be simplified as

$$\int \mathbf{x}'W(\mathbf{x}-\mathbf{x}',h)d\mathbf{x}' = \mathbf{x}\,. \tag{5.80}$$

Multiplying $\boldsymbol{x}$ to both side of equation (5.78), we have the following identity

$$\int \mathbf{x}W(\mathbf{x}-\mathbf{x}',h)d\mathbf{x}' = \mathbf{x}\,. \tag{5.81}$$

Subtracting equation (5.80) from the above identity yields

$$\int (\mathbf{x}-\mathbf{x}')W(\mathbf{x}-\mathbf{x}',h)d\mathbf{x}' = 0\,. \tag{5.82}$$

Equation (5.82) is just the previously described symmetric condition.

More generally, by performing Taylor series analyses on the kernel approximation of a function $f(\mathbf{x})$ $(= \int f(\mathbf{x}')W(\mathbf{x}-\mathbf{x}',h)d\mathbf{x}')$ in a one-dimensional space, we have already obtained a set of requirements of the smoothing function described previously in equation (5.47). Equations (5.78) (normalization condition) and (5.82) (symmetric condition) are actually components in equation (5.47), which describe the 0th and 1st moments.

For the integrations expressed in equations (5.47), the integration domain is assumed to be a full continuous support domain that is not truncated by the boundaries. Equation (5.47) states that the requirements on the moments of a smoothing function need to reproduce certain order of polynomials.

Equation (5.47) can thus be used as an approximation accuracy indicator. If a smoothing function satisfies equations (5.47), a function can be approximated to n-th order accuracy. Furthermore the 0th moment in equations (5.47) states the normalization condition, and the 1st moment states the symmetry property of the smoothing function.

Similar to the consistency concept in the traditional FEM, if an SPH approximation can reproduce a polynomial of up to n^{th} order exactly, the SPH approximation is said to have n^{th} order or C^n consistency. If the consistency of an SPH kernel approximation in continuous form is termed as *kernel consistency*, the kernel consistency of an SPH kernel approximation is of nth order when the smoothing function satisfies equation (5.47). Therefore the expressions in equations (5.47) are also the kernel consistency conditions of the smoothing function for an SPH kernel approximation.

Note that if the SPH kernel approximations are carried out for regions truncated by boundaries, constant and linear functions can not be reproduced exactly since equations (5.78) and (5.82) are not satisfied for these regions. Therefore we can conclude that, since a conventional smoothing function satisfies the normalization and symmetric conditions, the conventional SPH method has up to C^1 consistency for the interior regions. However, for the boundary regions, it even does not have C^0 kernel consistency.

5.5.2 Consistency in particle approximation (particle consistency)

Satisfying the consistency conditions at the kernel approximation stage does not necessarily mean that the discretized SPH model will have such a consistency. This is because such a consistency can be distorted by the particle approximation process in discrete SPH model. Therefore, the consistency analysis should be conducted for the discrete SPH model in the particle approximation process, and this consistency can be termed as *particle consistency*.

The discrete counterparts of the constant and linear consistency conditions as expressed in equations (5.78) and (5.82) are

$$\sum_{j=1}^{N} W(\mathbf{x}-\mathbf{x}_j, h)\Delta v_j = 1, \tag{5.83}$$

and

$$\sum_{j=1}^{N} (\mathbf{x}-\mathbf{x}_j) W(\mathbf{x}-\mathbf{x}_j, h)\Delta v_j = 0. \tag{5.84}$$

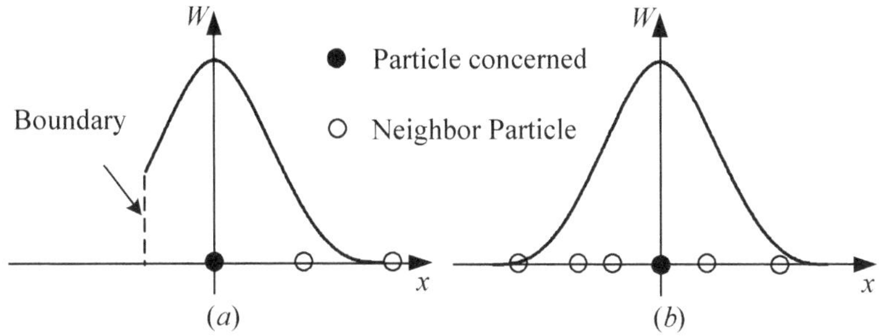

Figure 5.6 SPH particle approximations in one-dimensional cases. (a) Particle approximation for a particle whose support domain is truncated by the boundary. (b) Particle approximation for a particle with irregular particle distribution in its support domain.

These discretized consistency conditions are not satisfied in general. One obvious and simple example is the particle approximations at the boundary particles (**Figure 5.6**a). Even for uniform particle distribution, due to the unbalanced particles contributing to the discretized summation, the LHS of equation (5.83) is smaller than 1 and the LHS of equation (5.84) will not vanish, due to the truncation of the smoothing function by the boundary. For cases with irregularly distributed particles (**Figure 5.6**b), it is also easy to verify that even for the interior particles whose support domains are not truncated, the constant and linear consistency conditions in discretized forms may not be exactly satisfied. Therefore the original SPH method does not even have C0 consistency in the particle approximation. It is clear that the inconsistency caused by the particle approximation is closely related to the corresponding kernel approximation and particles involved in the approximation. Such an inconsistency problem results in directly the solution inaccuracy in the original SPH method.

Besides the particle approximation features associated with boundary particles or irregular distributed particles, the choice of the smoothing length is also important in the particle approximation process. In a one-dimensional domain with the cubic spline smoothing function, it is easy to verify that for uniformly distributed interior particles, the original SPH method has C^0 particle consistency if the smoothing length is taken exactly as the particle spacing ($h = \Delta x$) since equation (5.83) is satisfied. However, varying the smoothing length can results in a dissatisfaction of equation (5.83), leading to poor accuracy in the original SPH method. This is a reason why we often need to examine the influence of the smoothing length on the SPH approximation results.

In summary, the original SPH models, in general, do not have even C^0 consistency. Such an inconsistency originates from the discrepancy between the SPH kernel and particle approximations. Boundary particles, irregular distributed particles, and variable smoothing length can usually produce inconsistency in the particle approximation process. In the next section, we discuss ways to restore the consistency in SPH models.

5.5.3 Review on approaches for restoring consistency

It has been shown that the original SPH method even does not have 0th particle consistency. Different approaches have been proposed to improve the particle inconsistency and hence the SPH approximation accuracy. Some of them involve reconstruction of a new smoothing function so as to satisfy the discretized consistency conditions. However, these approaches are usually not preferred for hydrodynamic simulations because the reconstructed smoothing function can be partially negative, non-symmetric, and not monotonically decreasing. Approaches which improve the particle consistency without changing the conventional smoothing function are usually more preferable in simulating hydrodynamics.

One early approach (Monaghan, 1992; Randles and Libersky, 1996) is based on the anti-symmetric assumption of the derivative of a smoothing function

$$\sum_{j=1}^{N} W_{i,\alpha} \Delta v_j = 0, \tag{5.85}$$

where $W_{i,\alpha} = \partial W_i(\mathbf{x}) / \partial \mathbf{x}^{\alpha}$, in which α is the dimension index repeated from 1 to d (d is the number of dimensions). Therefore when approximating the derivative of a function f, the particle approximation can be rewritten as

$$f_{i,\alpha} = \sum_{j=1}^{N} \left(f_j - f_i \right) W_{i,\alpha} \Delta v_j , \tag{5.86}$$

or

$$f_{i,\alpha} = \sum_{j=1}^{N} \left(f_j + f_i \right) W_{i,\alpha} \Delta v_j . \tag{5.87}$$

It should also be noted that equation (5.85) is not necessarily valid, even if its corresponding continuous counterpart $\int W_{i,\alpha} d\mathbf{x} = 0$ is valid (for interior regions). This is also a manifestation of the particle inconsistency. Therefore equations (5.86) and (5.87) actually use the particle inconsistency in approximating the derivative of the smoothing function to offset or balance the particle inconsistency in approximating the derivatives of a field function, with a hope to improve the accuracy of the approximations.

Randles and Libersky (Randles and Libersky, 1996) derived a normalization formulation for the density approximation

$$\rho_i = \frac{\sum_{j=1}^{N} \rho_j W_{ij} \Delta v_j}{\sum_{j=1}^{N} W_{ij} \Delta v_j}, \tag{5.88}$$

and a normalization for the divergence of the stress tensor σ

$$(\nabla \cdot \sigma)_i = \frac{\sum_{j=1}^{N} (\sigma_j - \sigma_i) \otimes \nabla_i W_{ij} \Delta v_j}{\sum_{j=1}^{N} (\mathbf{x}_j - \mathbf{x}_i) \otimes \nabla_i W_{ij} \Delta v_j}, \tag{5.89}$$

where $\otimes$ is the tensor product. Again, equations (5.88) and (5.89) also use the inconsistency in approximating the smoothing function and its derivatives to offset the inconsistency in approximating a field function and its derivatives, also with an aim to improve the accuracy of the approximations.

Based on Taylor series expansion on the SPH approximation of a function, Chen et al. (Chen and Beraun, 2000) suggested a corrective smoothed particle method (CSPM). In one-dimensional space, the process of CSPM can be briefed as follows.

Performing Taylor series expansion at a nearby point x_i, a sufficiently smooth function $f(x)$ can be expressed as

$$f(x) = f_i + (x - x_i) f_{i,x} + \frac{(x - x_i)^2}{2!} f_{i,xx} + \dots . \tag{5.90}$$

Multiplying both sides of the equation (5.90) by the smoothing function W and integrating over the entire computational domain yield

$$\begin{aligned} \int f(x) W_i(x) dx = f_i \int W_i(x) dx + f_{i,x} \int (x - x_i) W_i(x) dx \\ + \frac{f_{i,xx}}{2} \int (x - x_i)^2 W_i(x) dx + \dots \end{aligned} . \tag{5.91}$$

If the terms involving *derivatives* in this equation are neglected, a corrective kernel approximation for function $f(x)$ at particle i is obtained as

$$f_i = \frac{\int f(x) W_i(x) dx}{\int W_i(x) dx}. \tag{5.92}$$

For a conventional smoothing function (non-negative and symmetric), the second term at the RHS of equation (5.91) is zero for interior region and not zero for boundary region. Therefore the corrective kernel approximation expressed in equation (5.92) is also of 2nd order accuracy for interior region and 1st order accuracy for boundary region. Comparing equation (5.92) with equation (5.5), it is found that for the interior regions, the kernel approximations in the original SPH and CSPM are actually the same due to the satisfaction of the normalization condition (in continuous form). For the boundary regions, since the integral of the smoothing function is truncated by the boundary, the normalization condition cannot be satisfied. By retaining the non-unity denominator, CSPM restores the C^0 kernel consistency.

The corresponding particle approximation for function $f(x)$ at particle i can be obtained using summation over nearest particles for each term in equation (5.91) and again neglecting the terms related to derivatives

$$f_i = \frac{\sum_{j=1}^{N} f_j W_{ij} \Delta v_j}{\sum_{j=1}^{N} W_{ij} \Delta v_j}. \tag{5.93}$$

It is noted that the particle approximation of the second term at the RHS of equation (5.91) is not necessarily zero even for the interior particles due to the irregularity of the particles. Therefore strictly speaking, the particle approximation expressed in equation (5.93) is of 1st order accuracy for both the interior and boundary particles. Only if the particles are uniformly distributed can the particle approximation of the second term at the RHS of equation (5.91) be zero. In this case, the particle approximation expressed in equation (5.93) is of 2nd order accuracy for the uniformly distributed interior particles.

If replacing $W_i(x)$ in equation (5.91) with $W_{i,x}$ and neglecting the *second and higher* derivatives, a corrective kernel approximation for the first derivative is generated as

$$f_{i,x} = \frac{\int [f(x) - f(x_i)] W_{i,x}(x) dx}{\int (x - x_i) W_{i,x}(x) dx}. \tag{5.94}$$

The particle approximations corresponding to equation (5.94) is

$$f_{i,x}=\frac{\sum_{j=1}^{N}(f_j-f_i)W_{i,x}\Delta v_j}{\sum_{j=1}^{N}(x_j-x_i)W_{i,x}\Delta v_j}. \tag{5.95}$$

Similarly, The CSPM kernel approximations for the derivatives are also of second order accuracy (or 1st order consistency) for interior regions, but 1st order accuracy (or 0th order consistency) for boundary regions. Except for cases with uniformly distributed interior particles, the CSPM particle approximations for the derivatives are of 1st order accuracy (or 0th order consistency) for both the interior and boundary particles.

5.5.4 A general approach to restore particle consistency

Liu et al gave a general approach to restore particle consistency through reconstructing the smoothing function (Liu and Liu, 2003). In general, a smoothing function can be written in the following form

$$\begin{aligned}W(\mathbf{x}-\mathbf{x}_j,h)&=b_0(\mathbf{x},h)+b_1(\mathbf{x},h)\left(\frac{\mathbf{x}-\mathbf{x}_j}{h}\right)+\\&\quad b_2(\mathbf{x},h)\left(\frac{\mathbf{x}-\mathbf{x}_j}{h}\right)^2+\ldots\ldots\\&=\sum_{l=0}^{k}b_l(\mathbf{x},h)\left(\frac{\mathbf{x}-\mathbf{x}_j}{h}\right)^l.\end{aligned} \tag{5.96}$$

By substituting the above smoothing function into equation (5.47), and after some trivial transformation, the discretized form of equation (5.47) can be written as

$$\left.\begin{aligned}&\sum_{l=0}^{k}b_l(\mathbf{x},h)\sum_{j=1}^{N}\left(\frac{\mathbf{x}-\mathbf{x}_j}{h}\right)^l\Delta\mathbf{x}_j=1\\&\sum_{l=0}^{k}b_l(\mathbf{x},h)\sum_{j=1}^{N}\left(\frac{\mathbf{x}-\mathbf{x}_j}{h}\right)^{l+1}\Delta\mathbf{x}_j=0\\&\vdots\\&\sum_{l=0}^{k}b_l(\mathbf{x},h)\sum_{j=1}^{N}\left(\frac{\mathbf{x}-\mathbf{x}_j}{h}\right)^{l+k}\Delta\mathbf{x}_j=0\end{aligned}\right\}. \tag{5.97}$$

The $k+1$ coefficients $b_l(\mathbf{x},h)$ can then be determined by solving the following matrix equation

$$\underbrace{\begin{bmatrix} m_0(\mathbf{x},h) & m_1(\mathbf{x},h) & \cdots & m_k(\mathbf{x},h) \\ m_1(\mathbf{x},h) & m_2(\mathbf{x},h) & \cdots & m_{1+k}(\mathbf{x},h) \\ \vdots & \vdots & \ddots & \vdots \\ m_k(\mathbf{x},h) & m_{k+1}(\mathbf{x},h) & \cdots & m_{k+k}(\mathbf{x},h) \end{bmatrix}}_{M} \underbrace{\begin{Bmatrix} b_0(\mathbf{x},h) \\ b_1(\mathbf{x},h) \\ \vdots \\ b_k(\mathbf{x},h) \end{Bmatrix}}_{b} = \underbrace{\begin{Bmatrix} 1 \\ 0 \\ \vdots \\ 0 \end{Bmatrix}}_{I}, \tag{5.98}$$

or

$$\mathbf{Mb} = \mathbf{I}, \tag{5.99}$$

where

$$m_k(\mathbf{x},h) = \sum_{j=1}^{N} \left(\frac{\mathbf{x}-\mathbf{x}_j}{h} \right)^k \Delta\mathbf{x}_j, \tag{5.100}$$

$\mathbf{M}$ is a moment matrix, $\mathbf{b}$ is a vector of coefficients, $\mathbf{I}$ is a vector of given constants.

After determining the coefficients $b_l(\mathbf{x},h)$, the smoothing function expressed in equation (5.96) can be calculated. The procedure ensures particle consistency to k^{th} order. Therefore, the particle consistency restoring process actually gives an approach to construct some kind of smoothing function for the SPH methods.

Comparing with the traditional smoothing function, which is only dependent on the particle distance and applicable for all the particles, the consistency restored smoothing function is particle-wise. It therefore depends on both the distance and positions of the interacting particles. The cost-effectiveness for this approach in constructing particle-wise smoothing functions needs to be considered since it will require additional CPU time to solve the particle-wise equation (5.100) for all the particles. Moreover, since all particles are moving, the particle location is changing as well. Hence, the particle-wise smoothing functions need to be computed for every time step. Another problem is that, to solve equation (5.100), the moment matrix $\mathbf{M}$ is required to be non-singular. Therefore, the particle distribution must satisfy certain conditions to avoid singular momentum matrix. This implies that when we enforce on consistency, we will face the stability problem shown as the bad-conditioned moment matrix in the SPH settings. In contrast, in the original SPH method, particles can be arbitrary distributed, though the obtained results may be less accurate.

As far as the approximation is concerned, restoring particle consistency is an improvement on the accuracy of the particle approximation, provided that the moment matrix **M** is not singular. However, it is noted that restoring the consistency in discrete form leads to some problems in simulating hydrodynamic problems. Firstly, the resultant smoothing function is negative in some parts of the region. Negative value of smoothing function can leads to unphysical representation of some field variables, such as negative density, negative energy that can lead to a breakdown of the entire computation. Secondly, the resultant smoothing function may not be monotonically decreasing with the increase of the particle (node) distance. Moreover, the constructed smoothing function may not be symmetric and using this non-symmetric smoothing function violates the equal mutual interaction in physics.

5.5.5 Finite particle method

Considering the disadvantages of the above-mentioned particle inconsistency restoring approach in constructing a point-wise smoothing function, Liu et al. devised another particle consistency restoring approach, which retains the conventional non-negative smoothing function instead of reconstructing a new smoothing function (Liu et al., 2005; Liu and Liu, 2006). This approach has been termed as *Finite Particle Method* (FPM), in which a set of basis functions can be used in the numerical approximation.

Performing Taylor series expansion at a nearby point $\mathbf{x}_i = \{x_i, y_i, z_i\}$ and retaining the second order derivatives, a sufficiently smooth function $f(\mathbf{x})$ at point $\mathbf{x} = \{x, y, z\}$ can be expressed as follows

$$f(x) = f_i + (x^\alpha - x_i^\alpha) f_{i,\alpha} + \frac{(x^\alpha - x_i^\alpha)(x^\beta - x_i^\beta)}{2!} f_{i,\alpha\beta} + r((x - x_i)^3), \tag{5.101}$$

where α, β are the dimension indices repeated from 1 to 3 (or from x to z). $r((\mathbf{x} - \mathbf{x}_i)^3)$ is the remainder of the expansion. f_i, $f_{i,\alpha}$ and $f_{i,\alpha\beta}$ are defined as

$$f_i = f(\mathbf{x}_i), \tag{5.102}$$

$$f_{i,\alpha} = f_\alpha(\mathbf{x}_i) = (\partial f / \partial x^\alpha)_i, \tag{5.103}$$

$$f_{i,\alpha\beta} = f_{\alpha\beta}(\mathbf{x}_i) = (\partial^2 f / \partial x^\alpha \partial x^\beta)_i. \tag{5.104}$$

Multiplying both sides of equation (5.101) with a function $\varphi_1(\mathbf{x}-\mathbf{x}_i)$ and integrating over the problem space Ω can yield the following equation

$$\int_\Omega f(\mathbf{x})\varphi_1(\mathbf{x}-\mathbf{x}_i)d\mathbf{x} = f_i\int_\Omega \varphi_1(\mathbf{x}-\mathbf{x}_i)d\mathbf{x} + f_{i,\alpha}\int_\Omega (x^\alpha - x_i^\alpha)$$
$$\varphi_1(\mathbf{x}-\mathbf{x}_i)d\mathbf{x} + \frac{f_{i,\alpha\beta}}{2}\int_\Omega (x^\alpha - x_i^\alpha)(x^\beta - x_i^\beta)$$
$$\varphi_1(\mathbf{x}-\mathbf{x}_i)d\mathbf{x} + r((\mathbf{x}-\mathbf{x}_i)^3). \tag{5.105}$$

It is seen that the integration is carried out over the entire problem space, and can be quite time-consuming. One usual assumption is that a field variable at point $\mathbf{x}_i$ is only strongly influenced by the field variables at nearby points and that the influence of the field variables at points far away from point $\mathbf{x}_i$ is very weak and hence can be neglected. Therefore, the global integration can be converted into a local integration by defining a *support domain* for point $\mathbf{x}_i$ in which the field variables at point $\mathbf{x}_i$ can be determined. The shape of the support domain can be conveniently taken as a circle (in 2D) or a sphere (in 3D) with a radius of κh, in which κ is a constant scalar factor, and h is a length characterizing the support domain. The function $\varphi_1(\mathbf{x}-\mathbf{x}_i)$ is also limited to the local support domain, and can be rewritten as $\varphi_1(\mathbf{x}-\mathbf{x}_i,h)$.

Since the points distributed in the problem space are actually particles, each occupying individual lumped volume, equation (5.105) can be numerically approximated by summation over the particles surrounding point $\mathbf{x}_i$ as follows

$$\sum_{j=1}^{N} f(\mathbf{x}_j)\varphi_1(\mathbf{x}_j-\mathbf{x}_i,h)\Delta V_j = f_i\sum_{j=1}^{N}\varphi_1(\mathbf{x}_j-\mathbf{x}_i,h)\Delta V_j$$
$$+f_{i,\alpha}\sum_{j=1}^{N}(x_j^\alpha - x_i^\alpha)\varphi_1(\mathbf{x}_j-\mathbf{x}_i,h)\Delta V_j$$
$$+\frac{f_{i,\alpha\beta}}{2}\sum_{j=1}^{N}(x_j^\alpha - x_i^\alpha)(x_j^\beta - x_i^\beta)\varphi_1(\mathbf{x}_j-\mathbf{x}_i,h)\Delta V_j, \tag{5.106}$$

where N is the number of particles within the support domain of particle i. The remainder term $r((\mathbf{x}-\mathbf{x}_i)^3)$ in equation (5.106) is omitted in equation (5.106) for the sake of conciseness..

Equation (5.106) can be further simplified as the following equation at point $\mathbf{x}_i$

$$B_{1i} = A_{1ki}F_{ki}, \tag{5.107}$$

where

$$F_{ki}=[f_i \quad f_{i,\alpha} \quad f_{i,\alpha\beta}]^T , \tag{5.108}$$

$$B_{1i}=\sum_{j=1}^{N} f(\mathbf{x}_j)\varphi_1(\mathbf{x}_j-\mathbf{x}_i,h)\Delta V_j , \tag{5.109}$$

$$A_{1ki}=\Bigg[\sum_{j=1}^{N}\varphi_1(\mathbf{x}_j-\mathbf{x}_i,h)\Delta V_j \quad \sum_{j=1}^{N}(x_j^{\alpha}-x_i^{\alpha})\varphi_1(\mathbf{x}_j-\mathbf{x}_i,h)\Delta V_j \quad \frac{1}{2}\sum_{j=1}^{N}(x_j^{\alpha}-x_i^{\alpha})(x_j^{\beta}-x_i^{\beta})\varphi_1(\mathbf{x}_j-\mathbf{x}_i,h)\Delta V_j\Bigg]. \tag{5.110}$$

Corresponding to 1, 2, and 3 dimensional cases, there are one function value, 1, 2 and 3 first derivatives, and 1, 3 and 6 second derivatives that will be approximated. It is clear that k in equations (5.108), (5.109) and (5.110) is 3, 6, and 10 respectively corresponding to 1, 2, and 3 dimensional cases. To calculate the function value, the first and the second derivatives at $\mathbf{x}_i$, 2, 5, and 9 other equations similar to equation (5.107) are required. Therefore in 1, 2, and 3 dimensional cases, totally 3, 6, and 10 functions ($\varphi_M(\mathbf{x}-\mathbf{x}_i,h)$, M = 3, 6, or 10) are required in order to approximate the function value, the first and second derivative. These functions form a set of basis functions used for approximating the function value, its first and second derivatives. A conventional SPH smoothing function and its first and second derivatives can form a set of basis function in the FPM. For example, in 2D space, a smoothing function W, its two first order derivatives, W_α and W_β, and its three second order derivatives, $W_{\alpha\alpha}$, $W_{\alpha\beta}$ and $W_{\beta\beta}$ forms a set of 6 basis functions.

In summary, multiplying a set of basis functions on both sides of equation (5.101), integrating over the problem domain, summing over the nearest particles within the local support domain of particle i, a set of matrix equation can be produced to approximate the function value as well as the first and second derivatives at particle i. The matrix equations at particle i at can be written as

$$B_{Mi}=A_{Mki}F_{ki} , \text{ or } \mathbf{B}=\mathbf{AF} , \tag{5.111}$$

where

$$B_{Mi} = \sum_{j=1}^{N} f(\mathbf{x}_j)\psi_M(\mathbf{x}_j - \mathbf{x}_i, h)\Delta V_j \,, \tag{5.112}$$

$$\begin{aligned} A_{Mki} = & \sum_{j=1}^{N} \varphi_M(\mathbf{x}_j - \mathbf{x}_i, h)\Delta V_j \\ & \sum_{j=1}^{N} (x_j^{\alpha} - x_i^{\alpha})\varphi_M(\mathbf{x}_j - \mathbf{x}_i, h)\Delta V_j \\ & \frac{1}{2}\sum_{j=1}^{N} (x_j^{\alpha} - x_i^{\alpha})(x_j^{\beta} - x_i^{\beta})\varphi_M(\mathbf{x}_j - \mathbf{x}_i, h)\Delta V_j \;. \end{aligned} \tag{5.113}$$

Equation (5.111) is the basis of the finite particle method and can be used to approximate a function value and its derivatives for a field variable. It is seen that only if the coefficient matrix **A** is not singular, can these M equations determine a unique set of solutions at particle i for the M unknowns in vector **F**. Solving the above pointwise matrix equations, the function value as well as the first and second derivatives at every particle can be simultaneously approximated. Note that the conditioning of **A** matrix reflects the stability of the FPM model.

Since the governing equations in CFD only involve the first and second derivatives, only the derivatives up to the second order are retained in equation (5.101). For problems in other areas such as computational solid mechanics, high order derivatives may be involved. If third or higher order derivatives are to be approximated, in expanding $f(\mathbf{x})$ at $\mathbf{x}_i$, the interested derivatives need to be retained in equation (5.101). To obtain the increased number of unknowns, more functions like $\varphi_M(\mathbf{x} - \mathbf{x}_i, h)$ are necessary to complete the matrix equation (5.111). Therefore, except for the increased number of unknowns, increased number of basis functions, and therefore more computational efforts, the solution procedure for the interested unknowns is the same.

Comparing conventional SPH and FPM, it is clear that both FPM and SPH are meshfree particle methods in which particles with lumped volumes are used to represent the state of a system. The particles form a frame for interpolation, differencing or integration in a certain approximation. Both FPM and SPH can be used as Lagrangian methods if allowing the particles to move in the problem space. However, the difference between FPM and SPH is obvious.

1. FPM uses a set of basis function to approximate the function value and its derivatives, whereas SPH employs a smoothing function and its derivatives to approximate a function value and the corresponding

derivatives. The smoothing functions in SPH should have some special properties as described in Section 3. However, the basis functions in FPM are more general. Any set of functions which do not lead to a singular coefficient matrix A can be used as basis functions. Therefore the smoothing function and its specific derivatives actually can be one possible choice as a suitable set of basis functions.

2. SPH can be regarded as a special case of FPM, whereas FPM is a generalized version of SPH with modifications. In equation (5.101), if all the terms related to derivatives are neglected, multiplying both sides of equation (5.101) with the smoothing function W, and integrating over the problem space can lead to the approximation in SPH. Summation over the nearest particles within the support domain of a particle further produces the particle approximation of the field variable at that particle (equation (5.106)).
3. FPM should have better accuracy than SPH. Since no derivative term is retained in equation (5.101), the SPH method actually is of first order accuracy. If a symmetric smoothing function is used, the terms related to the first order derivatives are actually zero for the interior particles in the problem domain. Therefore SPH is of second order accuracy in interior parts. In contrast, since up to second order derivatives are retained in the expansion process, the accuracy of FPM is of third order. Moreover, if higher derivatives are retained, better accuracy can be achieved. FPM should have a better accuracy than SPH both for the interior particles and boundary particles.
4. As the solution is to be obtained from solving the matrix equation (5.111), a good matrix inversion algorithm is necessary to prevent the co-efficient matrix to be negative.
5. The matrix equation (5.111) is to be solved at every particle, and every time step. It can be more computationally expensive than the conventional SPH method.

One simpler example of FPM is to only consider the first derivative in equation (5.101). Using the smoothing function and its first derivatives as the basis functions, the following equations can be obtained

$$\int f(\mathbf{x})W_i(\mathbf{x})dx = f_i \int W_i(\mathbf{x})dx + f_{i,\alpha} \int (x^\alpha - x_i^\alpha)W_i(\mathbf{x})d\mathbf{x}\,, \tag{5.114}$$

and

$$\int f(\mathbf{x})W_{i,\beta}d\mathbf{x} = f_i \int W_{i,\beta}d\mathbf{x} + f_{i,\alpha} \int (x^\alpha - x_i^\alpha)W_{i,\beta}(\mathbf{x})d\mathbf{x}\,. \tag{5.115}$$

Again β is the dimension index repeated from 1 to d. The corresponding discrete forms for equations (5.114) and (5.115) are

$$\sum_{j=1}^{N} f_j W_{ij} \Delta v_j = f_i \cdot \sum_{j=1}^{N} W_{ij} \Delta v_j + f_{i,\alpha} \cdot \sum_{j=1}^{N} (x_j^{\alpha} - x_i^{\alpha}) W_{ij} \Delta v_j \ , \tag{5.116}$$

and

$$\sum_{j=1}^{N} f_j W_{ij,\beta} \Delta v_j = f_i \cdot \sum_{j=1}^{N} W_{ij,\beta} \Delta v_j + f_{i,\alpha} \cdot \sum_{j=1}^{N} (x_j^{\alpha} - x_i^{\alpha}) W_{ij,\beta} \Delta v_j \ . \tag{5.117}$$

There are $d+1$ equations for $d+1$ unknowns (f_i and $f_{i,\alpha}$). Equations (5.116) and (5.117) are therefore complete for solving with respect to f_i and $f_{i,\alpha}$, and the solutions for f_i and $f_{i,\alpha}$ are

$$\begin{bmatrix} f_i \\ f_{i,\alpha} \end{bmatrix} = \begin{bmatrix} \sum_{j=1}^{N} W_{ij} \Delta v_j & \sum_{j=1}^{N} (x_j^{\alpha} - x_i^{\alpha}) W_{ij} \Delta v_j \\ \sum_{j=1}^{N} W_{ij,\beta} \Delta v_j & \sum_{j=1}^{N} (x_j^{\alpha} - x_i^{\alpha}) W_{ij,\beta} \Delta v_j \end{bmatrix}^{-1} \begin{bmatrix} \sum_{j=1}^{N} f_j W_{ij} \Delta v_j \\ \sum_{j=1}^{N} f_j W_{ij,\beta} \Delta v_j \end{bmatrix} . \tag{5.118}$$

In equations (5.116) and (5.117), the terms related to the function and the first order derivatives are all retained, only the terms related to second or high order derivatives are neglected. Therefore the resultant particle approximations for a function and its derivatives (equation (5.118)) are able to exactly reproduce a constant and a linear function (C^0 and C^1 consistency). Hence the algorithm shown in equation (5.118) actually restores the particle consistency that conventional SPH method does not have. Again, this particle consistency restoring approach is independent of the particle distribution (either regular or irregular), and the choices of the smoothing kernel and smoothing length. Another advantage is that this particle consistency restoring approach does not change the conventional smoothing function and should be preferable in simulating hydrodynamics.

One notable pointed is that the accuracy of FPM is not sensitive to the selection of smoothing length, and extremely irregular particle distribution. To verify this point, a number of numerical tests have been given. For example, the second order FPM (expressed in equation (5.118)) to approximate a constant, $f(x)=1$, and a linear function, $f(x)=x$. Since approximation of the derivatives follow the same procedure as the approximation of a function itself, only the approximation results for the functions are presented. To prevent the boundary features from being concealed by boundary treatment tricks, there is no treatment on the boundary conditions in these analyses. The function approximation results obtained using the original SPH, CSPM and the FPM are comparatively investigated. The cubic spline function (equation (5.37)) is chosen as the smoothing function in the study.

5.5.6 A comparative study of particle consistency

5.5.6.1 0th order consistencies (C^0)

C^0 consistency represents the capability of reproducing or approximating a constant function. Consider an arbitrarily selected constant function $f(x)=1$ in a one-dimensional domain of [0, 1].

Table 5.1 Approximation results for $f(x)=1$ with 10 evenly distributed particles and $h=\Delta v$.

Particle number	Exact	Errors (%) SPH	CSPM	FPM
1	1.	-16.67	0.00	0.00
2	1.	0.00	0.00	0.00
3	1.	0.00	0.00	0.00
4	1.	0.00	0.00	0.00
5	1.	0.00	0.00	0.00
6	1.	0.00	0.00	0.00
7	1.	0.00	0.00	0.00
8	1.	0.00	0.00	0.00
9	1.	0.00	0.00	0.00
10	1.	-16.67	0.00	0.00

Table 5.1 shows the approximation results for the constant with 10 particles evenly located from $x=0.05$ to $x=0.95$ with an even increment of $\Delta x=0.1$. The lumped volume associated with each particle is $\Delta v=0.1$, and the smoothing length is selected exactly the same as the particle spacing, i.e., $h=\Delta v$. It is seen that CSPM and the new approach can exactly reproduce the constant function, no matter the particles are boundary particles and interior particles. The original SPH can also correctly approximate the constant for the interior particles. However, the approximation results for the two boundary particles (particle 1 and 10) are obviously below the constant.

Different from the above case with a smoothing length $h=\Delta x$, **Table 5.2** shows the approximation results with a smoothing length $h=1.2\Delta v$. Note that increasing the smoothing length means two more particles turn to be boundary particles since the support domains of two more particles are truncated by the boundaries. It is seen that varying smoothing length can influence the approximation results. The original SPH cannot correctly approximate the constant no matter the particles are boundary and interior particles, though the errors for interior particles are smaller than those for boundary particles. Again CSPM and the new approach do exactly reproduce the constant.

Table 5.2 Approximation results for $f(x)=1$ with 10 evenly distributed particles and $h=1.2\Delta v$.

Particle number	Exact	Errors (%) SPH	CSPM	FPM
1	1.	-19.68	0.00	0.00
2	1.	-0.05	0.00	0.00
3	1.	+0.04	0.00	0.00
4	1.	+0.04	0.00	0.00
5	1.	+0.04	0.00	0.00
6	1.	+0.04	0.00	0.00
7	1.	+0.04	0.00	0.00
8	1.	+0.04	0.00	0.00
9	1.	-0.05	0.00	0.00
10	1.	-19.68	0.00	0.00

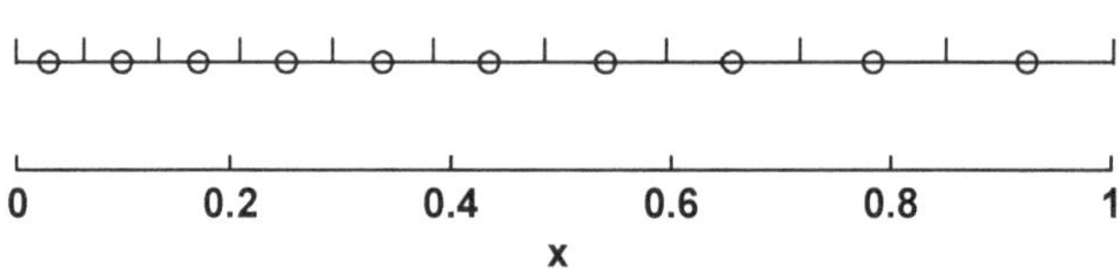

Figure 5.7 Approximation of a function with 10 non-uniformly distributed particles. The ratio of the lumped volumes for two neighboring particles is 1.1.

Table 5.3 shows the approximation results for the constant with 10 non-uniformly distributed particles with an even ratio of $\Delta v = 1.1$ (**Figure 5.7**). The smoothing length is selected exactly the same as the particle spacing, i.e., $h_i = \Delta v_i$. In determining interaction particle pair, a mean smoothing length $h_{ij} = (h_i + h_j)/2$ is used. It is also found that for irregular particle distribution, the original SPH cannot correctly represent the constant no matter the particles are boundary and interior particles. In contrast, CSPM and the new approach also do exactly represent the constant.

Table 5.3 Approximation results for $f(x)=1$ with 10 non-uniformly distributed particles and $h_i = \Delta v_i$.

Particle number	Exact	Errors (%)		
		SPH	CSPM	FPM
1	1.	-29.28	0.00	0.00
2	1.	-3.71	0.00	0.00
3	1.	+0.25	0.00	0.00
4	1.	+0.17	0.00	0.00
5	1.	-0.01	0.00	0.00
6	1.	-0.25	0.00	0.00
7	1.	-0.18	0.00	0.00
8	1.	+0.45	0.00	0.00
9	1.	+2.16	0.00	0.00
10	1.	+3.18	0.00	0.00

5.5.6.2 1st order consistencies (C^1)

C^1 consistency represents the capability of reproducing or approximating a linear function. Consider an arbitrarily selected linear function $f(x)=x$ in a one-dimensional domain of [0, 1].

Table 5.4 shows the approximation results for the linear function with 10 particles evenly located from $x=0.05$ to $x=0.95$ with an even increment of $\Delta x = 0.1$. The lumped volume associated with each particle is $\Delta v = 0.1$, and the smoothing length is selected exactly the same as the particle spacing, i.e.,

Table 5.4 Approximation results for $f(x)=x$ with 10 evenly distributed particles and $h=\Delta v$.

Particle number	Exact	Errors (%)		
		SPH	CSPM	FPM
1	0.0500	16.6000	40.0000	0.00
2	0.1500	0.00	0.00	0.00
3	0.2500	0.00	0.00	0.00
4	0.3500	0.00	0.00	0.00
5	0.4500	0.00	0.00	0.00
6	0.5500	0.00	0.00	0.00
7	0.6500	0.00	0.00	0.00
8	0.7500	0.00	0.00	0.00
9	0.8500	0.00	0.00	0.00
10	0.9500	-18.4211	-2.1053	0.00

$h = \Delta v$. It is seen that the new approach does exactly reproduce the linear function, no matter the particles are boundary particles and interior particles. Both the original SPH and CSPM fail to correctly approximate the linear function at boundary particles.

Different from the above case with a smoothing length $h = \Delta x$, **Table 5.5** shows the approximation results with a smoothing length $h = 1.2\Delta v$. Therefore two more particles turn to be boundary particles. The original SPH fail to correctly approximate the linear function no matter the particles are boundary and interior particles. Except for the boundary particles, CSPM can correctly approximate the linear function. In contrast, the new approach does exactly reproduce the linear function for both the boundary and interior particles.

Table 5.6 shows the approximation results for the linear function with 10 non-uniformly distributed particles with an even ratio of $\Delta v = 1.1$ (**Figure 5.7**). Similarly, the smoothing length is selected exactly the same as the particle spacing, i.e., $h_i = \Delta v_i$. In determining interaction particle pair, a mean smoothing length $h_{ij} = \left(h_i + h_j\right)/2$ is used. It is also found that for irregular particle distributions, the original SPH and CSPM cannot correctly approximate the linear function no matter the particles are boundary and interior particles. The new approach also can correctly approximate the function for both the boundary and interior particles.

Extending the linear function to a two-dimensional domain of [0, 1, 0, 1]. The linear function is $f(x, y) = x + y$. Similar conclusions can be made on the particle consistency for the original SPH, CSPM, and the new approach. Here only the approximation results using irregularly distributed particles are presented.

Table 5.5 Approximation results for $f(x) = x$ with 10 evenly distributed particles and $h = 1.2\Delta v$.

Particle number	Exact	Errors (%)		
		SPH	CSPM	FPM
1	0.0500	20.0000	49.4000	0.00
2	0.1500	0.0667	0.1333	0.00
3	0.2500	0.0400	0.00	0.00
4	0.3500	0.0286	0.00	0.00
5	0.4500	0.0444	0.00	0.00
6	0.5500	0.0364	0.00	0.00
7	0.6500	0.0462	0.00	0.00
8	0.7500	0.0400	0.00	0.00
9	0.8500	-0.0706	-0.0235	0.00
10	0.9500	-21.7579	-2.6000	0.00

Table 5.6 Approximation results for $f(x) = x$ with 10 non-uniformly distributed particles and $h_i = \Delta v_i$.

Particle number	Exact	Errors (%)		
		SPH	CSPM	FPM
1	0.0314	38.2166	95.2229	0.00
2	0.0973	1.4388	5.3443	0.1028
3	0.1697	0.1768	-0.0589	0.00
4	0.2494	0.1604	0.00	0.00
5	0.3371	0.0297	0.0297	0.00
6	0.4336	-0.2998	-0.0461	0.00
7	0.5397	-0.3706	-0.2038	-0.0185
8	0.6564	0.1523	-0.2895	-0.0152
9	0.7848	1.8349	-0.3186	-0.0127
10	0.9260	2.4946	-0.6695	-0.0648

Table 5.7 Approximation errors of the linear function $f(x, y) = x + y$.

	SPH	CSPM	FPM
Maximum error (%)	3.0288e-001	2.4840e-001	5.1768e-003
Minimum error (%)	1.1152e-004	1.7661e-006	0.
Root-mean-square error (%)	9.5534e-002	4.6493e-003	1.9500e-004

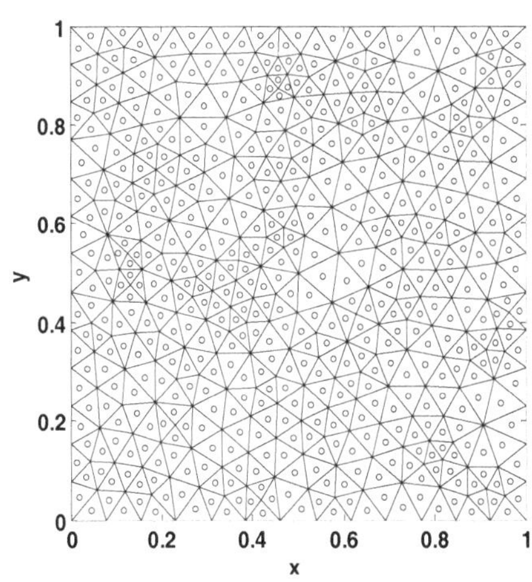

Figure 5.8 Particle distributions for the approximation of a linear function $f(x, y) = x + y$ in a two-dimensional domain of [0, 1, 0, 1] (Liu et al., 2005).

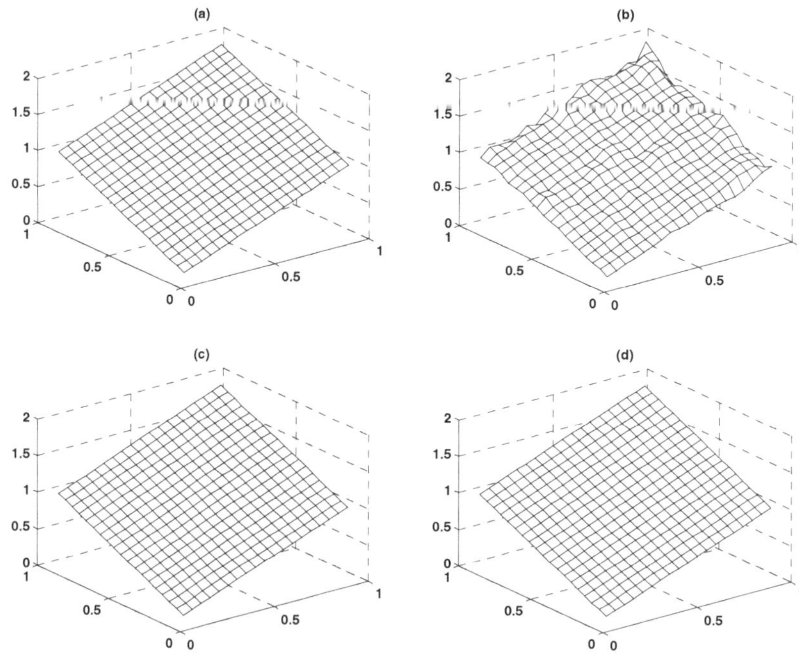

Figure 5.9 Approximation results of the linear function as well as the exact values. (a) Exact values, (b) results from the original SPH, (c) results from CSPM, and (d) results from the FPM (Liu et al., 2005).

As shown in **Figure 5.8**, the domain is irregular distributed with 528 particles. These particles are located at the mass centers of the triangular elements generated using the MATLAB partial differential toolbox (Matlab, 1995). The lumped volumes of the particles are exactly the same as the corresponding particles. The smoothing length for a particle i is taken as the diameter of an equivalent circle with the same area as the corresponding triangular elements. **Figure 5.9** shows the approximation results of the linear function using the original SPH, CSPM and the new approach as well as the exact values. It is clear that the results from the original SPH method obviously oscillate for both the boundary and interior particles. The results from CSPM and FPM are apparently better.

To more precisely compare the three methods in approximating the linear function, the maximum error, minimum error and root-mean-square error are calculated and listed in **Table 5.7**. It is clear that the errors resulting from the new approach are much smaller than those from the original SPH and CSPM.

Table 5.8 shows the results of the linear function along the diagonal line connecting the left-bottom and right-top corner. The results along the diagonal line (regular) are obtained through interpolation from the corresponding values on the irregular particles within the domain using a 21×21 grid. It is seen that the new approach also gives much better results than the original SPH and CSPM.

Table 5.8 Approximation results of the linear function $f(x,y)=x+y$ along the diagonal line connecting the left-bottom and right-top corner.

Points	Exact	Error (%)		
		SPH	CSPM	FPM
(2, 2)	1.2468e-001	-3.2243	2.3741	-0.1845
(3, 3)	2.2198e-001	1.1758	0.2388	0.0
(4, 4)	3.1927e-001	-0.2725	-0.0031	-0.0125
(5, 5)	4.1657e-001	6.6159	0.1104	-0.0072
(6, 6)	5.1386e-001	4.4720	0.0817	-0.0292
(7, 7)	6.1115e-001	1.4252	0.0589	0.0
(8, 8)	7.0845e-001	1.0234	-0.4376	0.0113
(9, 9)	8.0574e-001	2.9848	-0.0074	0.0012
(10, 10)	9.0304e-001	2.4816	0.2680	-0.0044
(11, 11)	1.0003e+000	1.6495	0.2999	-0.0520
(12, 12)	1.0976e+000	-1.3393	0.0638	0.0273
(13, 13)	1.1949e+000	1.1633	-0.0670	0.0
(14, 14)	1.2922e+000	0.9828	-0.0851	0.0310
(15, 15)	1.3895e+000	-0.8852	-0.1511	-0.0072
(16, 16)	1.4868e+000	2.0985	-0.0336	0.0
(17, 17)	1.5841e+000	-0.3346	-0.0505	0.0
(18, 18)	1.6814e+000	4.0799	0.0178	0.0119
(19, 19)	1.7787e+000	2.0352	-0.0787	0.0225
(20, 20)	1.8760e+000	2.8731	-0.0586	0.0213

5.5.6.3 Modeling Poiseuille flow

The above two numerical examples show the effectiveness of FPM in reproducing 0th and 1st order consistency, while the no solid boundary walls are involved. It is clear that FPM is not consistent to the selection of smoothing length, and extremely irregular particle distribution. Here FPM is used to model the Poiseuille flow, which is an incompressible flow with solid boundaries. In the work, the weakly compressible SPH (WCSPH) (will be discussed later in the next chapter) is used.

The classic Poiseuille flow is frequently used to validate the SPH algorithms in dealing with incompressible flow (Morris et al., 1997; Liu and Liu, 2003; Liu et al., 2005), due to its simplicity, and the existence of an analytical series solution (Liu and Liu, 2003). It involves flow between two infinite parallel plates driven by a body force F. In our test, the problem domain is a rectangle of 0.0005 m $\times$ 0.001 m, and is modeled with 20×40 particles, which are initialized regularly (**Figure 5.10**a). The fluid density, kinetic viscosity and body force are taken as $\rho = 10^3$ kg/m^3, $\upsilon = 10^{-6}$ m^2/s, $\mathbf{F} = 2 \times 10^{-4}$ m/s^2 respectively. The peak fluid velocity is $\mathbf{v}_0 = 2.5 \times 10^{-5}$ m/s, which corresponds to a Reynolds number of $Re = 2.5 \times 10^{-2}$. The infinite effects are modeled on a short region by applying the periodic boundary condition along the flow direction both in the particle movement and particle interaction process. The upper and lower plates are simulated using ghost or virtual particles. We first study the Poiseuille flow with 40 particles fixed on the upper and lower plates, each with 20 particles (**Figure 5.10**a). This Monaghan type of virtual particles are located right on the solid boundary to produce a highly repulsive force to the flow particles near the

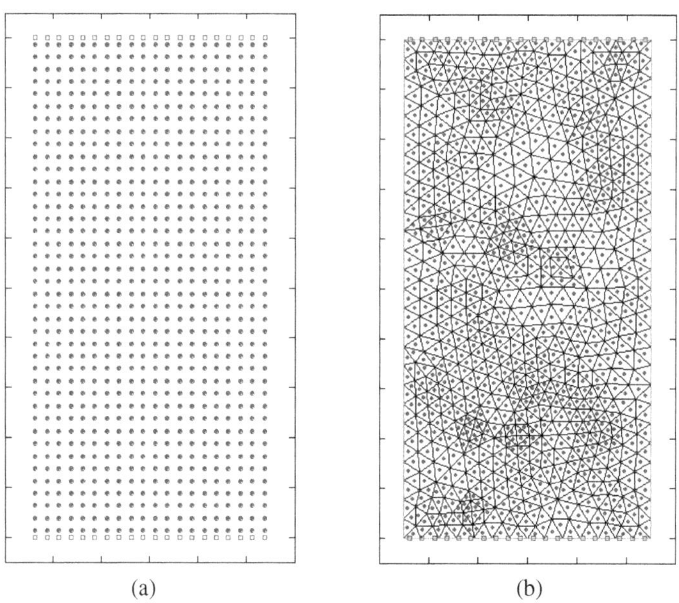

Figure 5.10 Initial particle distributions for the simulations of Poiseuille flow, (a) regular distribution with 800 flow particles, and (b) irregular distribution with 1040 flow particles (Liu and Chang, 2010).

boundary, and thus to prevent these particles from unphysical penetration through the boundary (Monaghan, 1994; Liu and Liu, 2003). This is an effective and commonly used approach to treat solid boundaries in SPH simulations. The smoothing length is slighter bigger than (e.g., 1.05 times) the initial particle spacing, and the time step is set to 10^{-4} s. The cubic spline smoothing function is used as the kernel in the simulation.

Figure 5.11 shows the velocity profiles at t = 0.01, 0.1, 1.0, 10.0, 10.5, 11.0, and 11.5 s obtained using the SPH method, in comparisons with the analytical solutions (circles). At t = 0.01, 0.1, and 1.0 s, the obtained SPH results agree well with the analytical results. At 1.0 s, both the SPH results and analytical solutions fully approach the steady state solution. When continuing to run the SPH simulations for more time steps till around 10.0 s, the obtained results oscillate around the steady state solution. However, after around 10.0 s, the simulation results tend to deviate from the steady state solution. Results at later stages deviate more from the steady state solution. This suggests that, even for regular particle distributions, the conventional SPH method can lead to oscillated or unstable results long-term simulations.

Figure 5.12 shows the velocity profiles at t = 0.01, 0.1, and 1.0 s obtained using the SPH method with irregular initial particle distribution and the analytical solutions. 1040 irregular distributed particles are located at the mass centers of the triangular elements generated using the MATLAB partial differential toolbox (Matlab, 1995) (**Figure 5.10**b). The lumped volumes of the particles are exactly the same as the corresponding particles. The smoothing length for a particle i is taken as the diameter of an equivalent circle with the

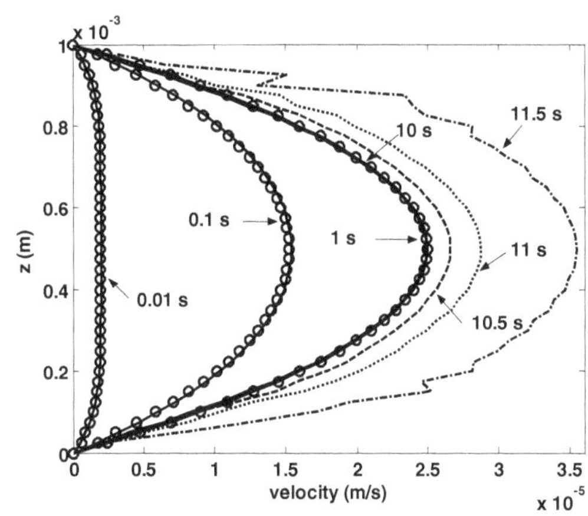

Figure 5.11 Velocity profiles at t = 0.01, 0.1, 1.0, 10.0, 10.5, 11.0, and 11.5 s obtained using the SPH method with regular initial particle distribution and the analytical solutions (circles) (Liu and Chang, 2010).

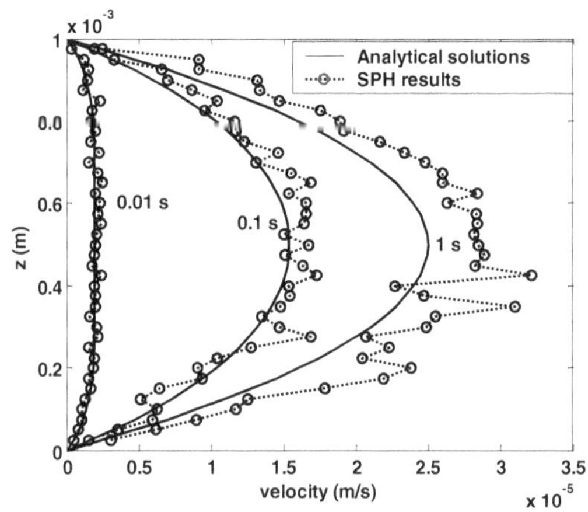

Figure 5.12 Velocity profiles at t = 0.01, 0.1, and 1.0 s obtained using the SPH method with irregular initial particle distribution and the analytical solutions (Liu and Chang, 2010).

same area as the corresponding triangular elements. It is clear that even at early stages, there are numerical oscillations in the SPH simulation. As time elapses, the numerical errors accumulate, the obtained SPH results deviates from the analytical solutions, and this leads to numerical instability.

To identify possible factors that may lead to this numerical oscillation in the conventional SPH method, different implementations of solid boundary treatment, different smoothing functions and smoothing lengths have been tested. It is found that this numerical instability is not related to solid boundary treatment algorithm and is also not dependent on different smoothing functions and smoothing lengths.

Figure 5.13 and **Figure 5.14** show the velocity profiles at t = 0.01, 0.1, and 1.0 s obtained from FPM simulation with regular and irregular initial particle distribution and the analytical solutions respectively. It is found that for both regular and irregular initial particle distribution, after around 1.0 s, the FPM results from both regular and irregular initial particle distribution reaches steady state. The obtained FPM results are very close to the analytical solutions at corresponding instants. In order to test the stability of the numerical results, the FPM simulations last for a long period of time to 10, 20, and even 30 s. The obtained FPM velocity profiles (for from both regular and irregular initial particle distribution) after 1 s nearly keep unchanged. **Figure 5.15** shows the particle distribution and velocity vectors at 15 s for FPM simulation with irregular initial particle distribution. It is seen that though the particles are highly disordered, the resultant flow pattern is accurate.

Table 5.9 shows the maximal velocities obtained from FPM simulation for regular and irregular initial particle distribution at t = 10, 20, and 30 s. It is seen for from both regular and irregular initial particle distribution, the relative errors are within 1%. It is demonstrated that FPM is able to obtain accurate and stable results, and the results are not sensitive to disordered particle distribution. Also the numerical oscillation in the conventional SPH results should be originated from the inconsistency inherent in the conventional SPH method.

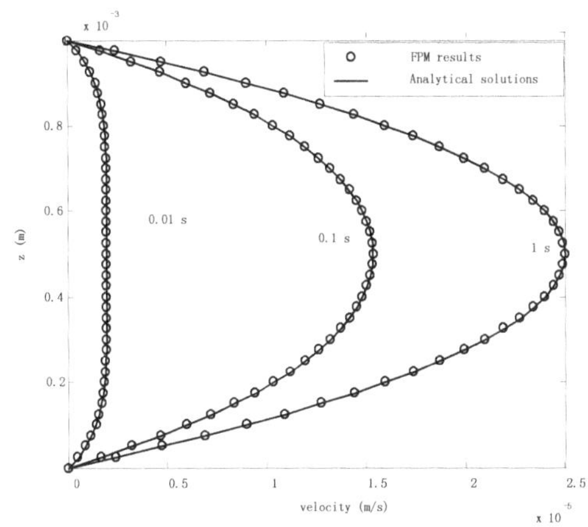

Figure 5.13 Velocity profiles at t = 0.01, 0.1, and 1.0 s obtained from FPM simulation with regular initial particle distribution and the analytical solutions (Liu and Chang, 2010).

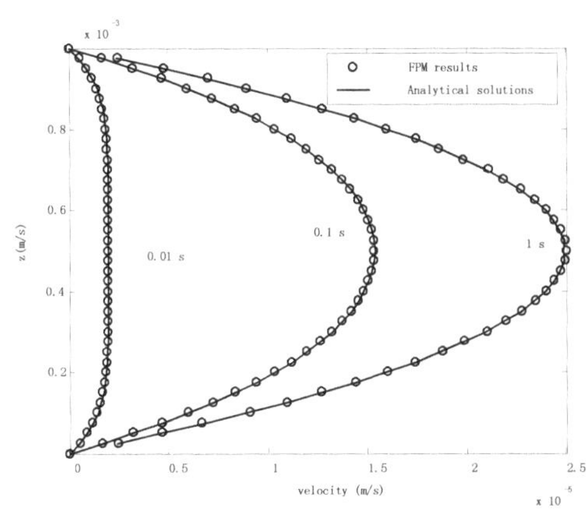

Figure 5.14 Velocity profiles at t = 0.01, 0.1, and 1.0 s obtained from FPM simulation with irregular initial particle distribution and the analytical solutions (Liu and Chang, 2010).

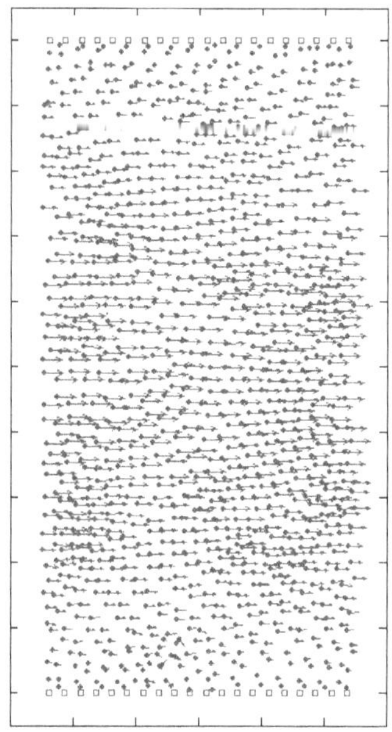

Figure 5.15 Particle distribution at 15 s for FPM simulation with irregular initial particle distribution (Liu and Chang, 2010).

Table 5.9 Maximal velocities obtained from FPM simulation for regular and irregular initial particle distribution at $t = 10$, 20, and 30 s.

Time instants (s)		Maximal velocity (m/s)	Relative errors (%)
10	Regular distribution	2.501360e-005	0.0544
	Irregular distribution	2.515600e-005	0.6240
20	Regular distribution	2.501400e-005	0.0560
	Irregular distribution	2.520722e-005	0.8289
30	Regular distribution	2.501320e-005	0.0528
	Irregular distribution	2.515600e-005	0.6240

5.5.6.4 Summary

In comparisons, the consistency of the conventional SPH method, CSPM, and FPM (if using equation (5.118), rather than equation (5.111), which is associated with higher order particle consistency than equation (5.118)) are described in **Table 5.10** and **Table 5.11**.

Table 5.10 Kernel consistency of SPH, CSPM, and FPM.

	Interior domain	Boundary area
SPH	1st order	less than 0th order
CSPM	1st order	0th order
FPM	1st order	1st order

Table 5.11 Particle consistency of SPH, CSPM, and FPM.

	Interior domain		Boundary area
	Regular distribution	Irregular distribution	
SPH	1st order	less than 0th order	less than 0th order
CSPM	1st order	0th order	0th order
FPM	1st order	1st order	1st order

5.5.7 Consistency vs. stability

We have seen that a typical dilemma exists for many numerical methods: consistency or stability. For a given "setting" of a numerical model such as a particle method, we may choose one over another, but probably a "balanced" one can be difficult to choose. The original SPH has clearly chosen the stability (and also flexibility) over the consistency, which gives the SPH a distinct feature of working well for many complicated problems with good efficiency, but less accuracy. It always tries to deliver some reasonably good results for the price paid. This seems to be a very practical choice for many practical engineering problems, as should be regarded as an advantage of the SPH method. Attempts to improve the accuracy of SPH via restoring the consistency can be helpful, provided that the stability and efficiency is not too much compromised.

The consistency restoring approaches, such as the FPM, put more emphases on the consistency (hence hopefully accuracy), but the stability (the conditioning of the **M** in equation (5.99) or **A** in equation (5.111)) can be in question for some types of problems. Hence proper measures are needed to establish the stability ensuring the accuracy of the solution.

The question is that can we have both the consistency and the stability at the same time? The answer is yes, provided we are willing to change the "setting" and pay the price. The recently proposed gradient smoothing method (GSM) (Liu and Xu, 2008; Xu et al., 2010; Li et al., 2011; Wang et al., 2013) is a typical example that guarantees both excellent stability for arbitrary grids and 2nd order accuracy. It uses also the gradient smoothed technique, but in a very carefully designed fashion. However, the GSM is not a particle method any more; the simplicity and efficiency features of the particle methods are lost. The GSM requires precise evaluation of the integrals for carefully chosen types of smoothing domains, and it works more like FVM. Therefore, the final question depends on what we want and at what cost: choosing a numerical method should be closely related to the nature of the problem, the requirement for the solutions and the resources we have.

5.6 Concluding remarks

In this chapter, the basic concepts, smoothing function and consistency of the smoothed particle hydrodynamics have been addressed. In summarizing the contents in this chapter, the following remarks can be made.

1. The SPH method employs particles to represent material and form the computational frame. There is no need for predefined connectivity between these particles. All one needs is the initial particle distribution.
2. The SPH approximation consists of kernel approximation and particle approximation. The kernel approximation of a function and its derivative are carried out in the continuum domain, and the particle approximations of a function and its derivative are carried out using discretized particles in the support domain at the current time step. By using the kernel and particle approximation techniques, it is possible to discretize partial differential equations governing concerned physics.
3. Smoothing function is very important in SPH as it determines the pattern to interpolate, and defines the cutoff distance of the influencing area of a particle. An SPH smoothing function should satisfy some special requirements — its moments determine the approximation accuracy of the SPH method. The frequently used SPH smoothing function generally satisfy the 0th and 1st moment equations (in continuous form) describing the normalization and symmetric requirements. By using the moment equations, it is feasible to construct new SPH smoothing function with special properties or high order accuracy.
4. The conventional SPH cannot exactly reproduce a constant and a linear function, and therefore it does not have 0th and 1st order particle consistency, though sometimes it can have 0th and 1st order kernel consistency. In order to improve the computational accuracy of the SPH method, particle consistency restoring approaches need to be used. The particle consistency restoring approaches usually involve reconstruction of a new smoothing function so as to satisfy the discretized consistency conditions or generating a correction matrix for the conventional SPH method without changing the conventional smoothing function.

5. As the conventional SPH method does not have 0th and 1st order particle consistency, it is sensitive to the selection of smoothing length and smoothing function as well as irregular particle distribution. Long time SPH simulation may lead to numerical oscillation and cause the numerical results deviating far away from the correction solution. In contrast, higher order SPH approximation schemes which restore particle consistency may not be sensitive to the selection of smoothing length and smoothing function as well as irregular particle distribution.

References

Allen, M. P. and Tildesley, D. J. (1987) Computer simulation of liquids. Oxford University Press, Oxford.

Balsara, D. S. (1995) von neumann stability analysis of smoothed particle hydrodynamics suggestions for optimal algorithms, *Journal of Computational Physics* **121**(2): 357–372.

Batra, R. C. and Zhang, G. M. (2004) Analysis of adiabatic shear bands in elasto-thermo-viscoplastic materials by modified smoothed-particle hydrodynamics (MSPH) method, *Journal of Computational Physics* **201**(1): 172–190.

Beissel, S. and Belytschko, T. (1996) Nodal integration of the element-free galerkin method, *Computer Methods in Applied Mechanics and Engineering* **139**(1–4): 49–74.

Belytschko, T., Guo, Y., Liu, W. K. and Xiao, S. P. (2000) A unified stability analysis of meshless particle methods, *International Journal for Numerical Methods in Engineering* **48**: 1359–1400.

Belytschko, T., Krongauz, Y., Dolbow, J. and Gerlach, C. (1998) On the completeness of the meshfree particle methods, *International Journal for Numerical Methods in Engineering* **43**(5): 785–819.

Belytschko, T., Krongauz, Y., Organ, D., Fleming, M. and Krysl, P. (1996) Meshless methods: An overview and recent developments, *Computer Methods in Applied Mechanics and Engineering* **139**(1–4): 3–47.

Belytschko, T. and Xiao, S. (2002) Stability analysis of particle methods with corrected derivatives, *Computers & Mathematics with Applications* **43**(3–5): 329–350.

Benz, W. (1988) Applications of smooth particle hydrodynamics (SPH) to astrophysical problems, *Computer Physics Communications* **48**(1): 97–105.

Benz, W. (1989) Applications of smoothed particle hydrodynamics (SPH) to astrophysical problems, *Computer Physics Communications* **48**: 97–105.

Benz, W. (1990). Smooth particle hydrodynamics: A review, the numerical modelling of nonlinear stellar pulsations, problems and prospects, 269–293, Boston, Kluwer Academic.

Berczik, P. (2000) Modeling the star formation in galaxies using the chemo-dynamical SPH code, *Astrophysics and Space Science* **271**(2): 103-126.

Bonet, J. and Kulasegaram, S. (2000) Correction and stabilization of smooth particle hydrodynamics methods with applications in metal forming simulations, *International Journal for Numerical Methods in Engineering* **47**(6): 1189–1214.

Bonet, J. and Kulasegaram, S. (2001) Remarks on tension instability of eulerian and lagrangian corrected smooth particle hydrodynamics (CSPH) methods, *International Journal for Numerical Methods in Engineering* **52**: 1203–1220.

Cabezon, R. M., Garcia-Senz, D. and Relano, A. (2008) A one-parameter family of interpolating kernels for smoothed particle hydrodynamics studies, *Journal of Computational Physics* **227**(19): 8523–8540.

Capuzzo-Dolcetta, R. and Di Lisio, R. (2000) A criterion for the choice of the interpolation kernel in smoothed particle hydrodynamics, *Applied Numerical Mathematics* **34**(4): 363–371.

Century Dynamics Incorporated (1997) Autodyn release notes version 3.1, AUTODYN™ Interactive Non-Linear Dynamic Analysis Software.

Chen, J. K. and Beraun, J. E. (2000) A generalized smoothed particle hydrodynamics method for nonlinear dynamic problems, *Computer Methods in Applied Mechanics and Engineering* **190**: 225–239.

Chen, J. K., Beraun, J. E. and Carney, T. C. (1999a) A corrective smoothed particle method for boundary value problems in heat conduction, *International Journal for Numerical methods in Engineering* **46**: 231–252.

Chen, J. K., Beraun, J. E. and Jih, C. J. (1999b) An improvement for tensile instability in smoothed particle hydrodynamics, *Computational Mechanics* **23**(4): 279–287.

Chen, J. S., Pan, C., Wu, C. T. and Liu, W. K. (1996) Reproducing kernel particle methods for large deformation analysis of nonlinear structures, *Computer Methods in Applied Mechanics and Engineering* **139**: 195–227.

Cleary, P. W., Prakash, M., Ha, J., Stokes, N. and Scott, C. (2007) Smooth particle hydrodynamics: Status and future potential, *Progress In Computational Fluid Dynamics* **7**(2–4): 70–90.

Dilts, G. A. (1999) Moving-least-squares-particle hydrodynamics. I: Consistency and stability, *International Journal for Numerical Methods in Engineering* **44**: 1115–1155.

Dilts, G. A. (2000) Moving least square particle hydrodynamics II: Conservation and boundaries, *International Journal for Numerical Methods in Engineering* **48**: 1503–1524.

Dyka, C. T. and Ingel, R. P. (1995) An approach for tension instability in smoothed particle hydrodynamics (SPH), *Computers & Structures* **57**(4): 573–580.

Dyka, C. T., Randles, P. W. and Ingel, R. P. (1997) Stress points for tension instability in SPH, *International Journal for Numerical Methods in Engineering* **40**(13): 2325–2341.

Evrard, A. E. (1988) Beyond n-body: 3D cosmological gas dynamics, *Monthly Notices of The Royal Astronomical Society* **235**: 911–934.

Fang, J. N., Owens, R. G., Tacher, L. and Parriaux, A. (2006) A numerical study of the SPH method for simulating transient viscoelastic free surface flows, *Journal of Non-Newtonian Fluid Mechanics* **139**(1–2): 68–84.

Fang, J. N. and Parriaux, A. (2008) A regularized lagrangian finite point method for the simulation of incompressible viscous flows, *Journal of Computational Physics* **227**(20): 8894–8908.

Fang, J. N., Parriaux, A., Rentschler, M. and Ancey, C. (2009) Improved SPH methods for simulating free surface flows of viscous fluids, *Applied Numerical Mathematics* **59**(2): 251–271.

Frederic, A. R. and James, C. L. (1999) Smoothed particle hydrodynamics calculations of stellar interactions, *Journal of Computational And Applied Mathematics* **109**: 213–230.

Frenkel, D. and Smit, B. (2002) Understanding molecular simulation: From algorithms to applications. Academic Press.

Fulk, D. A. (1994). A numerical analysis of smoothed particle hydrodynamics. PhD. Thesis, Air Force Institute of Technology.

Fulk, D. A. and Quinn, D. W. (1996) An analysis of 1-D smoothed particle hydrodynamics kernels, *Journal of Computational Physics* **126**(1): 165–180.

Gingold, R. A. and Monaghan, J. J. (1977) Smoothed particle hydrodynamics-theory and application to non-spherical stars, *Monthly Notices of The Royal Astronomical Society* **181**: 375–389.

Gingold, R. A. and Monaghan, J. J. (1982) Kernel estimates as a basis for general particle method in hydrodynamics, *Journal of Computational Physics* **46**: 429–453.

Gray, J. P., Monaghan, J. J. and Swift, R. P. (2001) Sph elastic dynamics, *Computer Methods in Applied Mechanics and Engineering* **190**(49): 6641–6662.

Hernquist, L. (1993) Some cautionary remarks about smoothed particle hydrodynamics, *Astrophysical Journal* **404**(2): 717–722.

Hernquist, L. and Katz, N. (1989) TREESPH — a unification of SPH with the hierarchical tree method, *Astrophysical Journal Supplement Series* **70**(2): 419–446.

Hoogerbrugge, P. J. and Koelman, J. (1992) Simulating microscopic hydrodynamic phenomena with dissipative particle dynamics, *Europhysics Letters* **19**: 155.

Hu, X. Y. and Adams, N. A. (2006) Angular-momentum conservative smoothed particle dynamics for incompressible viscous flows, *Physics of Fluids* **18**: 101702.

Hultman, J. and Pharayn, A. (1999) Hierarchical, dissipative formation of elliptical galaxies: Is thermal instability the key mechanism ? Hydrodynamic simulations including supernova feedback multi-phase gas and metal enrichment in cdm: Structure and dynamics of elliptical galaxies, *Astronomy and Astrophysics* **347**: 769–798.

Jiang, T., Ouyang, J., Li, Q., Ren, J. L. and Yang, B. X. (2011a) A corrected smoothed particle hydrodynamics method for solving transient viscoelastic fluid flows, *Applied Mathematical Modelling* **35**(8): 3833–3853.

Jiang, T., Ouyang, J., Li, X. J., Zhang, L. and Ren, J. L. (2011b) The first order symmetric SPH method for transient heat conduction problems, *Acta Physica Sinica* **60**(9): 10.

Jin, H. B. and Ding, X. (2005) On criterions for smoothed particle hydrodynamics kernels in stable field, *Journal of Computational Physics* **202**(2): 699–709.

Johnson, G. R. (1996) Artificial viscosity effects for SPH impact computations, *International Journal of Impact Engineering* **18**(5): 477–488.

Johnson, G. R. and Beissel, S. R. (1996) Normalized smoothing functions for SPH impact computations, *International Journal for Numerical Methods in Engineering* **39**(16): 2725–2741.

Johnson, G. R., Stryk, R. A. and Beissel, S. R. (1996) SPH for high velocity impact computations, *Computer Methods in Applied Mechanics and Engineering* **139**(1–4): 347–373.

Lanson, N. and Vila, J. P. (2007) Renormalized meshfree schemes I: Consistency, stability, and hybrid methods for conservation laws, *SIAM Journal on Numerical Analysis* **46**(4): 1912–1934.

Lattanzio, J. C., Monaghan, J. J., Pongracic, H. and Schwartz, M. P. (1986) Controlling penetration, *SIAM Journal on Scientific and Statistical Computing* **7**(2): 591–598.

Lee, W. H. (2000) Newtonian hydrodynamics of the coalescence of black holes with neutron stars — III. Irrotational binaries with a stiff equation of state, *Monthly Notices of The Royal Astronomical Society* **318**(2): 606–624.

Lee, W. H. and Kluzniak, W. (1999) Newtonian hydrodynamics of the coalescence of black holes with neutron stars II. Tidally locked binaries with a soft equation of state, *Monthly Notices of The Royal Astronomical Society* **308**(3): 780–794.

Li, E., Tan, V., Xu, G. X., Liu, G. R. and He, Z. C. (2011) A novel linearly-weighted gradient smoothing method (LW-GSM) in the simulation of fluid dynamics problems, *Computers & Fluids* **50**(1): 104–119.

Liu, G. R. (2002) Meshfree methods: Moving beyond the finite element method. CRC Press, Florida.

Liu, G. R. and Liu, M. B. (2003) Smoothed particle hydrodynamics: A meshfree particle method. World Scientific, Singapore.

Liu, G. R. and Xu, X. G. (2008) A gradient smoothing method (gsm) for fluid dynamics problems, *International Journal for Numerical Methods in Fluids* **58**(10): 1101–1133.

Liu, M. B. and Chang, J. Z. (2010) Particle distribution and numerical stability in the SPH method, *Acta Physica Sinica* **59**(6): 3654–3662.

Liu, M. B. and Liu, G. R. (2005) Meshfree particle simulation of micro channel flows with surface tension, *Computational Mechanics* **35**(5): 332–341.

Liu, M. B. and Liu, G. R. (2006) Restoring particle consistency in smoothed particle hydrodynamics, *Applied Numerical Mathematics* **56**(1): 19–36.

Liu, M. B. and Liu, G. R. (2010) Smoothed particle hydrodynamics (SPH): An overview and recent developments, *Archives of Computational Methods in Engineering* **17**(1): 25–76.

Liu, M. B., Liu, G. R. and Lam, K. Y. (2003a) Constructing smoothing functions in smoothed particle hydrodynamics with applications, *Journal Of Computational And Applied Mathematics* **155**(2): 263–284.

Liu, M. B., Liu, G. R. and Lam, K. Y. (2003b) A one-dimensional meshfree particle formulation for simulating shock waves, *Shock Waves* **13**(3): 201–211.

Liu, M. B., Liu, G. R. and Lam, K. Y. (2006) Adaptive smoothed particle hydrodynamics for high strain hydrodynamics with material strength, *Shock Waves* **15**(1): 21–29.

Liu, M. B., Liu, G. R., Zhou, L. V. and Chang, J. Z. (2014) Dissipative particle dynamics (DPD): An overview and recent developments, *Archives of Computational Methods in Engineering*.

Liu, M. B., Xie, W. P. and Liu, G. R. (2005) Modeling incompressible flows using a finite particle method, *Applied Mathematical Modelling* **29**(12): 1252–1270.

Liu, W. K., Chen, Y., Jun, S., Chen, J. S., Belytschko, T., Pan, C., Uras, R. A. and Chang, C. T. (1996) Overview and applications of the reproducing kernel particle methods, *Archives of Computational Methods in Engineering* **3**(1): 3–80.

Lucy, L. B. (1977) A numerical approach to the testing of the fission hypothesis, *Astronomical Journal* **82**(12): 1013–1024.

Matlab (1995) Partial differential equation toolbox. The MathWorks Inc.

Melean, Y., Sigalotti, L. D. G. and Hasmy, A. (2004) On the SPH tensile instability in forming viscous liquid drops, *Computer Physics Communications* **157**(3): 191–200.

Monaghan, J. (2012) Smoothed particle hydrodynamics and its diverse applications, *Annual Review Of Fluid Mechanics* **44**: 323–346.

Monaghan, J. J. (1982) Why particle methods work, *Siam Journal on Scientific and Statistical Computing* **3**(4): 422–433.

Monaghan, J. J. (1985) Particle methods for hydrodynamics, *Computer Physics Reports* **3**: 71–124.

Monaghan, J. J. (1987) Sph meets the shocks of noh, *Monash University Paper*.

Monaghan, J. J. (1988) An introduction to SPH, *Computer Physics Communications* **48**: 89–96.

Monaghan, J. J. (1989) On the problem of penetration in particle methods, *Journal of Computational Physics* **82**(1): 1–15.

Monaghan, J. J. (1990) Modelling the universe, *Astronomical Society of Australia Proceedings* **8**(3): 233–237.

Monaghan, J. J. (1992) Smooth particle hydrodynamics, *Annual Review of Astronomy and Astrophysics* **30**: 543–574.

Monaghan, J. J. (1994) Simulating free surface flows with SPH, *Journal of Computational Physics* **110**(2): 399–406.

Monaghan, J. J. (1995). Heat conduction with discontinuous conductivity. Applied Mathematics Reports and Preprints, Monash University. Melbourne.

Monaghan, J. J. (2000) SPH without a tensile instability, *Journal of Computational Physics* **159**(2): 290–311.

Monaghan, J. J. (2005) Smoothed particle hydrodynamics, *Reports on Progress In Physics* **68**(8): 1703–1759.

Monaghan, J. J. and Gingold, R. A. (1983) Shock simulation by the particle method SPH, *Journal of Computational Physics* **52**: 374–389.

Monaghan, J. J. and Lattanzio, J. C. (1985) A refined particle method for astrophysical problems, *Astronomy and Astrophysics* **149**(1): 135–143.

Monaghan, J. J. and Lattanzio, J. C. (1991) A simulation of the collapse and fragmentation of cooling molecular clouds, *Astrophysical Journal* **375**(1): 177–189.

Monaghan, J. J. and Poinracic, J. (1985) Artificial viscosity for particle methods, *Applied Numerical Mathematics* **1**: 187–194.

Morris, J. P. (1996a). Analysis of smoothed particle hydrodynamics with applications. PhD Thesis, Monash University.

Morris, J. P. (1996b) A study of the stability properties of smooth particle hydrodynamics, *Publications of the Astronomical Society Of Australia* **13**(1): 97–102.

Morris, J. P., Fox, P. J. and Zhu, Y. (1997) Modeling low reynolds number incompressible flows using SPH, *Journal of Computational Physics* **136**(1): 214–226.

Nelson, R. P. and Papaloizou, J. C. B. (1994) Variable smoothing lengths and energy-conservation in smoothed particle hydrodynamics, *Monthly Notices of the Royal Astronomical Society* **270**(1): 1–20.

Noh, W. F. (1987) Errors for calculations of strong shocks using an artificial viscosity and an artificial heat flux, *Journal of Computational Physics* **72**: 78–120.

Omang, M., Borve, S. and Trulsen, J. (2005) Alternative kernel functions for smoothed particle hydrodynamics in cylindrical symmetry, *Shock Waves* **14**(4): 293–298.

Owen, J. M., Villumsen, J. V., Shapiro, P. R. and Martel, H. (1998) Adaptive smoothed particle hydrodynamics: Methodology. II, *Astrophysical Journal Supplement Series* **116**(2): 155–209.

Rabczuk, T., Belvtschko, T. and Xiao, S. P. (2004) Stable particle methods based on lagrangian kernels, *Computer Methods in Applied Mechanics and Engineering* **193**(12): 1035–1063.

Randles, P. W. and Libersky, L. D. (1996) Smoothed particle hydrodynamics: Some recent improvements and applications, *Computer Methods in Applied Mechanics and Engineering* **139**(1): 375–408.

Randles, P. W. and Libersky, L. D. (2000) Normalized SPH with stress points, *International Journal for Numerical Methods in Engineering* **48**(10): 1445–1462.

Randles, P. W. and Libersky, L. D. (2005) Boundary conditions for a dual particle method, *Computers and Structures* **83**(17–18): 1476–1486.

Randles, P. W., Petschek, A. G., Libersky, L. D. and Dyka, C. T. (2003). Stability of DPD and SPH, Meshfree methods for partial differential equations, Springer, Berlin.

Ren, J., Ouyang, J., Yang, B., Jiang, T. and Mai, H. (2011) Simulation of container filling process with two inlets by improved smoothed particle hydrodynamics (SPH) method, *International Journal of Computational Fluid Dynamics* **25**(7): 365–386.

Senz, D. G., Bravo, E. and Woosley, S. E. (1999) Single and multiple detonations in white dwarfs, *Astronomy and Astrophysics* **349** 177–188.

Shapiro, P. R., Martel, H., Villumsen, J. V. and Owen, J. M. (1996) Adaptive smoothed particle hydrodynamics, with application to cosmology: Methodology, *Astrophysical Journal Supplement Series* **103**(2): 269–330.

Sigalotti, L. D. G. and Lopez, H. (2008) Adaptive kernel estimation and SPH tensile instability, *Computers & Mathematics with Applications* **55**(1): 23–50.

Steinmetz, M. and Mueller, E. (1993) On the capabilities and limits of smoothed particle hydrodynamics, *Astronomy and Astrophysics* **268**(1): 391–410.

Swegle, J. W., Attaway, S. W., Heinstein, M. W., Mello, F. J. and Hicks, D. L. (1994). An analysis of smoothed particle hydrodynamics. Albuquerque, NM, US, Sandia National Labs.

Swegle, J. W., Hicks, D. L. and Attaway, S. W. (1995) Smoothed particle hydrodynamics stability analysis, *Journal of Computational Physics* **116**(1): 123–134.

Thacker, R. J. and Couchman, H. M. P. (2001) Star formation, supernova feedback, and the angular momentum problem in numerical cold dark matter cosmogony: Halfway there, *Astrophysical Journal* **555**(1): L17–L20.

Vignjevic, R., Campbell, J. and Libersky, L. (2000) A treatment of zero-energy modes in the smoothed particle hydrodynamics method, *Computer Methods in Applied Mechanics and Engineering* **184**(1): 67–85.

von Neumann, J. and Richtmyer, R. D. (1950) A method for the numerical calculation of hydrodynamic shocks, *Journal of Applied Physics* **21**: 232–247.

Wang, S., Khoo, B. C., Liu, G. R. and Xu, G. X. (2013) An arbitrary lagrangian-eulerian gradient smoothing method (gsm/ale) for interaction of fluid and a moving rigid body, *Computers & Fluids* **71**: 327–347.

Xu, G. X., Liu, G. R. and Tani, A. (2010) An adaptive gradient smoothing method (gsm) for fluid dynamics problems, *International Journal for Numerical Methods in Fluids* **62**(5): 499–529.

Xu, X. Y., Ouyang, J., Yang, B. X. and Liu, Z. J. (2013) Sph simulations of three-dimensional non-newtonian free surface flows, *Computer Methods in Applied Mechanics and Engineering* **256**: 101–116.

Yang, X. F., Liu, M. B. and Peng, S. L. (2014a) Smoothed particle hydrodynamics modeling of viscous liquid drop without tensile instability, *Computers & Fluids* **92**(20): 199–208.

Yang, X. F., Peng, S. L. and Liu, M. B. (2014b) A new kernel function for SPH with applications to free surface flows, *Applied Mathematical Modelling* **38**: 3822–3813.

Zhang, G. M. and Batra, R. C. (2004) Modified smoothed particle hydrodynamics method and its application to transient problems, *Computational Mechanics* **34**(2): 137–146.

Chapter 6

Smoothed Particle Hydrodynamics — Applications

In Chapter 5, the basic concepts, smoothing function and consistency of the SPH have been addressed. As a comparatively new computational method, SPH combines the advantages of meshfree, Lagrangian and particle methods. Particles are used to represent the state of a system and these particles can freely move according to internal particle interactions and external forces. The discretization of the governing equations is based on these discrete particles, and a variety of particle-based formulations have been used to calculate the local density, velocity and acceleration of the fluid. In SPH, there is no explicit interface tracking for multiphase flows – the motion of the fluid is represented by the motion of the particles, and fluid surfaces or fluid-fluid interfaces move with particles representing their phase defined at the initial stage. Therefore it can naturally obtain history of fluid motion, and can easily track material interfaces, free surfaces and moving boundaries. The meshfree nature of SPH method remove the difficulties due to large deformations since SPH uses particles rather than mesh as a computational frame to approximate related governing equations. These features of SPH make it fairly attractive in modeling problems with free surfaces, moving interfaces, deformable boundaries and large deformations. Therefore since its invention, the SPH method has been applied to different areas of interest, just as reviewed in literature (Liu and Liu, 2003; Cleary et al., 2007; Liu and Liu, 2010; Monaghan, 2012).

This chapter is outlined as follows.

- In Section 6.1, applications of the SPH method are briefly reviewed, with a current focus on viscous incompressible free surface flows.

- In Section 6.2, the governing equations for viscous incompressible fluid flow and moving rigid bodies are described.
- In Section 6.3, two approaches, the weakly compressible SPH (WCSPH) model and incompressible SPH (ISPH) model, for modeling incompressible flows are described. The advantages and disadvantages of WCSPH and ISPH are compared with a number of numerical examples.
- In Section 6.4, the SPH method is applied to modeling free surface flows with a number of examples including dam-breaking against a vertical wall and a sharp-edged obstacle, and the movement of an elliptical cylinder near water free surface.
- In Section 6.5, the SPH method is applied to modeling liquid sloshing with a number of examples including liquid sloshing in a rectangular container with different excitations and geometries, and the movement of water in a ballast water tank.
- In Section 6.6, the SPH method is applied to modeling the entry and exit of water with a number of examples including the sinking, free falling and exit of water of a cylinder, and underwater launch of a projectile.
- In Section 6.7, the SPH method is applied to modeling oil spill and boom movement considering effects of oil type, boom velocity, skirt angle, and water waves.
- In Section 6.8, the SPH method is applied to modeling hydro-elastic problems.
- In Section 6.9, some remarks and conclusion are given.

6.1 Introduction

6.1.1 Review on SPH applications

The original applications of the SPH method is in astrophysical phenomena, such as the simulations of binary stars and stellar collisions (Benz, 1988; Monaghan, 1992; Frederic and James, 1999), supernova (Hultman and Pharayn, 1999; Thacker and Couchman, 2001), collapse as well as the formation of galaxies (Monaghan and Lattanzio, 1991; Berczik, 2000), coalescence of black holes with neutron stars (Lee and Kluzniak, 1999; Lee, 2000), single and multiple detonation of white dwarfs (Senz et al., 1999), and even the evolution of the universe (Monaghan, 1990). It also has been extended to a vast range of problems in both fluid and solid mechanics because of the strong ability to incorporate complex physics into the SPH formulations (Liu and Liu, 2003). The applications of SPH to many other engineering applications include

- multi-phase flows (Monaghan and Kocharyan, 1995; Ritchie and Thomas, 2001; Colagrossi and Landrini, 2003; Password, 2003; Liu et al., 2005a; Hu and Adams, 2006; Garg et al., 2007; Ge et al., 2007; Hu and Adams, 2007; Hong et al., 2008; Zhou et al., 2008; Hu and Adams, 2009; Zhou et al., 2010),
- coastal hydrodynamics including water wave impact, dam break, sloshing and overtopping (Monaghan, 1994; Monaghan et al., 2003; Gomez-Gesteira and Dalrymple, 2004; Gotoh et al., 2004; Iglesias et al., 2004; Shao and Gotoh, 2004; Bulgarelli, 2005; Panizzo, 2005; Rhee and Engineer, 2005; Rogers and Dalrymple, 2005; Zou and Dalrymple, 2005; Dalrymple and Rogers, 2006; Gotoh and Sakai, 2006; Lohner et al., 2006; Shao, 2006; Shao et al., 2006; Souto-Iglesias et al., 2006; Crespo et al., 2007; Kim, 2007; Violeau et al., 2007; Crespo et al., 2008a; Crespo et al., 2008b; Issa and Violeau, 2008; Khayyer et al., 2008; Qiu, 2008; Yim et al., 2008; Shao, 2009),
- environmental and geophysical flows including flood and river dynamics, landslide, flow in fractures and porous media, seepage, soil mechanics and mudflow (Morris et al., 1999; Zhu et al., 1999; Moresi et al., 2001; Zhu and Fox, 2002; Bursik et al., 2003; Cleary and Prakash, 2004; McDougall and Hungr, 2004; Gallati et al., 2005; McDougall and Hungr, 2005; Tartakovsky and Meakin, 2005b; Bui et al., 2006; Hui et al., 2006; Kipfer and Westermann, 2006; Sakai and Maeda, 2006; Tartakovsky and Meakin, 2006; Bui et al., 2007; Laigle et al., 2007; Maeda and Sakai, 2007; Bui et al., 2008; Ghazali and Kamsin, 2008; Herrera et al., 2009; Pastor et al., 2009),
- heat and/or mass conduction (Cleary, 1998; Chen et al., 1999; Jeong et al., 2003; Jiang and Sousa, 2006a; Jiang and Sousa, 2006b; Rook et al., 2007; Sousa and Jiang, 2007),
- ice and cohesive grains (Gutfraind and Savage, 1998; Oger and Savage, 1999; Shen et al., 2000; Ji et al., 2005; Wang et al., 2005; Ji et al., 2007; Schafer et al., 2007),
- microfluidics and/or liquid drop dynamics (Apfel et al., 1997; Nugent and Posch, 2000; Liu and Liu, 2005; Melean and Sigalotti, 2005; Tartakovsky and Meakin, 2005a; Lopez et al., 2006; Garg et al., 2007; Zhang et al., 2007; Sigalotti and Lopez, 2008; Wang et al., 2008; Zhang et al., 2008; Zhou et al., 2008; Fang et al., 2009; Zhang et al., 2009),

- high explosive detonation and explosion (Swegle and Attaway, 1995; Mair, 1999; Liu et al., 2002; Bromm et al., 2003; Liu et al., 2003b; Liu et al., 2003a; Liu et al., 2003e, d, c; Liu et al., 2003f; Liu and Liu, 2004; Alia and Souli, 2006, Busegnies et al., 2007; Yang et al., 2013),
- underwater explosions and water mitigation (Swegle and Attaway, 1995; Liu et al., 2002; Liu et al., 2003a; Liu et al., 2003e, d; Liu et al., 2003f; Zhang et al., 2012; Zhang et al., 2013a; Zhang et al., 2013c; Ming et al., 2014),
- elastic and/or plastic flow (Libersky and Petschek, 1991; Swegle, 1992; Libersky et al., 1993; Randles and Libersky, 1996; Zhou et al., 2007),
- fracture of brittle solids (Benz and Asphaug, 1995),
- metal forming and high pressure die casting (Chen et al., 1998; Bonet and Kulasegaram, 2000; Cleary and Ha, 2000; Cleary et al., 2000; Ha and Cleary, 2000; Cleary et al., 2002; Ha and Cleary, 2005; Cleary et al., 2006; Cleary et al., 2007; Hu et al., 2007; Prakash et al., 2007a),
- magneto-hydrodynamics and magnetic field simulation (Meglicki, 1994; Dolag et al., 1999; Borve et al., 2001; Price and Monaghan, 2004a, b, c; Ala et al., 2006; Jiang et al., 2006; Ala et al., 2007a; Ala et al., 2007b; Pimenta et al., 2007; Francomano et al., 2009),
- problems with fluid-solid interactions (Chikazawa et al., 2001; Iglesias et al., 2004; Anghileri et al., 2005; Guilcher et al., 2006; Antoci et al., 2007; Bui et al., 2007; Crespo et al., 2007; Hosseini and Amanifard, 2007; Prakash et al., 2007b; Zhang et al., 2013b; Hu et al., 2014), and
- many other problems like blood flow (Hieber, 2004; Muller, 2004; Tanaka and Takano, 2005; Tsubota et al., 2006; Hosseini and Feng, 2009), traffic flow (Rosswog and Wagner, 2002).

Some review articles have provided applications of SPH to problems in different areas in engineering and sciences (Liu and Liu, 2003; Cleary et al., 2007; Liu and Liu, 2010; Monaghan, 2012). This chapter mainly focuses on the SPH applications to incompressible fluid flows in hydrodynamics and ocean engineering.

6.1.2 Applications to hydrodynamics and ocean engineering

Flow phenomena in hydrodynamics and ocean (including coastal and offshore) engineering are significantly important as they can greatly influence the nearly personnel and structures. The flow phenomena include

- wave dynamics (waver generation, wave breaking, and wave interaction with other structures),
- dam breaking,
- water filling and water discharge (to and from a water tank or reservoir),
- shallow water flows,
- entry of water, sloshing phenomena with fluid-solid interaction, and
- different other problems.

These phenomena involve special features, which make it difficult for numerical simulation. For example, water waves can propagate shoreward where they undergo changes induced by the near-shore topography and increase in height. Upon reaching the shoreline, they can break into pieces, and travel inland for large distances with potential damage of property and loss of life. Experimental setups for fluid flow in coast hydrodynamics and offshore engineering are expensive and only limited to laboratory applications. Numerical simulation has thus become a great tool to predicting fluid flow in ocean and coast hydrodynamics and offshore engineering.

However, numerical simulation of fluid flow in these related areas is a formidable task as it involves not only complex geometries and free surfaces, but also fluid-solid interaction as well as other complex physics in a comparably very large scale. In many circumstances, violent fluid–structure interactions lead to air entrapment and multi-phase flows, where the dynamics of the entrapped air at the impact may play a dominant role during the process and contribute to the high pressure maxima and pressure oscillations. Though conventional grid based methods like FDM, FVM and FEM have achieved greatly in simulating fluid flow in coast hydrodynamics and offshore engineering, there is still a long way to go for practical engineering applications.

Smoothed particle hydrodynamics, due to its meshfree, Lagrangian and particle nature, has been attracting more and more researchers in coast hydrodynamics and offshore engineering. From the very early simulation of a simple dam break problem (Monaghan, 1994), there have been a lot of literatures addressing the applications of SPH method in related areas.

This chapter shall focus on SPH modeling of viscous, incompressible free surface flows interacting with moving solid objects. The governing equations, turbulence model, two different modeling approaches in simulating incompressible flows, and typical applications shall be provided.

6.2 Governing equations

6.2.1 Governing equation for viscous incompressible fluid flow

In chapter 5, the SPH equations of motion for general fluid flows from N-S equation are given. For incompressible viscous hydrodynamic problems, the energy equation can be ignored. Fluid flows in hydrodynamics and ocean engineering usually involves strong turbulence and vortex, and a turbulence model is necessary to be incorporated into SPH equations of motion. Early SPH works did not include turbulence models. Recently a number of researchers gradually incorporated turbulence models into SPH equations of motion (Shao and Gotoh, 2005; Shao, 2006). It is demonstrated that turbulence modeling is important for problems or areas with strong turbulence and vortex. Here the RANS (Reynolds-averaged Navier-Stokes) turbulence model shall be incorporated into the SPH method.

The Lagrangian form of the N-S equation can be written as

$$\frac{d\rho}{dt} = -\rho \cdot \nabla \mathbf{v}, \tag{6.1}$$

$$\frac{d\mathbf{v}}{dt} = -\frac{1}{\rho}\nabla P + \frac{\mu}{\rho}\nabla^2 \mathbf{v} + \mathbf{F}, \tag{6.2}$$

where ρ is fluid density, $\mathbf{v}$ is the velocity vector, P is pressure, μ is the dynamic viscosity, $\mathbf{F}$ is external force (e.g., the gravitational acceleration, $\mathbf{g}$).

If considering RANS turbulence model, a term describing the Reynolds stress tensor should be added into the right hand side of equation (6.2) as

$$\frac{d\bar{\mathbf{v}}}{dt} = -\frac{1}{\rho}\nabla \bar{P} + \frac{\mu}{\rho}\nabla^2 \bar{\mathbf{v}} + \mathbf{F} + \frac{1}{\rho}\nabla(\rho \mathbf{R}), \tag{6.3}$$

where $\mathbf{R}$ is the Reynolds stress tensor whose elements are $R_{ij} = -\overline{v_i' v_j'}$. The eddy viscosity assumption is used to model the Reynolds stress tensor as

$$\mathbf{R} = 2\nu_t \mathbf{S} - \frac{2}{3}k\mathbf{I}, \tag{6.4}$$

where ν_t is the eddy viscosity, $\mathbf{S}$ is the mean rate-of-strain tensor, k is turbulence kinetic energy and $\mathbf{I}$ is a unit tensor. The Smagorinsky model of eddy viscosity $\nu_t = (C_s \Delta l)^2 \sqrt{2S_{ij}S_{ij}}$ is widely used (Smagorinsky, 1963), where C_s is the Smagorinsky constant usually taken as 0.1~0.24 and Δl is a mixing length which is assumed to be the initial particle spacing in SPH. S_{ij} are the elements of $\mathbf{S}$ given by:

$$S_{ij} = \frac{1}{2}\left(\frac{\partial v_j}{\partial x_i} + \frac{\partial v_i}{\partial x_j}\right). \tag{6.5}$$

Considering $k = \overline{v_i' v_i'}/2$ and $R_{ii} = -\overline{v_i' v_i'}$, the relation between k and the Reynolds stress tensor can be written as $k = -R_{ii}/2$.

6.2.2 Governing equation for moving rigid body

For a moving rigid body, the equation of motion is simply the Newton's law of motion. The centre of mass can be written as follows

$$\frac{d\mathbf{u}_o}{dt} = \frac{\mathbf{F}}{M} + \mathbf{g}, \tag{6.6}$$

where $\mathbf{u}_o$ is the velocity of the centre of mass, $\mathbf{F}$ is the summation of forces without gravity around the rigid body, M is the mass of the rigid body. For two-dimensional problems, the equation for the angular velocity is

$$\frac{d\omega}{dt} = \frac{\mathbf{J}}{I}, \tag{6.7}$$

where ω is the angular velocity, I is the moment of inertia, and $\mathbf{J}$ is the total moment of force on the rigid body about the centre of mass. The velocity of a point a on the rigid body is

$$\mathbf{u}_a = \mathbf{u}_o + \omega \times \mathbf{r}_{oa}, \tag{6.8}$$

where $\mathbf{r}_{oa}$ is the vector from the centre of mass o to point a.

6.2.3 SPH equations of motion

As described in Chapter 5, in SPH, A field function and its derivative can then be written in the following forms

$$< f(\mathbf{x}_i) > = \sum_{j=1}^{N} \frac{m_j}{\rho_j} f(\mathbf{x}_j) W(\mathbf{x}_i - \mathbf{x}_j, h), \tag{6.9}$$

$$< \nabla f(\mathbf{x}_i) > = \sum_{j=1}^{N} \frac{m_j}{\rho_j} f(\mathbf{x}_j) \nabla_i W_{ij}, \tag{6.10}$$

where $< f(\mathbf{x}_i) >$ is the approximated value of particle i; $f(\mathbf{x}_j)$ is the value of $f(\mathbf{x})$ associated with particle j; $\mathbf{x}_i$ and $\mathbf{x}_j$ are the positions of corresponding particles; m denotes mass; h is the smooth length; N is the number of the particles in the support domain; W is the smoothing function representing a weighted contribution of particle j to particle i.

Substituting the SPH approximations for a function and its derivative to the RANS equations, after some trivial transformation, the SPH equations of motion for viscous fluid flows can be obtained as

$$\frac{d\rho_i}{dt} = \sum_{j=1}^{N} m_j \mathbf{v}_{ij} \nabla_i W_{ij}, \tag{6.11}$$

$$\frac{d\mathbf{v}_i}{dt} = -\sum_{j=1}^{N} m_j \left(\frac{p_i}{\rho_i^2} + \frac{p_j}{\rho_j^2} \right) \nabla_i W_{ij} + \sum_{j=1}^{N} \frac{4m_j \left(\mu_i + \mu_j \right) \mathbf{x}_{ij} \cdot \nabla_i W_{ij}}{\left(\rho_i + \rho_j \right)^2 \left(x_{ij}^{\ 2} + 0.01h^2 \right)} \mathbf{v}_{ij} + \mathbf{F}, \tag{6.12}$$

where the first, second and third terms on the LHS of equation (11) describe the SPH approximations for pressure and viscosity respectively.

If considering RANS turbulence model, the SPH equations of motion with the approximation of Reynolds stress tensor is

$$\frac{d\mathbf{v}_i}{dt} = -\sum_{j=1}^{N} m_j \left(\frac{p_i}{\rho_i^2} + \frac{p_j}{\rho_j^2} \right) \nabla_i W_{ij} + \sum_{j=1}^{N} \frac{4m_j \left(\mu_i + \mu_j \right) \mathbf{x}_{ij} \cdot \nabla_i W_{ij}}{\left(\rho_i + \rho_j \right)^2 \left(x_{ij}^{\ 2} + 0.01h^2 \right)} \mathbf{v}_{ij}$$
$$+ \sum_{j} m_j \left(\frac{\mathbf{R}_i}{\rho_i} + \frac{\mathbf{R}_j}{\rho_j} \right) \nabla_i W_{ij} + \mathbf{F}. \tag{6.13}$$

Another approach to approximate density is substitute density directly into equation (6.9), and the resultant density approximation is as follows

$$\rho_i = \sum_{j=1}^{N} m_j W_{ij} \ . \tag{6.14}$$

For the motion of moving rigid body, $\mathbf{F}$ and $\mathbf{J}$ can be expressed in SPH as follows

$$\mathbf{F} = \sum_b \mathbf{f}_b \ , \tag{6.15}$$

$$\mathbf{J} = \sum_b \mathbf{r}_{ob} \times \mathbf{f}_b \ , \tag{6.16}$$

where $\mathbf{f}_b$ denotes the force on particle b due to fluids. The summation is over all rigid body particles.

6.2.4 Density and kernel gradient correction

It is descried in Chapter 5 that the conventional SPH method has been hindered with low accuracy as it cannot exactly reproduce quadratic and linear functions, and even cannot exactly reproduce a constant. The accuracy of the conventional SPH method is also closely related to the distribution of particles, selection of smoothing function and the support domain (described by the smoothing length h multiplied by a scalar factor). During the last decade, different approaches have been proposed to improve the particle inconsistency and hence the SPH approximation accuracy. Some of them involve reconstruction of a new smoothing function so as to satisfy the discretized consistency conditions. However, these approaches are usually not preferred for hydrodynamic simulations because the reconstructed smoothing function can be partially negative, non-symmetric, and not monotonically decreasing. Recently, one popular way is to construct improved SPH approximation schemes based on Taylor series expansion on the SPH approximation of a function and/or its derivatives. Typical examples include the corrective smoothed particle method (CSPM) by Chen et al. (Chen and Beraun, 2000b) and the finite particle method (FPM) by Liu et al. (Liu et al., 2005b; Liu and Liu, 2006). Both CSPM and FPM do not need to reconstruct smoothing function. It is noted that in the conventional SPH method, a field function and its derivatives are approximated separately. Instead, in CSPM, the derivatives are approximated through solving a coupled matrix equation while the field function is approximated separately. In FPM, both the field function and its derivatives are coupled together and can be approximated simultaneously through solving a general matrix equation.

Fluid flows in hydrodynamics and ocean engineering is usually associated with changing and breakup of free surfaces. When wave front violently impacts onto solid walls of the container, water particles can first be splashed away from bulky fluid, and then fall onto the bulky fluid. The changing and breakup of free surfaces as well as splashing and fall of water particles lead to highly disordered particle distribution, which can seriously influence computational accuracy of SPH approximations. Hence an SPH approximation scheme, which is of higher order accuracy and is insensitive to disordered particle distribution, is necessary for modeling liquid sloshing dynamics.

In this work, we used two modified schemes for approximating density (density correction) and kernel gradient (kernel gradient correction, or KGC). As to the density correction, a simple re-normalization approach is used for (6.14), and the density can be approximated as

$$\rho_i^{new} = \sum_{j=1}^{N} \rho_j W_{ij}^{new} \frac{m_j}{\rho_j} = \sum_{j=1}^{N} m_j W_{ij}^{new}, \tag{6.17}$$

$$W_{ij}^{new} = \frac{W_{ij}}{\sum_{j=1}^{N} W_{ij} \frac{m_j}{\rho_j}}. \tag{6.18}$$

Another approach is to use the moving least square (MLS) density re-initialization algorithm developed by Dilts (Dilts, 1999, 2000) as follows

$$\rho_a^{new} = \sum_b \rho_b W_{ab}^{MLS} \frac{m_b}{\rho_b} = \sum_b m_b W_{ab}^{MLS}, \tag{6.19}$$

$$W_{ab}^{MLS} = \left[\beta_0 + \beta_x (x_a - x_b) + \beta_z (y_a - y_b)\right] W_{ab}, \tag{6.20}$$

$$\begin{bmatrix} \beta_0 \\ \beta_x \\ \beta_z \end{bmatrix} = \left(\sum_b W_{ab} \mathbf{A} \frac{m_b}{\rho_b} \right)^{-1} \begin{bmatrix} 1 \\ 0 \\ 0 \end{bmatrix}, \tag{6.21}$$

$$\mathbf{A} = \begin{bmatrix} 1 & x_a - x_b & y_a - y_b \\ x_a - x_b & (x_a - x_b)^2 & (x_a - x_b)(y_a - y_b) \\ y_a - y_b & (x_a - x_b)(y_a - y_b) & (y_a - y_b)^2 \end{bmatrix}. \tag{6.22}$$

In above equations, x and y are components of a position vector. The density re-initialization can be conducted every a number of time steps (e.g., 20 or 30 time steps).

As to kernel gradient correction, from equation (6.10), it is known that the approximation accuracy of the derivatives is closely related to the accuracy of the gradient of the smoothing function (or kernel gradient). It is possible to use a corrective kernel gradient rather than the conventional kernel gradient in (6.10) to obtain better approximation accuracy. For example, in a two-dimensional space, based on Taylor series expansion on the SPH approximation of a function, it is possible to get the following formulation

$$\begin{aligned}\int_{\Omega} f(\mathbf{x}')\nabla W d\mathbf{x}' &= f(\mathbf{x})\int_{\Omega}\nabla W d\mathbf{x}' + \frac{\partial f(\mathbf{x})}{\partial x}\int_{\Omega}(\mathbf{x}'-\mathbf{x})\nabla W d\mathbf{x}' \\ &+ \frac{\partial f(\mathbf{x})}{\partial y}\int_{\Omega}(y'-y)\nabla W d\mathbf{x}' + O(h^2).\end{aligned} \tag{6.23}$$

Considering the following equation

$$\langle \nabla f(\mathbf{x})\rangle = \int_{\Omega} f(\mathbf{x}')\nabla W d\mathbf{x}' - f(\mathbf{x})\int_{\Omega}\nabla W d\mathbf{x}', \tag{6.24}$$

a new formulation for approximating derivatives can be obtained as

$$\begin{aligned}\langle \nabla f(\mathbf{x})\rangle &= \frac{\partial f(\mathbf{x})}{\partial x}\int_{\Omega}(x'-x)\nabla W d\mathbf{x}' \\ &+ \frac{\partial f(\mathbf{x})}{\partial y}\int_{\Omega}(y'-y)\nabla W d\mathbf{x}' + O(h^2).\end{aligned} \tag{6.25}$$

It is can be further written as follows in terms of particle approximation

$$\begin{aligned}\langle \nabla f(\mathbf{x}_i)\rangle &= \frac{\partial f(\mathbf{x}_i)}{\partial x_i}\underbrace{\sum_j (x_j - x_i)\nabla_i W_{ij} V_j}_{X} \\ &+ \frac{\partial f(\mathbf{x}_i)}{\partial y_i}\underbrace{\sum_j (y_j - y_i)\nabla_i W_{ij} V_j}_{Y} + O(h^2),\end{aligned} \tag{6.26}$$

where $V_j\ (=m_j/\rho_j)$ is the volume of particle j. From (6.26), it can be concluded that if $X=\begin{pmatrix}1\\0\end{pmatrix}$ and $Y=\begin{pmatrix}0\\1\end{pmatrix}$, the SPH particle approximation scheme for a gradient (expressed in (6.26)) is of second order accuracy. However, for general cases (e.g., irregular particle distribution, variable smoothing length, and/or truncated boundary areas), these two requirements cannot be satisfied, and therefore the accuracy of (6.26) can be seriously reduced. It is possible to restore the accuracy for general cases with the following correction on the kernel gradient,

$$\nabla_i^{new} W_{ij} = L(\mathbf{x}_i)\nabla_i W_{ij} \tag{6.27}$$

$$L(\mathbf{x}_i)=\left(\sum_j \begin{pmatrix} x_{ji}\dfrac{\partial W_{ij}}{\partial x_i} & y_{ji}\dfrac{\partial W_{ij}}{\partial x_i} \\ x_{ji}\dfrac{\partial W_{ij}}{\partial y_i} & y_{ji}\dfrac{\partial W_{ij}}{\partial y_i}\end{pmatrix} V_j\right)^{-1}, \tag{6.28}$$

where $x_{ji}=x_j-x_i$, $y_{ji}=y_j-y_i$. If replacing $\nabla_i W_{ij}$ on the RHS of (6.26) with $\nabla_i^{new} W_{ij}$, it is easy to verify that $X=\begin{pmatrix}1\\0\end{pmatrix}$ and $Y=\begin{pmatrix}0\\1\end{pmatrix}$. Therefore, for general cases with irregular particle distribution, variable smoothing length, and/or truncated boundary areas, the SPH particle approximation scheme for a gradient based on kernel gradient correction is of second order accuracy. It is noted that for both density correction, and gradient correction, since only kernel and its gradient are corrected, there is no need to significantly change the structure of SPH computer programs and procedure of SPH simulations. It is therefore convenient to implement SPH equations of motion.

6.3 Modeling incompressible flows in SPH

In the standard SPH method for solving compressible flows, the particle motion is driven by the pressure gradient, while the particle pressure is calculated by the local particle density and internal energy through the equation of state. However, for incompressible flows, the actual equation of state of the fluid will lead to prohibitive time steps that are extremely small. How to effectively calculate the pressure term in the momentum equation is a major task for simulation of incompressible flows. This is also true for other numerical methods like the FDM, not just for the SPH method. Though it is possible to include the constraint of the constant density into the SPH formulations, the resultant equations are usually too cumbersome.

In modeling incompressible flows, early SPH models are usually based on the assumption that the modeling fluid can be regarded as weakly compressible, and this approach is usually referred to as weakly compressible SPH (WCSPH). WCSPH is able to obtain reasonable flow patterns, but it encounters a number of difficulties especially pressure oscillations. However, during recent years, great progresses have been made in WCSPH methods and the previous defections such as pressure oscillations have been resolved quite well either by using high order SPH approximation schemes such as the corrective smoothed particle method (CSPM) by Chen et al. (Chen and Beraun, 2000a) and the finite particle method (FPM) by Liu et al. (Liu et al., 2005b; Liu and Liu, 2006) or by using a well-tuned artificial viscosity (Colagrossi and Landrini, 2003). Density re-initialization (Dilts, 1999, 2000) is another effective approach to reduce large oscillation of the particle pressure and can be applied every several time steps without too much computational cost. Furthermore, the improvements in treating solid boundaries (Liu et al., 2012a) and free surfaces (Zheng et al., 2012) also contribute to WCSPH in better modeling free surface incompressible flows.

Another approach in resolving pressure is to model the incompressible flows as rigorously incompressible rather than weakly compressible as in WCSPH (Shao et al., 2006). This leads to the projection-based incompressible smoothed particle hydrodynamics (ISPH) method (Cummins and Rudman, 1999; Rafiee and Thiagarajan, 2009). This is similar to the Moving Particle Semi-implicit (MPS) method (Koshizuka and Oka, 1996) for simulating incompressible flows, in which the pressure term is implicitly calculated using particle interaction with the constant particle number density, while the source term is explicitly calculated. Unlike WCSPH, the particle density in ISPH remains unchanged, ensuring the incompressibility of the modeling fluid. The pressure is implicitly obtained from solving the pressure Poisson's equation (PPE) rather than from an artificial equation of state as in WCSPH. The PPE is only related on the relative positions and relative velocities between particles without artificial parameters. Meanwhile, the solution of PPE in ISPH is based on the entire computational domain. Therefore compared with traditional WCSPH methods, the ISPH methods are usually able to obtain much smoother pressure fields (Liu et al., 2013).

6.3.1 Weakly compressible SPH (WCSPH) model

In early SPH modeling of incompressible flow, an artificial compressibility technique is usually used to model the incompressible flow as a slightly compressible flow. The artificial compressibility considers that every theoretically incompressible fluid is actually compressible. Therefore, it is feasible to use a quasi-incompressible equation of state to model the incompressible flow. This is therefore usually referred to as *weakly compressible SPH* (WCSPH) for incompressible flows.

In WCSPH for modeling incompressible fluid flow, the density and momentum equations can be approximated using equations (6.11) and (6.12), with possible density and kernel corrections. Pressure can be obtained from the artificial equation of state (EOS). A frequently used artificial EOS is

$$p = c^2 \rho , \tag{6.29}$$

where c is the sound speed which is a key factor that deserves careful consideration. If the actual sound speed is employed, the real fluid is approximated as an artificial fluid, which is ideally incompressible. Monaghan argued that the relative density variation δ is related to the fluid bulk velocity and sound speed in the following way (Monaghan, 1994)

$$\delta = \frac{\Delta \rho}{\rho_0} = \frac{|\rho - \rho_0|}{\rho_0} = \frac{V_b^2}{c^2} = M^2 , \tag{6.30}$$

where ρ_0, $\Delta\rho$, V_b and M are the initial density, absolute density variation, fluid bulk velocity and Mach number respectively.

Morris, through considering the balance of pressure, viscous force and body force, proposed an estimate for the sound speed (Morris et al., 1997). He argued that the square of the sound speed should be comparable with the largest value of V_b^2/δ, $\upsilon V_b/\delta l$ and Fl/δ, i.e.

$$c^2 = \max\left(\frac{V_b^2}{\delta}, \frac{\upsilon V_b}{\delta l}, \frac{Fl}{\delta}\right), \tag{6.31}$$

where $\upsilon (\upsilon = \mu/\rho)$ is the kinetic viscosity, F is the magnitude of the external body force, and l is the characteristic length scale.

It is clear that WCSPH is an explicit method and is similar to conventional SPH modeling of compressible flow. The difference is that in WCSPH, an artificial rather than true equation of state is used. It is also noted that in WCSPH, an artificial equation of state should be used in modeling incompressible flow, and empirical values such as artificial sound need to be tuned. Moreover the resultant pressure in conventional WCSPH is usually rough and oscillatory. Therefore, artificial viscosity and density corrections are usually used to obtain smooth pressure field (Colagrossi and Landrini, 2003).

For incompressible flows with free surfaces, theoretically the pressure on free surfaces should rigorously be zero. With the SPH approximation of density (equation (6.14)), density approaches zero when particles approach free surfaces. As such, in WCSPH, the free surface boundary conditions are usually ignored as pressure calculated from an equation of state also approaches zero and the free surface boundary condition is implicitly implemented. However, in SPH simulation, as free surface particles are not easy to be exactly identified, it is natural to render the free surface condition by setting a criterion value of density ratio to identify the free surface particles. For example, for a specific particle i, if $\rho_i/\rho_0 < 0.9$ (where ρ_0 is the reference or initial density and ρ_i is calculated from equation (6.14) or improved with possible corrections), i can be regarded as a surface particle, and the p_i can be set to be zero to implement the free surface condition.

6.3.2 Incompressible SPH (ISPH) algorithm

As WCSPH may involve empirical parameters and oscillation pressure field, Some researchers try to develop rigorously *incompressible SPH* (ISPH) models for incompressible flows. In ISPH, the governing equations are the same as those in WCSPH (equations (6.1) and (6.2)), the only difference is the calculation of pressure. In WCSPH, the pressure is obtained from an artificial equation of state. Whereas in ISPH, the pressure is obtained from solving a Poisson pressure equation, somewhat similar to that in the moving particle semi-implicit (MPS) method (Koshizuka and Oka, 1996).

One typical approach of ISPH is to apply the pressure projection method to SPH method (Chen et al., 2013b). In this ISPH model, the time integration is accomplished through two loops in one time step, ensuring second order accuracy.

In the first loop, the position and the velocity of a fluid particle are firstly predicted as

$$\mathbf{x}^* = \mathbf{x}^n + \Delta t\left(\frac{3}{2}\mathbf{v}^n - \frac{1}{2}\mathbf{v}^{n-1}\right), \tag{6.32}$$

$$\mathbf{v}^* = \frac{4}{3}\mathbf{v}^n - \frac{1}{3}\mathbf{v}^{n-1} + \frac{2}{3}\Delta t\left(\mathbf{F}^{n+1} + \frac{\mu}{\rho}\nabla^2\mathbf{v}^n\right), \tag{6.33}$$

where superscript * denotes the estimated value, n means the number of time step.

Then a correction step is applied based on the pressure force. Since the incompressibility ensures that the density does not change, the pressure of fluid particles and boundary particles are obtained by solving the pressure Poisson equation (PPE)

$$\nabla\cdot\left(\frac{\nabla P^{n+1}}{\rho}\right)=\frac{3}{2\Delta t}\nabla\cdot\mathbf{v}^{*}, \tag{6.34}$$

where the left-hand-side (LHS) can be expressed in SPH form as

$$\left(\nabla\cdot\frac{1}{\rho}\nabla P\right)_a\approx\sum_b m_b\frac{8}{\left(\rho_a+\rho_b\right)^2}\frac{\left(P_a-P_b\right)\mathbf{x}_{ab}\cdot\nabla W_{ab}}{\left(\left|\mathbf{x}_{ab}\right|^2+0.01h^2\right)}, \tag{6.35}$$

The corrected velocity is expressed as

$$\mathbf{v}^{n+1}=\mathbf{v}^{*}-\frac{2}{3}\frac{\Delta t}{\rho}\nabla P^{n+1}. \tag{6.36}$$

The velocity and the position in half time step are obtained by

$$\mathbf{v}^{n+1/2}=\frac{1}{8}\left(3\mathbf{v}^{n+1}+6\mathbf{v}^{n}-\mathbf{v}^{n-1}\right), \tag{6.37}$$

$$\mathbf{x}^{n+1/2}=\mathbf{x}^{n}+\frac{\Delta t}{24}\left(2\mathbf{v}^{n+1}+11\mathbf{v}^{n}-\mathbf{v}^{n-1}\right). \tag{6.38}$$

The second loop in this numerical scheme advances another half time step to obtain the velocity and position at time step $n+1$.

In ISPH, a common solid boundary treatment is the combinational use of edge particles and dummy particles. Based on the work in literature (Lee et al., 2008), one layer of edge particles (named here as E*) and three layers of dummy particles (named here as D*) are placed on the solid boundary. The edge particles are involved in the solving of Poisson pressure equation. Dummy particles are used here to avoid the truncation of support domain for fluid particles near solid walls. The pressures of these dummy particles are given by the following formula to meet the Neumann boundary condition:

$$P_i=P_j+\rho\mathbf{g}\cdot\left(\mathbf{x}_i-\mathbf{x}_j\right), \tag{6.39}$$

where i and j denote to the dummy particle and its reference edge particle, respectively, $\mathbf{g}$ is the acceleration of gravity. The nearest edge particle is chosen as the reference particle for a dummy particle, as shown in **Figure 6.1**.

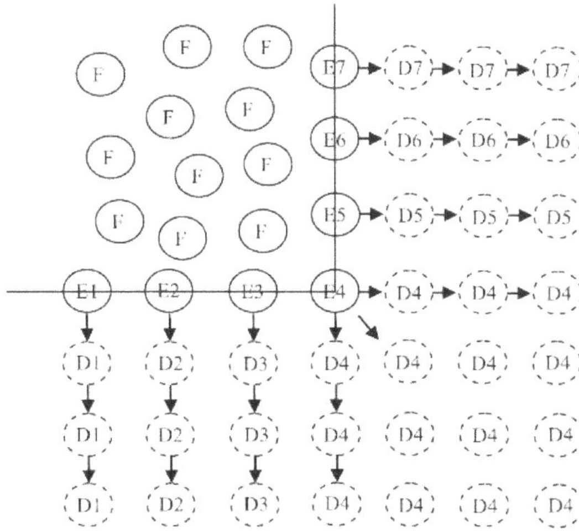

Figure 6.1 Illustration of boundary treatment in ISPH.

In ISPH, the widely used divergence of the position vectors is applied in the identification of free surface particles

$$\nabla \cdot \mathbf{x}_a = \sum_b \frac{m_b}{\rho_b} \mathbf{x}_{ab} \cdot \vec{\nabla}_a W\left(\mathbf{x}_a - \mathbf{x}_b, h\right). \tag{6.40}$$

For particles far from the surface, this divergence has a theoretical value of 2 in 2D. However, due to the loss of particles in the support domains of free surface particles, their divergence values of the position vectors shall be much less than 2. In general, a threshold value of 1.5 can be used to determine the free surface particles.

There are different ways to treat the detected free surface particles. One is to apply accurate Dirichlet boundary condition on the free surface particles, which means the pressures of all free surface particles are strictly set to zero. However, this treatment may lead to particle aggregation on free surfaces because of the lack of repulsive forces between free surface particles. Bøckmann et al. (Bøckmann et al., 2012) use another method to keep small pressure divergence between free surface particles. In the approach, for a free surface particle, (6.35) is replaced as follows

$$\left(\nabla \cdot \frac{1}{\rho} \nabla P\right)_a = \sum_b m_b \frac{8}{\left(\rho_a + \rho_b\right)^2} \frac{\left(2P_a - P_b\right)\mathbf{x}_{ab} \cdot \vec{\nabla} W_{ab}}{\left(\left|\mathbf{x}_{ab}\right|^2 + 0.01h^2\right)}. \tag{6.41}$$

6.3.3 Comparisons of WCSPH and ISPH

There are several papers offering comparisons between ISPH and WCSPH. These papers compared the traditional WCSPH method and ISPH method and concluded that ISPH method has overall advantages over WCSPH method (Lee et al., 2008). However, these works did not consider the recent improvements in the SPH method especially in correction of particle approximation, enhancements in treatments of solid boundaries and free surfaces, and therefore the obtained conclusions are insufficient. Shadloo et al. (Shadloo et al., 2012) considered a robust WCSPH model and offered its comparison with ISPH method. But the presented numerical examples are not associated with changing and breakup of free surfaces, and violent fluid-solid interactions with liquid impact, sloshing and splashing. These are usually the most important features for violent free surface flows with moving objects.

As existing results and conclusions from ISPH and WCSPH are usually inconsistent and even controversial, Chen et al. performed a comparative study of a truly incompressible and an improved weakly compressible SPH (IWCSPH) method for free surface incompressible flows (Chen et al., 2013b). The WCSPH model and ISPH model are described in Section 6.3.1 and 6.3.2. The difference between IWCSPH and conventional WCSPH is that IWCSPH uses a simple moving least square approach to re-initialize density shown in equation (6.19). It is noted that to make ISPH and IWCSPH comparable, high order SPH approximation schemes and artificial viscosity are not used in both models. Three representative numerical examples, including a benchmark test for hydrostatic pressure, a dam breaking problem and a liquid sloshing problem, are comparatively analyzed with ISPH and IWCSPH.

6.3.3.1 Water tank with hydrostatic pressure

The examination of hydrostatic pressure in a tank partially filled with still water is a benchmark problem and it is quite a challenge for traditional SPH method to obtain a stable hydrostatic pressure (Oger et al., 2007). In this section, the hydrostatic pressures obtained through the ISPH method and the improved WCSPH model are compared and discussed.

As shown in **Figure 6.2**, the breadth and height of the water tank are 1 m and 1 m, respectively. The depth of water column is 0.3 m, and a pressure probe point (point P in **Figure 6.2**) is placed 0.1 m below the initial free surface on the left wall. Theoretically, the pressure value of this point should be a constant one of $\rho gd = 981$ Pa. The total number of fluid particles used in this simulation is $N = 100\times 30$, with an initial particle spacing of 0.01 m. The time step taken in the ISPH simulation is 10^{-4} s, and in the IWCSPH simulation the time step is 10^{-5} s.

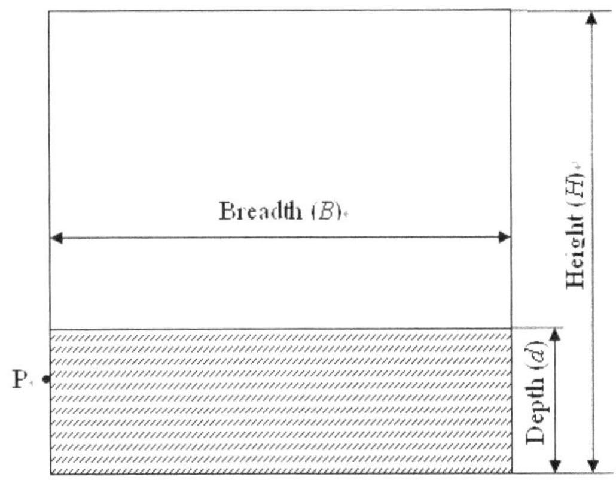

Figure 6.2 Illustration of the numerical water tank (Chen et al., 2013b).

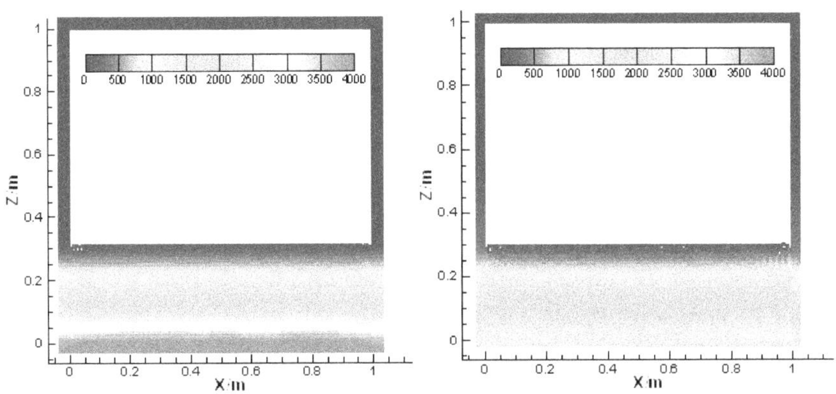

Figure 6.3 Pressure snapshots at $t = 15$ s, left: ISPH; right: improved WCSPH (Chen et al., 2013b).

Figure 6.3 shows the pressure snapshots obtained by using ISPH and IWCSPH at t = 15 s. This time instant is chosen to stay away from initial disturbance and examine the numerical stability of the two numerical models discussed above. For the convenience of comparisons, the calibration scale is kept uniform in the pressure snapshots obtained by ISPH and IWCSPH. It is observed that both models give smooth pressure fields but the average pressure in the ISPH snapshot is higher than that in the IWCSPH snapshot.

Figure 6.4 demonstrates the comparison of pressure profiles along the middle section of the water tank (x = 0.5 m) at t = 15 s. It is observed that the pressure profile obtained by ISPH is significantly higher than the analytical solution. And the discrepancy becomes larger as the depth increases. The pressure profile obtained by IWCSPH is very close to the analytical pressure profile, except for some small discrepancies near the solid boundary area.

Figure 6.5 shows the pressure histories of the probe point *P* obtained by ISPH and IWCSPH. It is observed that both models give stable numerical results for this still water tank with hydrostatic pressure. However, the pressure history obtained by ISPH keeps rising as time marches, which may be due to ISPH's sensitivity to small disturbances. In contrast, the pressure curve from IWCSPH is very close to the analytical solution and slightly keeps going down due to the small numerical diffusion in this model.

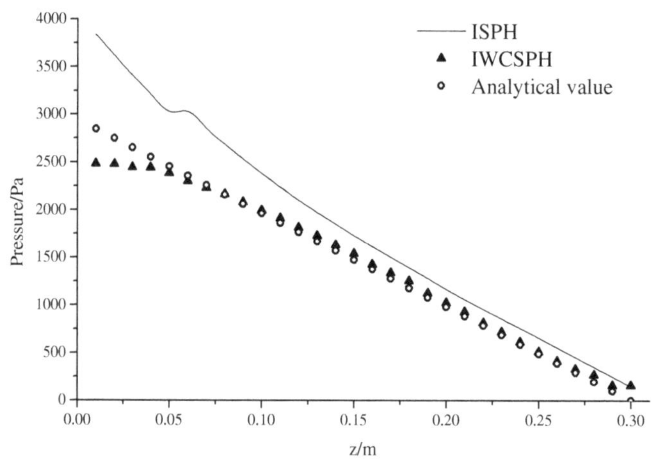

Figure 6.4 Pressure profiles along the middle section of the water tank (Chen et al., 2013b).

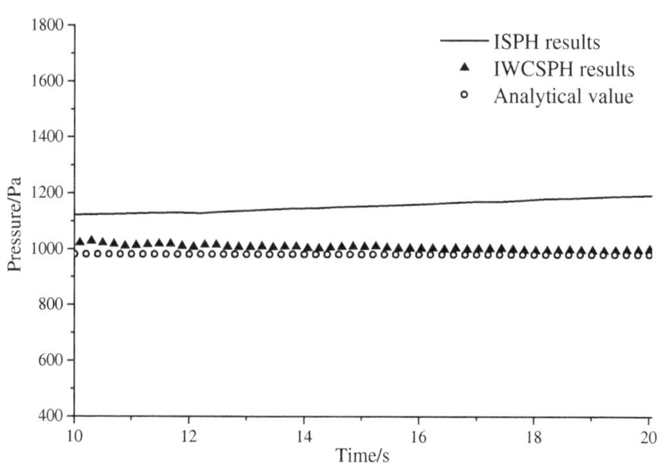

Figure 6.5 Pressure histories of the probe point *P* (Chen et al., 2013b).

In general, both ISPH and IWCSPH give acceptable results for this hydrostatic test. The results of IWCSPH are about 10% lower than those of ISPH, and are closer to the theoretical values. Hence it can be concluded that IWCSPH is more accurate than ISPH in spatial and temporal pressure distributions for this benchmark problem with hydrostatic pressure.

6.3.3.2 Dam breaking problem with large impact pressure

Dam breaking is a common and important benchmark problem for numerical simulations of free surface incompressible flows (Shao and Lo, 2003). It usually comes with lots of complex phenomenon, including changing and breakup of free surfaces, splashing and strong impact against solid boundaries. In this section, two cases of 2D dam breaking are presented and discussed to comparatively investigate the reliability of the ISPH method and the IWCSPH model. Meanwhile, some comparisons and analyses of the computational costs are offered in this section.

In the first case of dam breaking, the experimental domain is $W = 0.4$ m long, and the initial dimension of water column is $L \times H = 0.1\ \text{m} \times 0.2\ \text{m}$, as shown in **Figure 6.6**. The total number of fluid particles is $N = 40 \times 80$, with the initial particle spacing of $d_0 = 0.0025$ m. In ISPH simulation, the time step is taken as 10^{-4} s, and in IWCSPH simulation it is 10^{-5} s.

The numerical results at two different moments are presented in **Figure 6.7**. The results obtained by ISPH and IWCSPH are compared at dimensionless time moments of $t(g/H)^{0.5} = 1.96$ and 4.55, respectively. It is observed that both methods can obtain reasonable flow pattern and smooth pressure fields for areas far from the solid boundaries and free surfaces. IWCSPH can obtain better

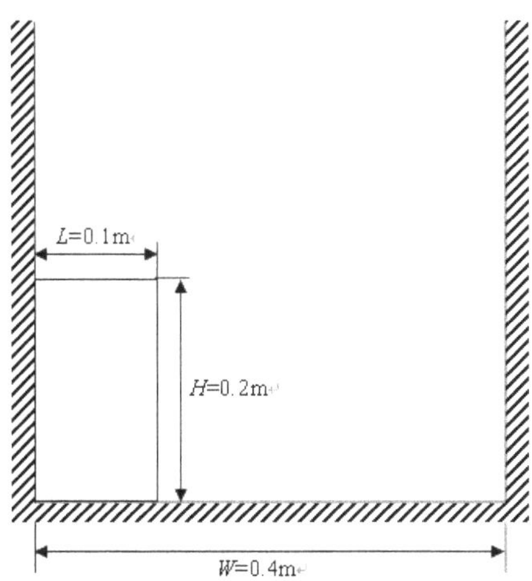

Figure 6.6 Initial set up of the first case of dam breaking (Chen et al., 2013b).

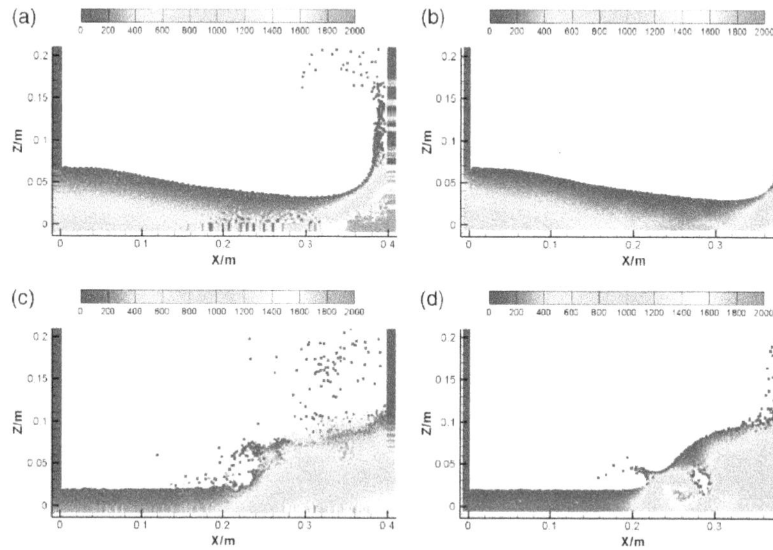

Figure 6.7 Comparison (pressure field) of the first case of dam breaking, (a) ISPH at $t(g/H)^{0.5} = 1.96$; (b) IWCSPH at $t(g/H)^{0.5} = 1.96$; (c) ISPH at $t(g/H)^{0.5} = 4.55$; (d) IWCSPH at $t(g/H)^{0.5} = 4.55$ (Chen et al., 2013b).

details in the flow jet and cavity. Most importantly, ISPH is associated with bigger oscillation near the solid boundaries and free surfaces and small disturbance can lead to numerical instability. This is due to the different schemes in calculating pressure in ISPH and WCSPH. In ISPH, pressure is obtained through solving a system matrix equation (describing PPE) on the entire computational domain, small perturbations in boundary or free surface areas can simultaneously propagate across the entire computational domain, and can further lead to bigger oscillations and instability in the numerical simulation. In contrast, in WCSPH, the pressure is obtained from the equation of state and is related to the approximation of density, which is conducted only within the support domain of a concerned particle. Numerical oscillations or small errors at a concerned particle only influence neighboring particles within the support domain rather than the entire computational domain at the same step. Their influences to particles far away from the concerned particle are effective only at later time steps. In other words, in ISPH, the influence of numerical oscillation is more direct and global. Any numerical disturbance can influence the entire computational domain at the same time step. While in WCSPH, the influence of numerical oscillation is indirect and local. The propagation of possible numerical disturbance can be smeared and vanished over a number time steps. Hence, ISPH is more sensitive to numerical oscillations and easier to induce numerical instability, while WCSPH is more flexible.

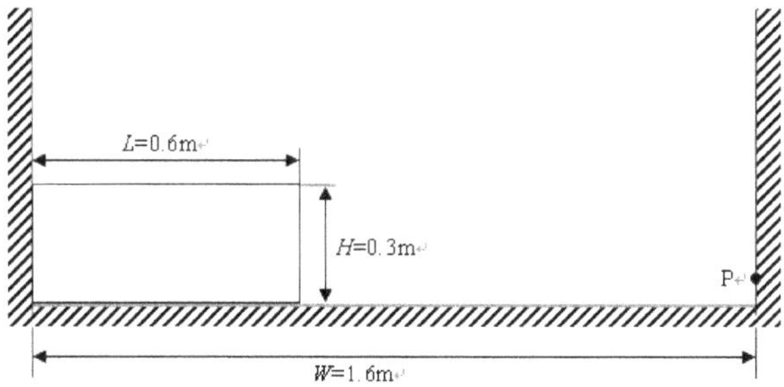

Figure 6.8 Initial set up of second case of dam breaking (Chen et al., 2013b).

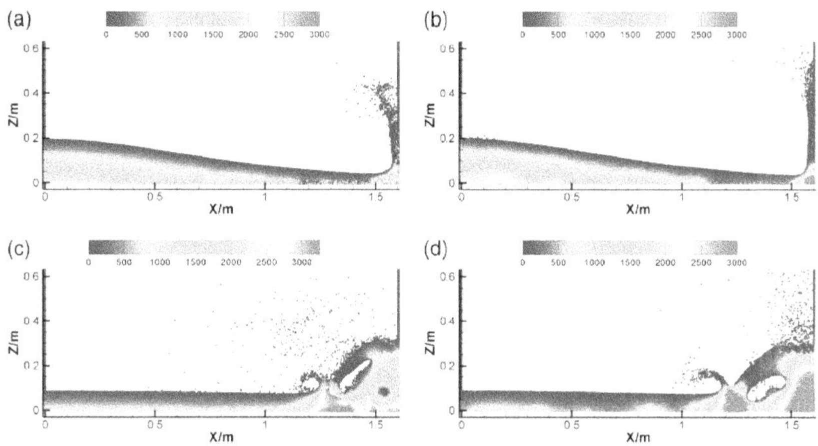

Figure 6.9 Comparison (pressure field) of the second case of dam breaking, (a) ISPH at $t(g/H)^{0.5} = 3.2$; (b) IWCSPH at $t(g/H)^{0.5} = 3.2$; (c) ISPH at $t(g/H)^{0.5} = 6.4$; (d) IWCSPH at $t(g/H)^{0.5} = 6.4$ (Chen et al., 2013b).

Another case of dam breaking is also simulated and the case offers some quantitative comparison between the results obtained by ISPH and IWCSPH. In this example of dam breaking, the experimental domain is W = 1.6 m long, and the initial dimension of water column is $L \times H$ = 0.6 m × 0.3 m, as shown in **Figure 6.8**. The total number of fluid particles is N = 300 × 150, with the initial particle spacing of d_0 = 0.002 m. In ISPH simulation, the time step is taken as 5×10^{-5} s, and in IWCSPH simulation it is 5×10^{-6} s. To track and analyze the time variations of pressure, a pressure measurement point P is located on the right wall and 0.06 m above the bottom.

The numerical results at two special instants are presented in **Figure 6.9**. The results obtained by ISPH and IWCSPH are compared at dimensionless time instants of $t(g/H)^{0.5}$ = 3.2 and 6.4, respectively. In this example, both ISPH and

IWCSPH give generally smooth pressure fields and present those important details such as flow cavity and flow jet. However, it is observed that, in the results obtained by ISPH, the pressure field inside the flow still suffers from some incorrect oscillations and discontinuities, especially in places where strong impacts happen. Meanwhile, the water splash in ISPH simulation is much more severe than that in IWCSPH simulation, which indicates that the IWCSPH is more flexible and stable than ISPH.

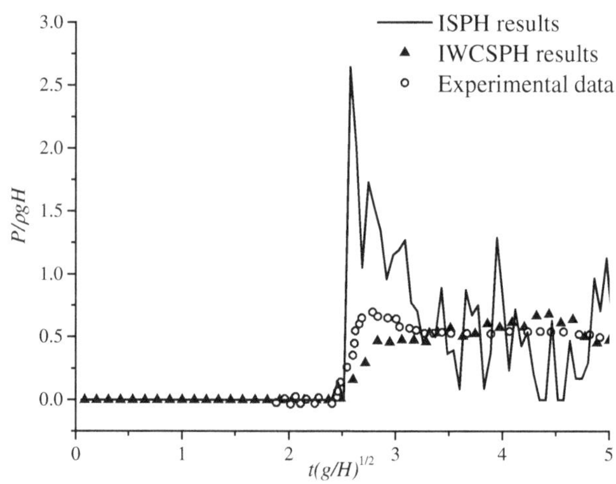

Figure 6.10 Comparison of time variations of dimensionless pressure on point *P* (Chen et al., 2013b).

A detailed comparison of the pressure history on point *P* is shown in **Figure 6.10**. The pressure curves obtained by ISPH and IWCSPH are compared with experimental data (Lee et al., 2002). It is observed that the pressure curve obtained by ISPH simulation oscillates more greatly than that obtained by IWCSPH simulation. And the oscillation amplitude is so big that it has concealed the valuable information about the pressure history on point P. Specifically, when strong impact happens, the instability of ISPH becomes much more severe, which is shown as the exorbitant pressure during dimensionless time from 2.6 to 3.0 in **Figure 6.10**. In contrast, the pressure history obtained in IWCSPH simulation is generally smooth. During the occurrence of strong impact, the IWCSPH captured the impact pressure very well. Although the pressure value is smaller than the experimental data, the discrepancy is acceptable for engineering applications.

To sum up, for the two dam breaking cases, both ISPH and IWCSPH can obtain reasonable flow patterns with smooth pressure fields in bulky flow region. However, IWCSPH is more accurate, more flexible and can capture more flow details such as jets and cavity. ISPH is more sensitive to numerical oscillations and it is easier to cause instability.

6.3.3.3 Liquid sloshing problem with periodically changing hydrodynamic pressure

Sloshing refers to the periodic liquid movement in a partially-filled container due to external excitations (Faltinsen and Timokha, 2009). It is a common and important phenomenon in engineering field. In this section, an example of liquid sloshing is presented to test the adaptabilities of ISPH and IWCSPH in periodic hydrodynamic problems.

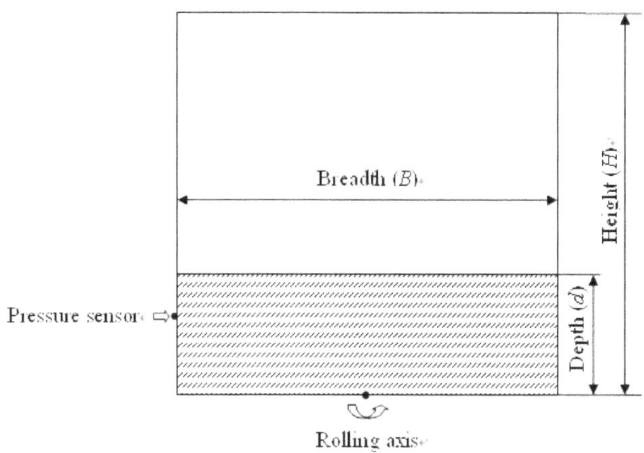

Figure 6.11 Illustration of the sloshing system (Chen et al., 2013b).

As shown in **Figure 6.11**, the geometry of the water tank is $B \times H$ = 1 m × 1 m, and the water depth is d = 0.3 m. The total number of fluid particles is N = 100 × 30, with the initial particle spacing of d_0 = 0.01 m. A pressure sensor is located on the left wall and is 0.1 m below the initial free surface. The tank rolls around the axis at the center of its bottom, and the rolling angle α follows motion law expressed as

$$\alpha = A\sin(\omega t), \tag{6.42}$$

where A, ω and t denote the rolling amplitude, the excitation circular frequency and time, respectively. In this example, A = 5°, ω = 0.95 rad/s.

Figure 6.12 and **Figure 6.13** show the pressure fields obtained by using ISPH and IWCSPH at different time instants. It is observed that both two models give smooth pressure pictures, but the pressure obtained by ISPH is a little bit higher than that obtained by IWCSPH. To find out which result is more reliable, the pressure history on the pressure sensor point is recorded and presented in **Figure 6.14**. Comparing the curves obtained in numerical simulations with the experimental data (Chen et al., 2013a), it is found that although the average pressure value decreases a little as the time marches, the pressure curve obtained by IWCSPH model is in better agreement with the real condition. For the pressure curve obtained by ISPH, although a similar shape is retained, the average pressure keeps rising and deviates from the experimental data more greatly compared with that obtained by IWCSPH. The greater discrepancies

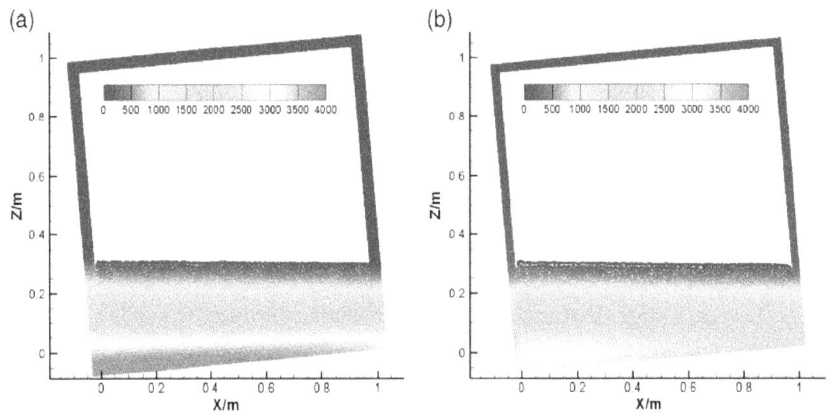

Figure 6.12 Comparison of the pressure fields, $t = 1.25$T, left: ISPH; right: IWCSPH (Chen et al., 2013b).

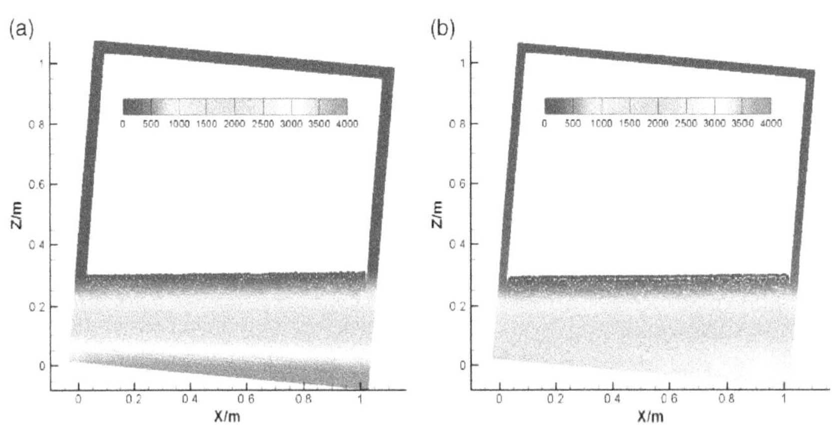

Figure 6.13 Comparison of the pressure fields, $t = 1.75$T, left: ISPH; right: IWCSPH (Chen et al., 2013b).

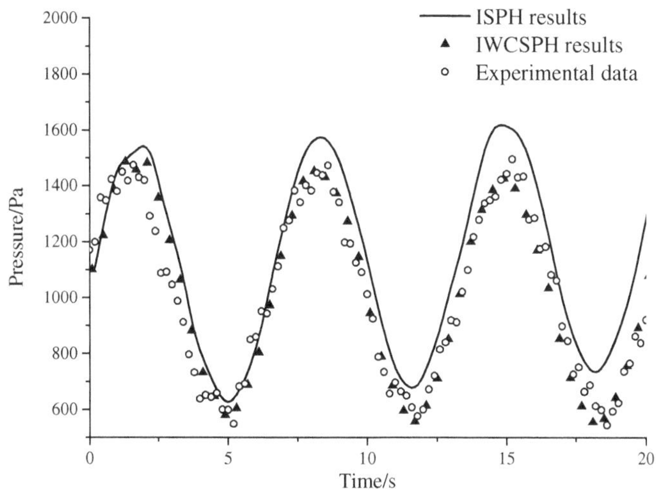

Figure 6.14 Comparison of the pressure histories on point *P* (Chen et al., 2013b).

between ISPH results and the experimental data may still come from the instability of ISPH algorithm. Therefore, it can be concluded that for this liquid sloshing problem with periodically changing hydrodynamic pressure, the IWCSPH model also gives more accurate results than ISPH.

6.3.3.4 Summary

ISPH and WCSPH are two approaches in modeling incompressible flows within the frame of SPH. ISPH is a semi-implicit method which needs to solve the pressure Poisson's equation, and WCSPH is an explicit method which is based on the weakly compressible assumption of the incompressible fluid.

As ISPH and WCSPH are very popular, while reported results are usually inconsistent, in this work, a comparative study of the ISPH method and an improved weakly compressible SPH method is conducted to model incompressible free surface flows. Both methods solve the Navier-Stokes equations in Lagrangian form and no artificial viscosity is used. The ISPH algorithm presented here is based on the classical SPH projection method with some common treatments on solid boundaries and free surfaces. The IWCSPH model includes some advanced corrective algorithms in density approximation and solid boundary treatment.

Three typical numerical examples, including a benchmark problem with hydrostatic pressure, a dam breaking problem with large impact pressure, and a liquid sloshing problem with periodically changing hydrodynamic pressure, are comparatively analyzed with ISPH and IWCSPH. For these three representative examples, it is found that, in general, both ISPH and IWCSPH can well capture the flow patterns with smooth pressure fields. For the benchmark problem with hydrostatic pressure, IWCSPH is more accurate than ISPH in spatial and temporal pressure distributions. For dam breaking problem and a liquid sloshing problem, IWCSPH is more accurate and more stable while ISPH is more sensitive to numerical oscillations and is easier to cause instability. And therefore the IWCSPH is more suitable for problems with violent water impact and fluid-structure interactions.

Another point is that although ISPH can use much bigger time steps, it needs to solve the time-consuming sparse matrix equation. Therefore ISPH may not be superior to IWCSPH in computational cost. For problems with a large number of particles, IWCSPH can even be more efficient than ISPH. Moreover, as the IWCSPH code in this work is serial and the ISPH code is partially parallel, the IWCSPH may be more competitive after parallelization.

It is known that the conventional WCSPH usually suffers from low accuracy with high pressure oscillations. Though using numerical techniques such as higher order SPH approximation schemes and artificial viscosity can greatly improve WCSPH, in the presented IWCSPH, only a simple MLS approach for re-initializing density is used. It is straightforward to implement on existing WCSPH without introducing inherent factors unsuitable for comparisons with ISPH.

In summary, although the ISPH method may have some advantages over traditional WCSPH method, the presented IWCSPH is more attractive than ISPH as it seems more accurate and more stable with comparable or even less computational efforts.

6.4 Free surface flows

Free surface flow motions, especially violent wave motions are common phenomena in coastal hydrodynamics and offshore engineering. When the violent wave encounters obstacles, very large impact and pressure oscillating can occur together with wave breaking and overturning. This can destroy the structure of the nearby obstacles. If the violent waves occur in coast areas, they may cause potential damages of property and loss of life. Therefore, studying the flow mechanism of the free surface flows is of great significance in theory and practice (Mei, 1989; Peregrine, 2003; Faltinsen et al., 2004).

In the past several decades, many researchers have already conducted some experiments and numerical simulations on violent free surface problem. However, experimental studies can be expensive, or sometimes can not be implemented. Numerical simulation has become an attractive tool to predicting fluid in coast hydrodynamics and offshore engineering.

Conventional grid-based numerical methods such as FDM (finite difference method) and FVM (finite volume method) have been widely applied to various areas of CFD (Computational Fluid Dynamics), and currently are the dominant methods in numerical simulations of domain discretization and numerical discretization. And many numerical techniques have been proposed to track the free surface, such as the volume of fluid (VOF) method (Hirt and Nichols, 1981), the level set method (Sethian, 2003) and the CIP-based method (Hu and Kashiwagi, 2004), and etc. Despite the great success, these numerical methods suffer from inherent difficulties in many aspects. Constructing a regular grid for irregular or complex geometry has never been an easy task, and usually requires additional complex mathematical transformation that can be even more expensive than solving the problem itself. Moreover, these methods have difficulties in avoiding numerical dissipation.

A recent strong interest has been focused on the development of the next generation computational methods, mesh free method, such as smoothed particle hydrodynamics (SPH). In the SPH method, the state of a system is represented by a set of particles, which possess individual material properties and move according to the governing conservation equations. SPH feature a remarkable flexibility in handling complex flow fields and in including physical effects. The special advantage of the SPH is the adaptive nature, therefore, the formulation of SPH is not affected by the arbitrariness of the particles, and it can naturally handle problems with extremely large deformation (Liu and Liu, 2003). Moreover, some improved SPH methods have been developed to restore the consistency in kernel and particle approximation. The improved SPH methods combined with suitable turbulence models and solid boundary treatment algorithms can well simulate free surface flows.

In this section, the SPH method shall be applied to modeling incompressible free surface flows. The SPH method is improved with kernel gradient correction (KGC) and density correction to improve the computational accuracy in particle approximation and with Reynolds Averaged Navier-Stokes turbulence model to capture the inherent physics of flow turbulence. Three numerical examples are modeled using this improved SPH method, and the obtained numerical results are compared with experimental observations and results from other sources.

6.4.1 Dam breaking against a vertical wall

Shao et al. ever provided an improved smoothed particle hydrodynamics with RANS for free surface flow problems (Shao et al., 2012b). In the improved SPH method, involves two major modifications on the traditional SPH method, 1) kernel gradient correction (KGC) and density correction to improve the computational accuracy in particle approximation, 2) RANS (Reynolds Averaged Navier-Stokes) turbulence model to capture the inherent physics of flow turbulence.

The improved SPH method is used to simulate the problem of dam breaking problem. Although it is not a new attempt for SPH to model such problems, the purpose of the work is to validate the effects of the kernel gradient correction and the RANS turbulence model, which can lead to in-depth results and findings. In the simulation, two dam break models are simulated. In this first case, the model of dam-break flow against a vertical wall is constructed, as shown in **Figure 6.15**. The length and height of the fluid are L and H, and the length of the container is d. The second test case is a model of dam-break flow with a gate, as shown in **Figure 6.16**. The length of container is 1.14 m, the fluid on the left of the gate has a length of 0.38 m and a height of 0.15 m. the other part has a length of 0.76 m and a height of 0.018 m.

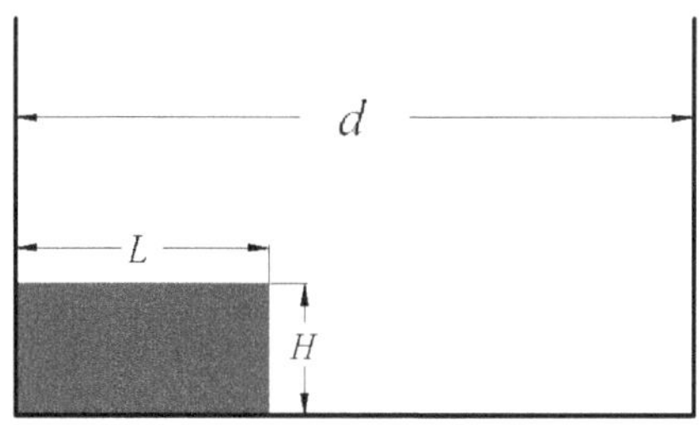

Figure 6.15 The model of dam-break flow against a vertical wall (Shao et al., 2012b).

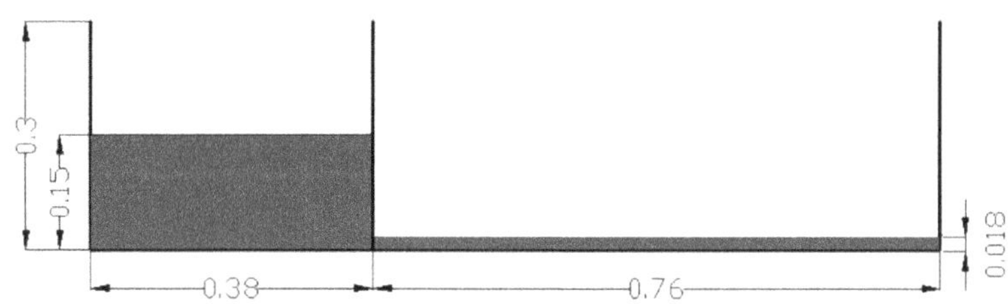

Figure 6.16 The model of dam-break flow with a controlling gate (unit: m) (Shao et al., 2012b).

For the first case, the time step is 10^{-5} s, the speed of sound is 40 m/s, and the coefficient of artificial viscosity is 0.08. About 15000 particles are used to represent the fluid. At first, an initial pressure field is given, based on the height ($P_i = \rho g h_i$), as shown in **Figure 6.17**.

After several time steps, the pressure field will be similar with the theoretical values, as shown in **Figure 6.18**. It also validates the adaptive nature of SPH.

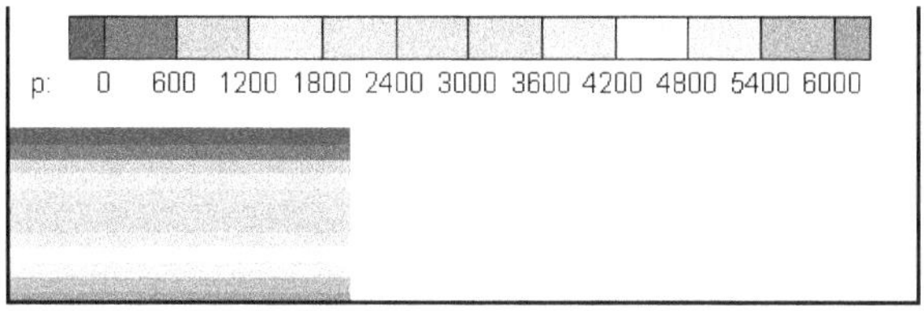

Figure 6.17 Initial pressure field of the first case (Shao et al., 2012b).

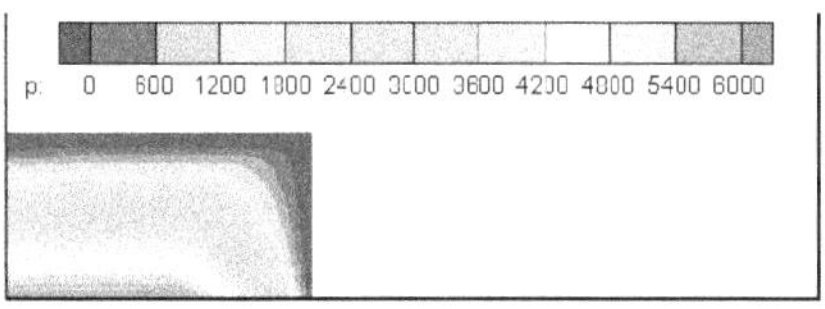

(a)

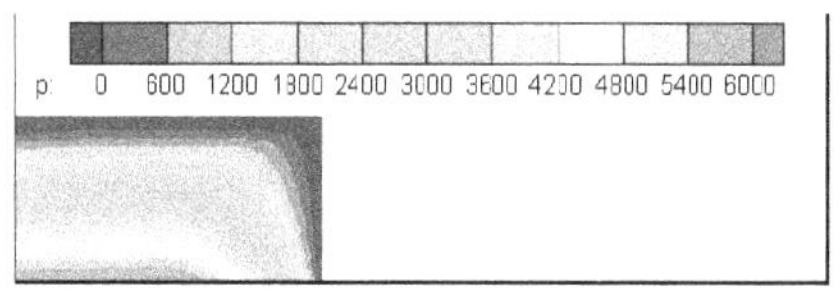

(b)

Figure 6.18 Pressure field of the dam-break problem at 0.01 s without (a) and with (b) KGC and density correction (Shao et al., 2012b).

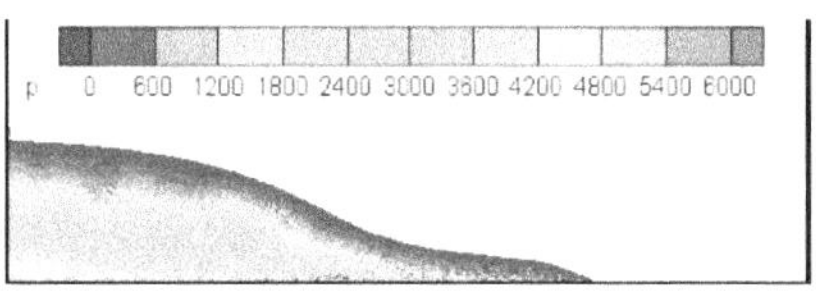

(a)

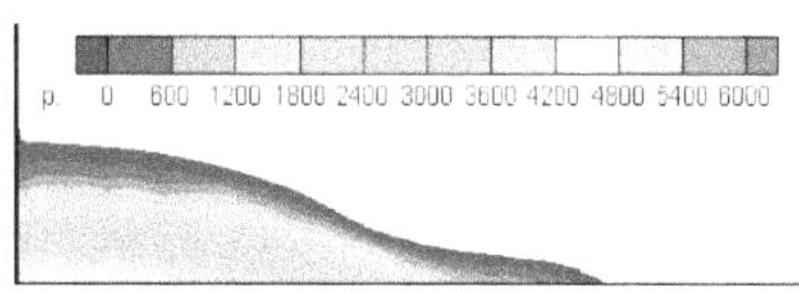

(d)

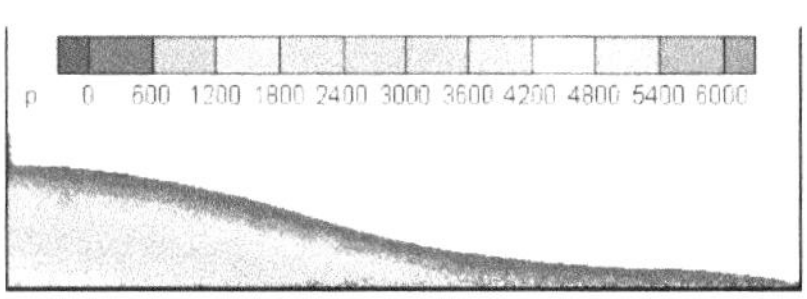

(b)

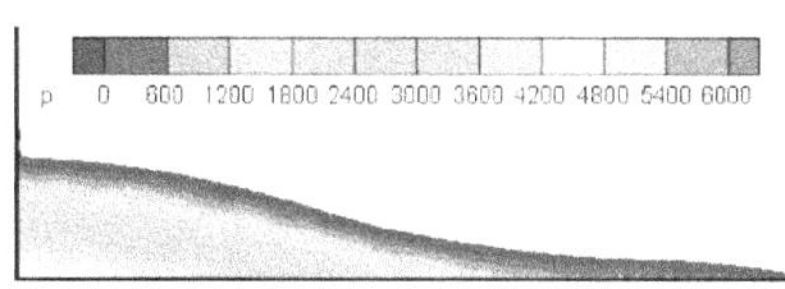

(e)

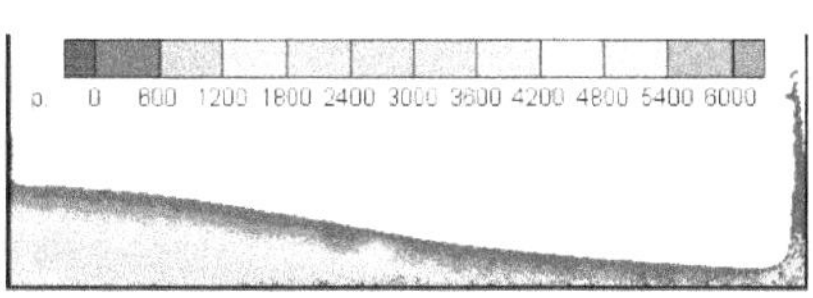

(c)

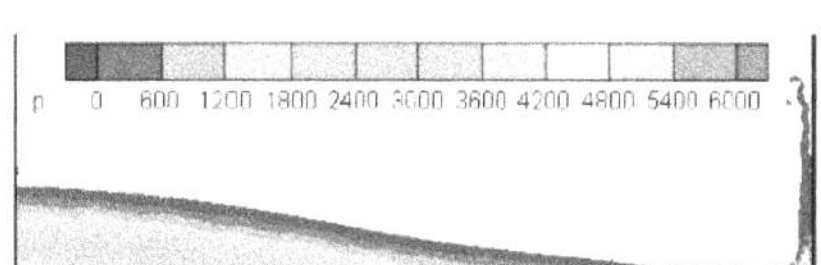

(f)

Figure 6.19 Pressure field of the dam-break problem at 0.2, 0.4 and 0.8 s without (a, b and c) and with (d, e and f) KGC and density correction (Shao et al., 2012b).

To show the effect of the correction method, three flow forms in different time are given in **Figure 6.19**. The left parts show the results from conventional SPH simulation without KGC and density correction, and the right parts show the results from the improved SPH method with KGC-model with density correction is used. It is seen that the conventional SPH method leads to strong pressure oscillations in the pressure field. At beginning, the pressure shocks only appear in the front of the fluid, with the increase of the time, the shocks transmit toward the back of the flow until the whole pressure field is destroyed. In contrast, the improved SPH model with KGC and density correction produces much smoother pressure field. Using the kernel gradient correction and density correction, the consistency of the kernel function is restored, and the accuracy of the kernel gradient is improved. Therefore, there are little oscillations in the whole pressure field. The layers of the pressure field are obviously.

To validate the roles of the RANS in describing the turbulence effects, a new case like the model in **Figure 6.16** is simulated; the difference is that no gate is used. Here, about 70000 particles are used, $H = 0.18$ m, $h = 0.15$ m, $l = 0.018$ m, $a = 0.38$ m, *and* $b = 0.76$ m. Beside the RANS, KGC with MLS correction are also be used. The time step is 5×10^{-6} s, the speed of sound is 20 m/s, and the coefficient of artificial viscosity is 0.08.

Similar with first case, the initial pressure field is given based on the height. **Figure 6.20** shows the pressure field at 0.01 s. **Figure 6.21** shows the flow patterns of the dam-break problem with a gate at 0.156, 0.219, and 0.281 s from experimental observation and SPH simulation. **Figure 6.22** shows the development of the cavities of the dam-break problem with a gate at 0.343, 0.406, 0.468, and 0.531 s. It is obvious that the flow patterns and turbulance effects obtained from SPH simulation agree very well with those in experimental observations. The change and evolution of the free surface with cavity generation, development and disapperance are very close. It is demostrated that SPH method with kernel gradient and density corrections and RANS turbulence model can well simulate violent free surface flows.

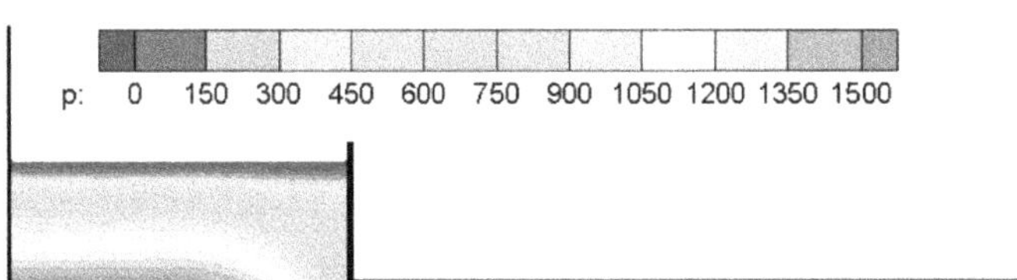

Figure 6.20 Pressure field at 0.01 s for the second case (Shao et al., 2012b).

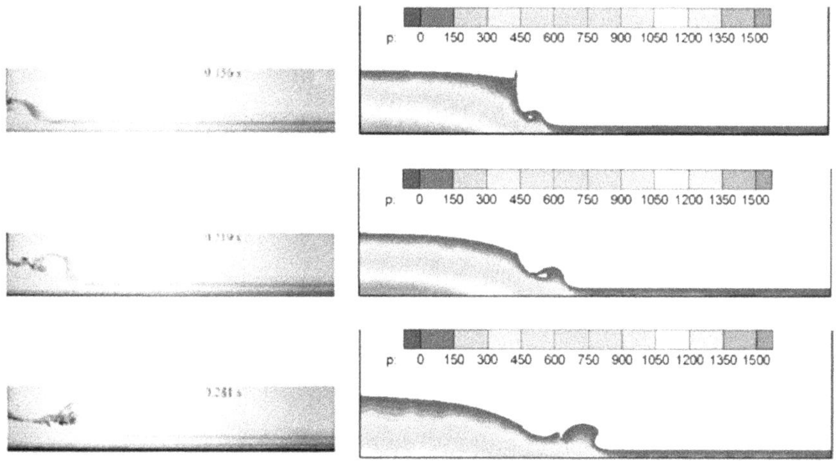

Figure 6.21 Flow patterns of the dam-break problem with a gate at 0.156, 0.219, and 0.281 s. Left column: experimental observations; Right column: SPH simulation (Shao et al., 2012b).

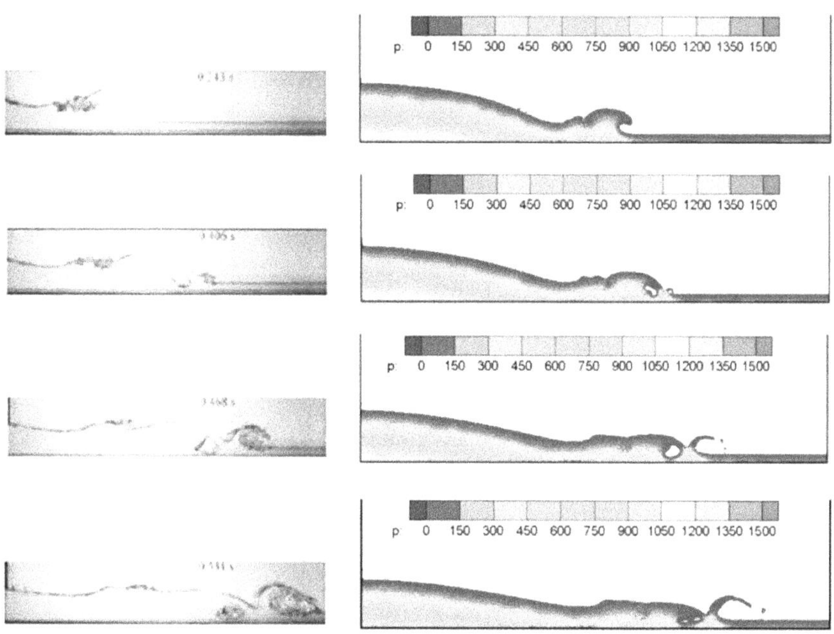

Figure 6.22 Development of the cavities of the dam-break problem with a gateat 0.343, 0.406, 0.468, and 0.531 s. Left column: experimental observations; Right column: SPH simulation (Shao et al., 2012b).

6.4.2 Dam breaking against a sharp-edged obstacle

This example is similar to the above-mentioned dam breaking except that a sharp-edged obstacle is placed before the right vertical wall, as shown in **Figure 6.23**. The sharp-edged obstacle can generate big water impact and it is difficult to precisely predict the pressure load. To track the value of the pressure, two probe points P1 and P2 are set in the sharp-edged obstacle, where OP2 = 2OP1 = 35.35 mm. The initial particle spacing is 0.01 m, about 20000 particles are used in the simulation, and the sound speed is 50 m/s.

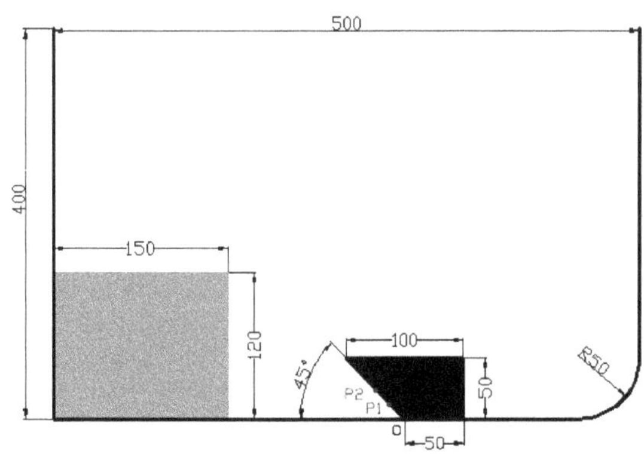

Figure 6.23 Numerical model of dam break flow against a sharp-edged obstacle (unit: mm) (Liu et al., 2012a).

Figure 6.24 shows the pressure evolution of the dam break against a sharp-edged obstacle. It is clear that when water front meets the sharp edge, a big pressure impact produces. After that, water particles spread away from the edge to generate a long strip of water. Most importantly, some particles splash away from the water strip and then fall onto the bulky water, leading to transient heavy pressure in some areas of the bulky water.

6.4.3 The movement of an elliptical cylinder near free surface

In this section, the SPH method is used to simulate the complicated movement of an elliptical cylinder near the free surface with prescribed acceleration and velocity. Early approaches for investigating waves generated by an accelerating rigid body are generally based on potential flow theory, which is not valid for long time movement with vortices. Lin provided a numerical simulation of this problem by solving the RANS equation with a cut-cell technique with a fixed-grid (Lin, 2007). The present SPH model also incorporate RANS turbulence model. The effectiveness of the SPH method as well as the RANS turbulence model will be examined.

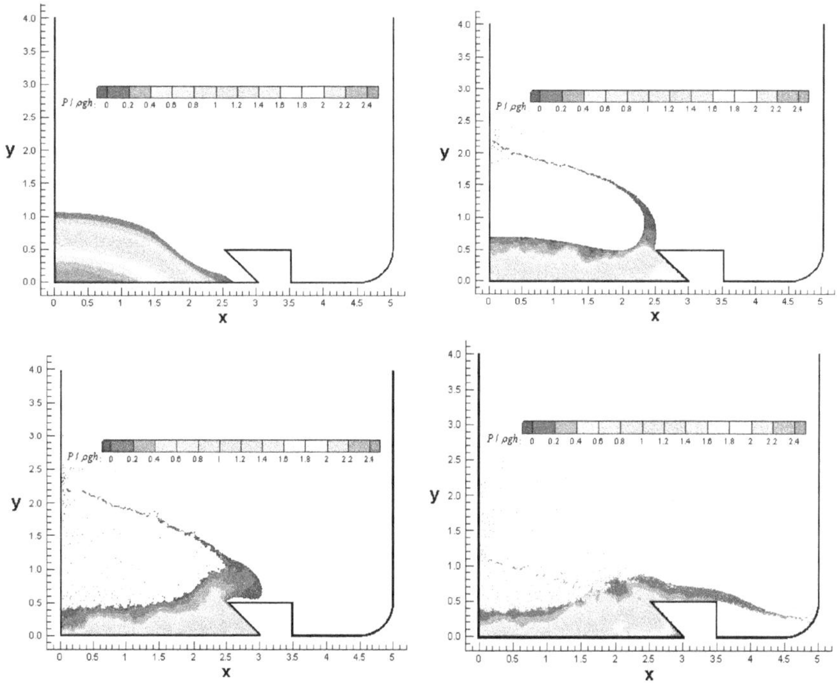

Figure 6.24 Pressure evolution at 0.37, 0.8, 1.31, 1.67 s of the dam break flow against a sharp-edged obstacle (Liu et al., 2012a).

In the simulation, approximately 100,000 SPH particles are regularly distributed in a computational domain of 10 m long and 1.0 m high with initial particle spacing of 0.01 m. The time step is taken as 0.2e-3 s. If taking the top-right corner as origin, the elliptical surface of the cylinder can be defined by $(x-8)^2/0.2^2+(y+0.2)^2/0.1^2=1$ with the center of the cylinder located at (8, -0.2). The elliptical cylinder moves leftwards from stationary with a periodic acceleration and deceleration as $a_x=-\sin\omega t$, where $\omega=-2\pi/T_p$ with $T_p=6$ s. The impulsive horizontal and vertical velocities are therefore $V_x=\omega^{-1}\cos wt-w^{-1}$ and $V_y=0.16\cos\omega t$ respectively. According to the periodically changing horizontal and vertical speed, the cylinder will move up for around 0.15 m and then exit from the free surface and then dives into the water, producing complicated flow pattern with pronounced vortices. The problem setup is the same as that in Lin's work except that the water height is 2.0 m smaller than that (3 m) in Lin's work in order to save computational effort.

Figure 6.25 shows the particle distribution of movement of an elliptical cylinder near free surface at 0.0, 1.0, 2.0, 3.0, 4.0, 5.0 and 6.0 s. The evolution of free surface and position of the elliptical cylinder are very close to the numerical

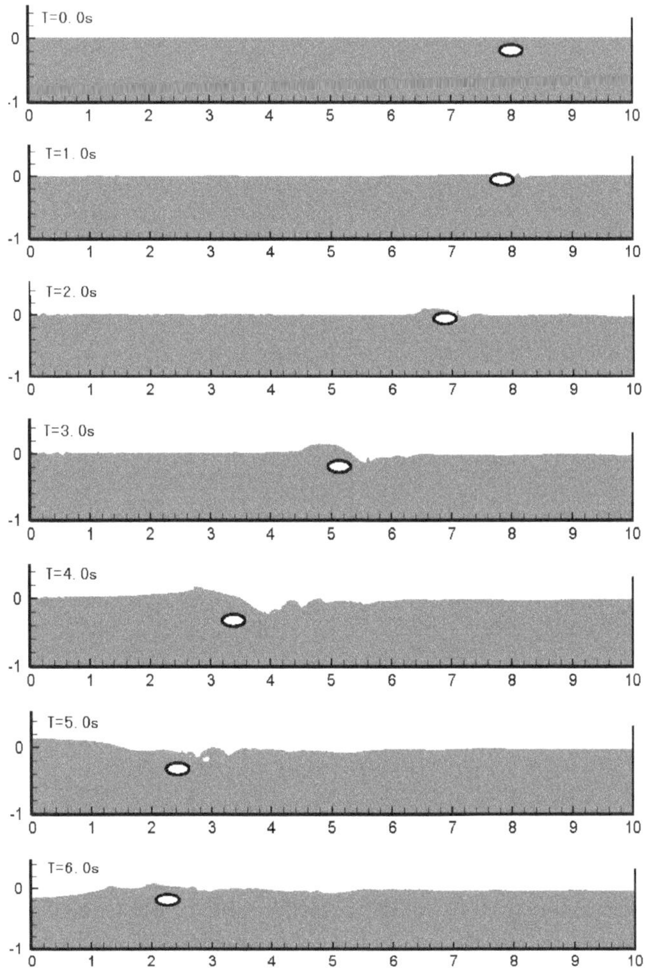

Figure 6.25 Particle distribution of the movement of an elliptical cylinder near free surface at 0.0, 1.0, 2.0, 3.0, 4.0, 5.0 and 6.0 s.

results provided in (Lin, 2007). The cylinder start to move with gradually increasing speed which reaches maximum at $T = 3$ s. During this period, waves are generated in front and behind the cylinder due to the interaction of the body motion and fluid flow. The front water wave can overtop the cylinder, impact onto water behind the cylinder, and produce violent free surface deformation with water splash. The interaction of the elliptical cylinder with the fluid flows induces vortices with obvious vertex shedding. The long period of movement, together with periodical acceleration and deceleration and elliptical shape makes the vortices generation and shedding in a complicated manner. After $T > 3$ s, the cylinder gradually slow down due to the deceleration and the wave front still keeps moving with a flattened crest. The overtopping and impacting of front water onto behind water will further produce more complicated flow pattern with cavity. **Figure 6.26** shows the zoomed in plots of velocity vector at 3.0, 4.0, 5.0 and 6.0 s. It is clear that a larger number of vortices with different strength are produced. **Figure 6.27** shows the quantitative comparisons of the present SPH results (vorticity) with the numerical results provided in (Lin, 2007). In general, the vorticity fields obtained from two different approaches are very close. The SPH vorticity field is not smooth in some areas, and this is due to the immature post-processing techniques in meshfree particle methods.

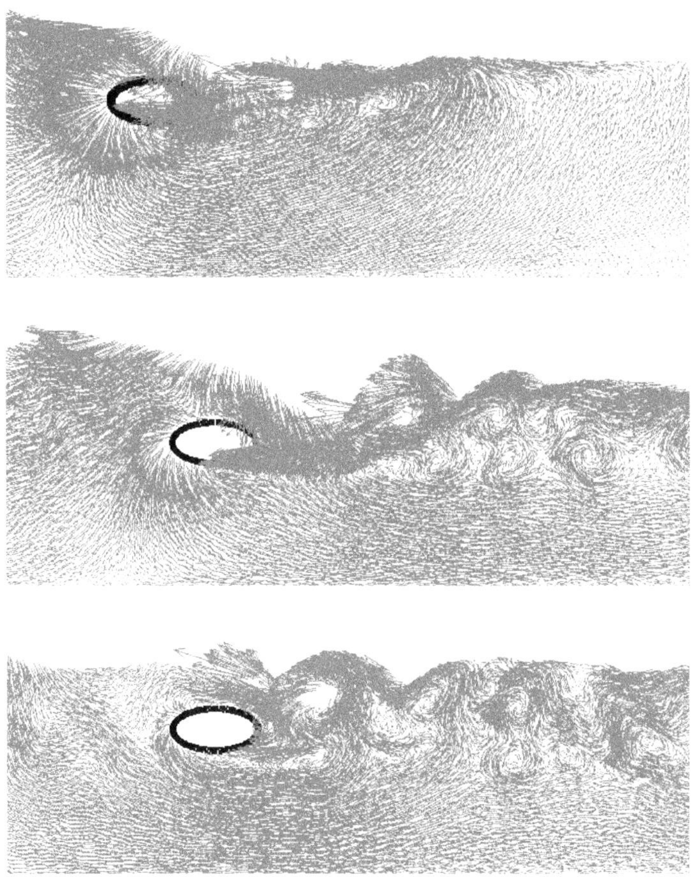

Figure 6.26 Velocity vector at 3.0, 4.0, 5.0 and 6.0 s.

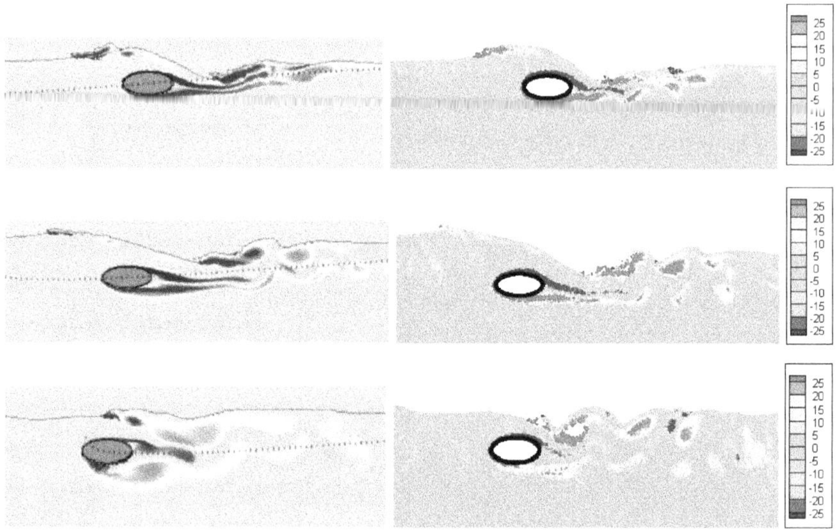

Figure 6.27 Numerical results (vorticity) of the moving elliptical cylinder. Left: results (Lin, 2007), Right: SPH results.

6.5 Liquid sloshing

Sloshing refers to the movement of liquid inside a partially-filled container due to external excitations. When the amplitude of an external excitation is very large or its frequency is close to the natural frequency of the liquid sloshing system, the liquid inside the container can exhibit violent oscillations, and exert strong impact load on the container (Ibrahim, 2005). Liquid sloshing can be frequently observed in daily life and in engineering and sciences, and can be of great importance both in theory and practices. For example, large liquid sloshing in an oil or liquefied natural gas (LNG) ship can result in local breakages and global instability to the ship, and can then further lead to leakage of oil, and capsizing of ship. The movement of water in a reservoir when experiencing an earthquake can produce tremendous impact pressure on the dam of the reservoir. The sloshing of liquefied fuel inside the fuel tank in an aeronautic or astronautic craft can disturb or even breakdown normal navigation of the craft.

Due to the ever-increasing interests on liquid sloshing dynamics, many researchers have conducted theoretical, experimental and numerical simulation works in this area. Theoretical researches are usually valid for simple cases with linear or weakly nonlinear liquid sloshing dynamics. Experimental works are generally expensive and sometimes certain physical phenomena related to liquid sloshing cannot be scaled in a practical experimental setup. Recently more and more researches on liquid sloshing are focused on numerical simulations with the advancement of the computer hardware and computational techniques. A number of researchers have provided comprehensive reviews on the problem of liquid sloshing, and the related numerical simulation methods (Cariou and Casella, 1999; Ibrahim et al., 2001; Faltinsen and Timokha, 2009). Most of the numerical simulations are focused on grid-based methods, such as finite difference method (FDM) (Chen, 2005; Chen and Nokes, 2005), finite element method (FEM) (Wu et al., 1998; Mitra et al., 2008), and boundary element method (BEM) (Faltinsen and Timokha; Faltinsen, 1978; Faltinsen et al., 2000; Faltinsen and Timokha, 2001). As a complex fluid motion, sloshing usually involves changing and breakup of free surfaces, strong turbulence and vortex, and violent fluid-solid interaction. It is therefore difficult for traditional grid-based numerical methods to model liquid sloshing problems. For example, traditional FEM cannot treat large fluid deformation very well, and mesh adjustment or rezoning can be necessary in model liquid sloshing. Traditional FDM requires special algorithms such as volume-of-fluid (VOF) (Veldman et al., 2007; Liu and Lin, 2008) and Level Set (Fang et al., 2007) to track changing free surfaces or moving interfaces when modeling liquid sloshing.

Recent developments in so-called meshfree and particle methods provide alternatives for traditional numerical methods in modeling free surface flows such as liquid sloshing dynamics (Koshizuka et al., 1998; Pan et al., 2008). For example, there are a few literatures addressing the application of SPH method to liquid sloshing dynamics. For example, Iglesias et al. simulated the anti-roll tanks and sloshing type problems (Iglesias et al., 2004). Rhee and Engineer studied liquid tank sloshing with Reynolds-averaged Navier-Stokes (Rhee and Engineer, 2005). Souto-Iglesias et al. assessed the liquid moment amplitude in sloshing type problems with smooth particle hydrodynamics (Souto-Iglesias et al., 2006). Anghileri investigated the fluid-structure interaction of water filled tanks during the impact with the ground (Anghileri et al., 2005). Delorme et al. simulated the sloshing loads in LNG tankers with SPH (Delorme et al., 2005). These works have demonstrated the feasibility of SPH method in modeling liquid sloshing dynamics. It is noted that previous works are generally based on traditional SPH method, which is believed to have poor computational accuracy. Also previous works usually did not incorporate turbulence models into SPH equations of motion. While when modeling turbulence and vortex in liquid sloshing, turbulence models can be very important.

In this section, the SPH method shall be applied to modeling liquid sloshing problems (Shao et al., 2012a). Similarly, the SPH method is also improved with kernel gradient correction (KGC) and density correction to improve the computational accuracy in particle approximation and with Reynolds Averaged Navier-Stokes turbulence model to capture the inherent physics of flow turbulence. Four numerical examples are modeled using this improved SPH method, and the obtained numerical results are compared with experimental observations and results from other sources.

6.5.1 Liquid sloshing under horizontal excitation

In this case, the liquid motion in a rectangular tank under a periodic horizontal excitation is simulated (Shao et al., 2012a). **Figure 6.28** shows the geometry of the liquid sloshing system, which is similar to what Faltinsen et al. provided (Faltinsen et al., 2000), i.e., L = 1.73 m, h_w = 0.6 m, and H = 1.15 m. For liquid sloshing with small amplitudes, there is no water impact on the top of the water tank. Under external excitation, the water tank moves according to $S = A\cos(2\pi t / T)$, where S is the location of the tank, A and T are amplitude and period of the external excitation, and are taken as 0.032 m and 1.5 s. The corresponding Reynold's number ($\mathrm{Re} = V_C L / \upsilon$, where V_C is the characteristic velocity from the movement of the tank, and υ is the kinetic viscosity of water) is approximately 0.23×10^6. A probe is placed on the initial free surface, and is 0.05 m away from the left wall.

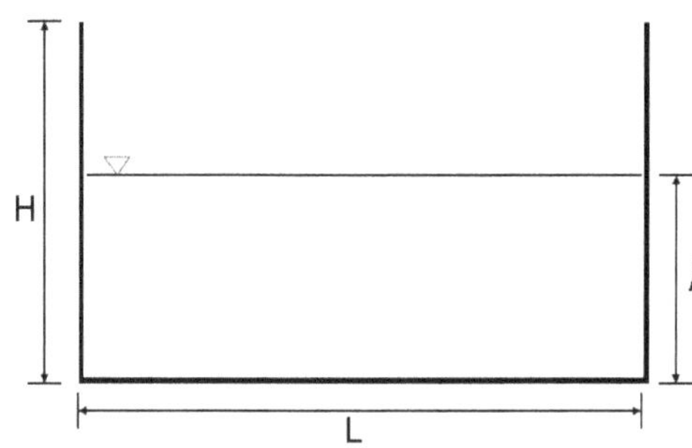

Figure 6.28 Illustration of the liquid sloshing system (Shao et al., 2012a).

In the SPH simulation, the time step is taken as 5×10^{-6} s, and around 40000 particles are used. The reciprocal movement of the water tank drives the contained water to and fro in the tank. **Figure 6.29** shows the flow pattern of sloshing at nine typical time instants within one period. As the water tank begins to move rightward, the contained water also starts to move rightward. Due to the blocking effect of the right wall, water particles will gradually aggregate and tends to run up along the right wall until a maximum water height is reached. After then water particles near the top right corner will then gradually fall down, move leftward, gradually aggregate and tends to run up along the left wall, and finally reach a maximum water height followed by new period.

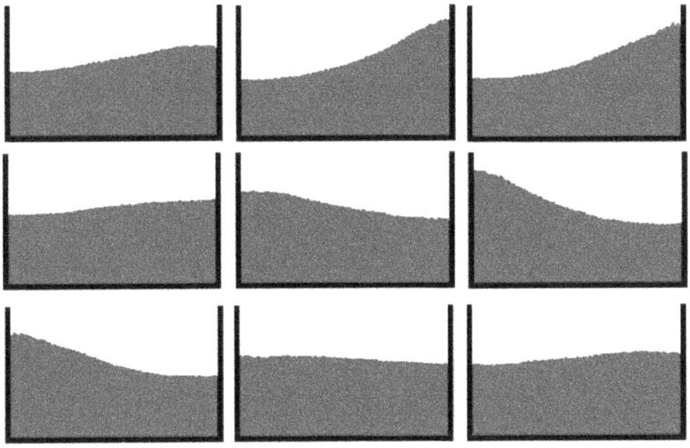

Figure 6.29 Particle distributions at 20, 20.2, 20.4, 20.6, 20.8, 21, 21.2, 21.4 and 21.5 s (Shao et al., 2012a).

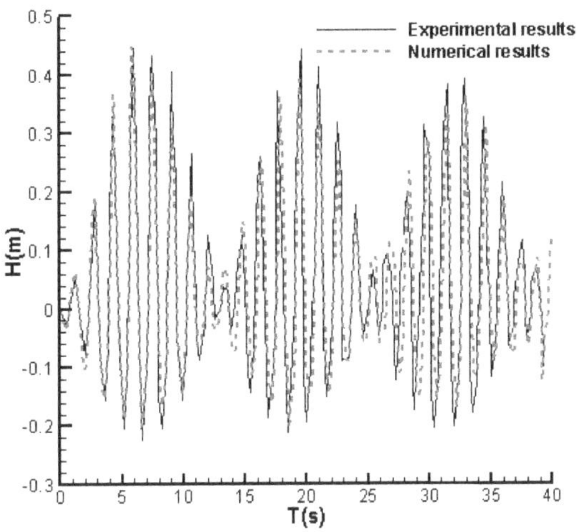

Figure 6.30 Wave heights obtained by SPH simulation and experiment (Shao et al., 2012a).

Figure 6.30 shows the wave heights obtained by SPH simulation and experiment (Faltinsen et al., 2000) at the probe point in a 40 s time course. The SPH results agree well with experimental observations, both in pattern, period and amplitude. There are some discrepancies in wave trough areas, which may be due to the sparser particle distribution when water reaches the lowest point.

Changing initial water height and period of the external excitation can lead to different liquid sloshing phenomena. **Figure 6.31** shows the wave heights for two scenarios with $h_w = 0.6$ m, $T = 1.3$ s, and $h_w = 0.5$ m, $T = 1.875$ s. Again for both scenarios, the obtained SPH results are comparable with experimental observations. Numerical results obtained using traditional SPH model (with simple repulsive boundary treatment, without density and kernel gradient correction) are also provided for these two scenarios. It is seen that the improved SPH model can get more agreeable results with experimental observations.

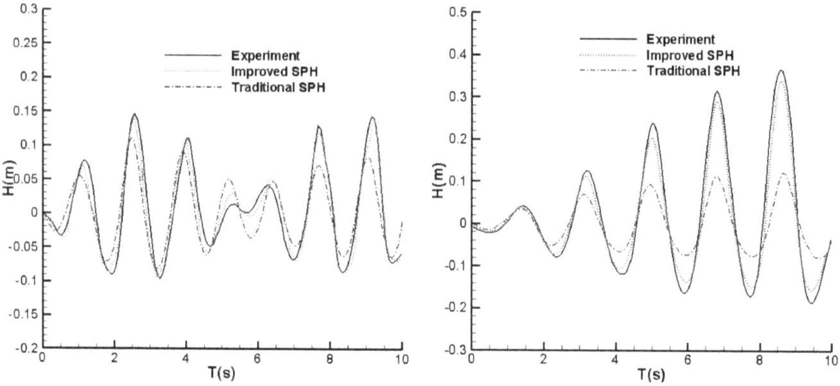

Figure 6.31 Wave heights for two scenarios with $h_w = 0.6$ m, $T = 1.3$ s (a), and $h_w = 0.5$ m, $T = 1.875$ s (b) (Shao et al., 2012a).

6.5.2 Liquid sloshing with a middle baffle

This example involves liquid sloshing in a rectangular tank with a middle baffle. The baffle can change the behavior of liquid sloshing, reduce the wave heights, and therefore help to suppress the impact load of water onto the tank. The geometry and external excitation are the same as the above case (i.e. $h_w = 0.032$ m and $T = 1.5$ s) except for the use of the baffle. In this simulation, the time step is taken as 5×10^{-6} s, the Reynold's number is approximately $Re \approx 0.23 \times 10^6$, and around 24000 particles are used.

Figure 6.32 shows the velocity vector distribution at 6.0, 6.3, 6.6, 6.9, 7.2, and 7.5 s while the height of the middle baffle (d) is 0.3 m. It is observed that a number of vortices have been reciprocally generated and disappeared. There is a large eddy around the top of the baffle, and it evolves quickly with the movement of the water particles. This reveals that the improved SPH method with the RANS turbulence model can effectively describe the inherent turbulence physics. These vortices and eddies complicate the liquid sloshing behavior, dissipate energy, and help to alleviate impact and vibration effects due to liquid sloshing.

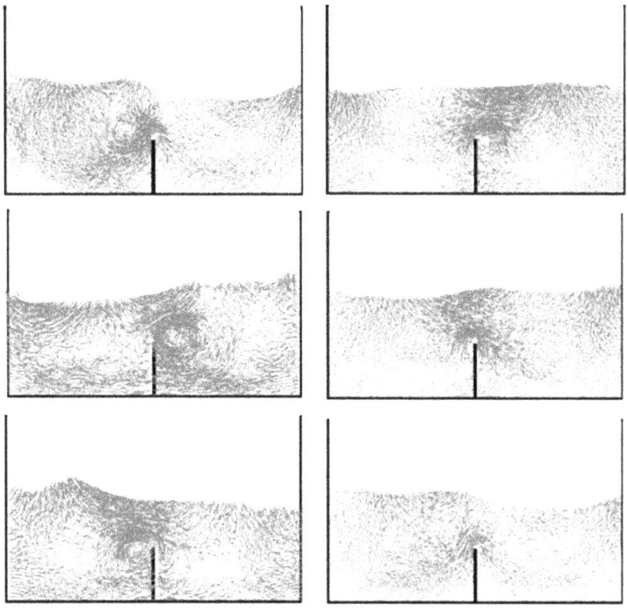

Figure 6.32 Velocity vector distributions at 6.0, 6.3, 6.6, 6.9, 7.2 and 7.5 s while the height of the middle baffle is 0.3 m (Shao et al., 2012a).

To investigate the influence of different heights of middle baffle, a probe is also placed on the initial free surface, and is 0.05 m away from the left wall to track the wave height. **Figure 6.33** shows the wave heights with different height of baffle (a: d = 0.3 m; b, d = 0.4 m) obtained using SPH and VOF simulations. The comparative work was done using commercial software, Fluent, with the VOF model to track free surface, while in the VOF model, the geometric reconstruction scheme is used to calculate the face fluxes (Krill and Chen, 2002). In the VOF simulation, about 28000 mesh cells (about 24000 particles are used in the corresponding SPH model) are used, and the time step 0.001 s. Experimental observations of liquid sloshing in the same rectangular tank without middle baffle are also provided. It is clear that the obtained SPH results agree well with the results obtained using VOF to track free surfaces. Compared with experimental data without middle baffle, it is seen that using a middle baffle greatly reduces the wave height. Also changing the height of the baffle can lead to different behavior of liquid sloshing including the period and amplitude of the wave heights. When d = 0.3 m, the maximum wave height is around 0.1 m. Increasing the baffle height to d = 0.4 m can reduce the maximum wave height to around 0.05 m.

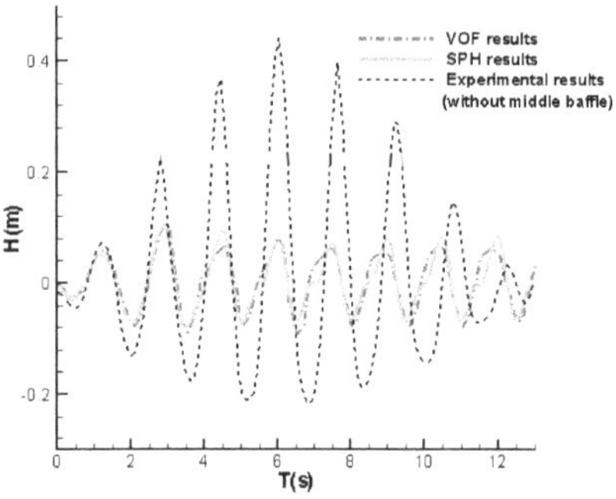

(a)

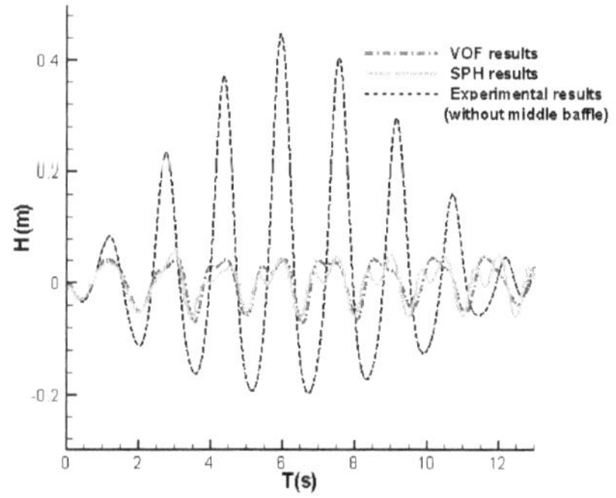

(b)

Figure 6.33 Wave heights with different heights of baffle (a: $d = 0.3$ m; b, $d = 0.4$ m) (Shao et al., 2012a).

It should be noted that though SPH and VOF results are close, there are some discrepancies. For both cases with the middle baffle, SPH results clearly demonstrate double wave troughs in a period, a main wave trough with an additional one. The double wave trough is closely related to the middle baffle. With a middle baffle, water wave in the liquid tank can be influenced, with a number of major or small vertices as shown in **Figure 6.32**. The middle baffle can even lead to second water waves on both sides. The accumulation of the major wave and the second wave produces a water wave with double wave trough. While in VOF simulation, there are no (**Figure 6.33**a) or only ambiguous (**Figure 6.33**b) double wave troughs in a period. This shows that the improved SPH method is more effective in resolving inherent complex flow patterns due to liquid sloshing with separated baffles.

6.5.3 Liquid sloshing due to the pitch motion of a rectangular tank

In this case, the tank is allowed to rotate around the transverse axis, and hence the pitch motion of a rectangular tank is studied. The external excitation can be described as $\theta = \theta_0 \sin(\omega_r t + \xi_0)$, where θ_0 is the angular displacement, ω_r is the circular frequency of the pitch motion, and ξ_0 is the initial phase. **Figure 6.34** shows an illustration of the problem geometry.

We firstly used following parameters to model the liquid sloshing due to the pitch motion of a rectangular tank, i.e. $\theta_0 = 6^o$, $\omega_r = 4.34$ rad/s, $\xi_0 = 0$, $L = 0.64$ m, $H = 0.14$ m. The water depth (h_w) is 0.03 m, and the center of rotation is 0.1 m below the baseline. In this case, about 24000 particles were used, and the Reynold's number is 0.16×10^6 approximately.

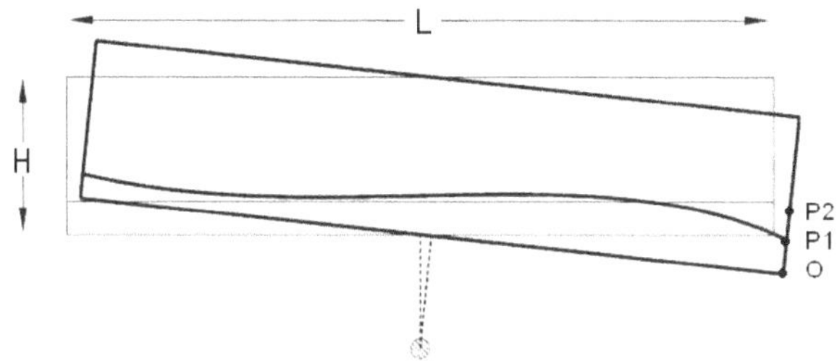

Figure 6.34 Liquid sloshing due to the pitch motion of a rectangular tank (Shao et al., 2012a).

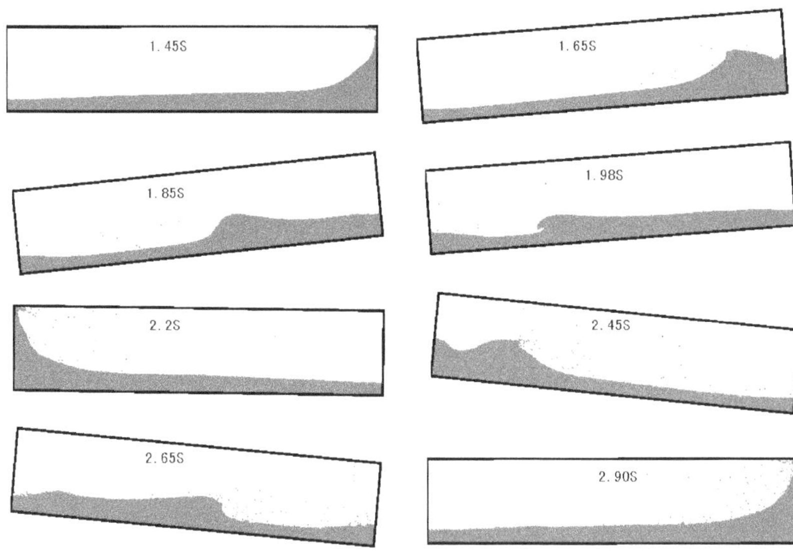

Figure 6.35 Particle distributions at 1.45, 1.65, 1.85, 1.98, 2.2, 2.45, 2.65 and 2.90 s (Shao et al., 2012a).

Figure 6.35 shows the particle distribution due to the pitch motion of a rectangular tank. The improved SPH method can effectively capture the flow dynamics associated with changing and breaking free surfaces. At around 1.45 s, water particles run up along the right wall and then impact onto the right and top side of the tank, with some particles splashed away from the bulky water. After then, water particles fall downwards, and propagate leftwards, with some splashed water particles falling onto the bulky water. With the leftward propagating movement of water, a number of surge fronts appear, depending on the geometry and angular velocity of the water tank as well as water height. Later, water particles aggregate and run up along the left wall and then impact onto the left and top side of the tank, also with some particles splashed away from the bulky water. As such, a half period ends, followed by symmetric flow pattern during the next half period.

We also studied another liquid sloshing problem with the same setup as Akyildiz and Unal (2006), in which H=0.62 m, L=0.92 m, $\theta = \theta_0 \cos(\omega_r t + \xi_0)$, ω_r = 2 rad/s, $\xi_0 = \pi / 2$, the filled depth is 75%, and the model rotates around the tank center. On the right wall, two probes are set to track the pressure values, as shown in **Figure 6.34**, OP1 = 0.06 m, P1P2 = 0.11 m. In this case, about 20000 particles were used, and the Reynold's number is 0.13×10^6 approximately. Two external excitations with different angular displacement ($\theta_0 = 4^o$ or 8°) are simulated.

Figure 6.36 shows the obtained pressure field with $\theta_0 = 8^o$ at 7.2 s (a) when the peak pressure values can be observed at the two probe points P1 and P2, and at 8.8 s (b), when the minimum pressure value can be observed. It is noted that different from traditional SPH models, which usually lead to rough and oscillated pressure distribution, the improved SPH method can produce smooth pressure distribution with clear pressure layers. Even near the solid boundary area, there are no obvious pressure oscillations.

Figure 6.38 shows the pressure values at two probes, P1 and P2, with two different angular displacement, $\theta_0 = 4^o$ and $\theta_0 = 8^o$. It is seen that with the movement of the water tank, the measured pressure values rise and fall periodically. The obtained SPH results agree in general with the results provided by Akyildiz and Erdem (2006), who used a VOF model to track free surfaces. This reveals that the improved SPH method with kernel gradient and density correction and coupled dynamic solid boundary treatment algorithm are effective in resolving pressure field, and in treating solid boundaries.

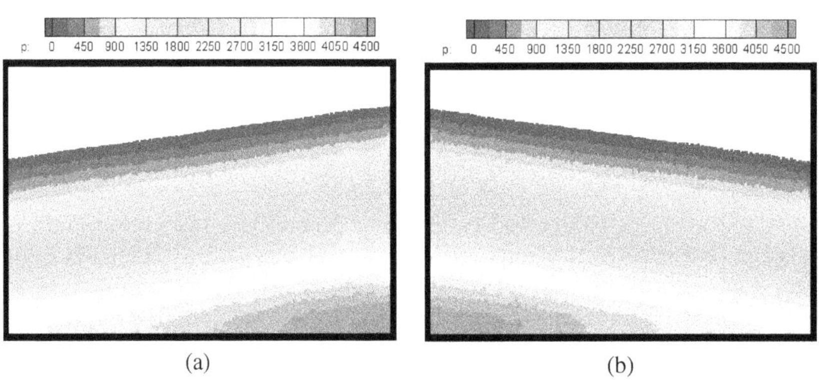

Figure 6.36 Pressure field with $\theta_0 = 8^o$ at 7.2 s (a) and 8.8 s (b) (Shao et al., 2012a).

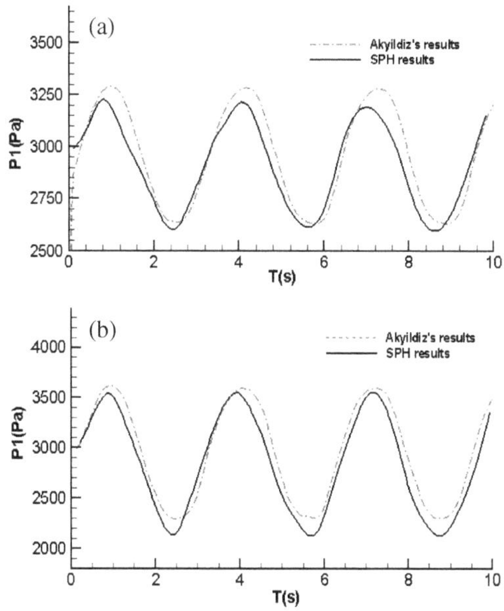

Figure 6.37 Pressure values at probe P1 with different angular displacement (a: $\theta_0 = 4^o$; b: $\theta_0 = 8^o$) (Shao et al., 2012a).

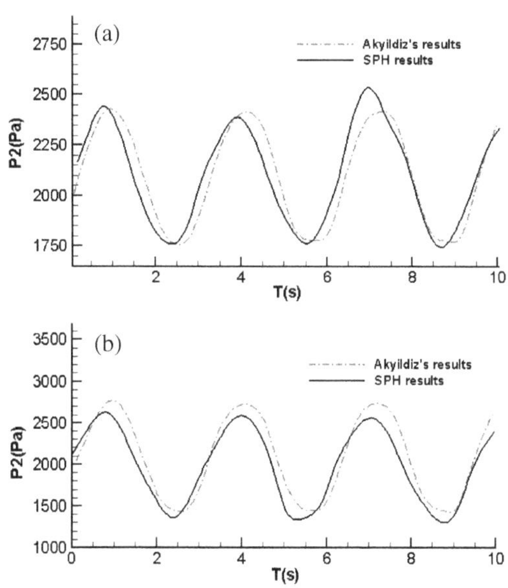

Figure 6.38 Pressure values at probe P2 with different angular displacement (a: $\theta_0 = 4^o$; b: $\theta_0 = 8^o$) (Shao et al., 2012a).

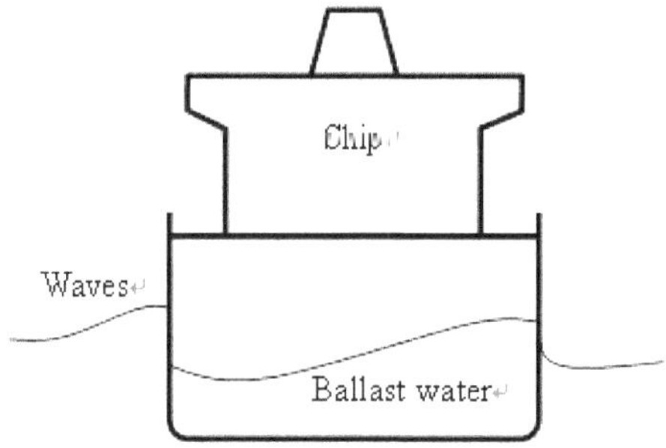

Figure 6.39 Sketch of a ship with ballast water in ocean environment with water waves.

6.5.4 Ballast water

Ballast water has frequently been used in ships to provide stability and adjust trim, stress, and torsion for optimal steering and propulsion (see **Figure 6.39**). The movement of ballast water in a ship is a special form of liquid sloshing. A ship must be a stable platform, whether at rest or moving under the influence of wind, currents, and waves. For a period of long time, people are trying to stabilize ships. It is usually taken into ballast tanks when cargo is being offloaded, and discharged when cargo is being loaded. Ships have carried solid ballast for thousands of years. In modern times, ships use water as ballast. The tanks on ships where in the ballast water is filled are known as ballast tanks or anti-roll tanks. The ballast tanks are located at the lowermost region of a ship. The movement of the free surface of ballast water is a typical phenomenon of sloshing. The rolling movement of a ship can be modeled as a one-degree harmonic movement. The ballast water follows the ship's rolling and acquires a phase lag with respect to the ship movement. If this lag is 90°, the moment generated by the ballast water at its maximum can partially counteract the sea wave moment outside the ship.

A free ship has six degrees of freedom (surge, sway, heave, roll, pitch, yaw, see **Figure 6.40**). Out of the six motions, the critical motion leading to capsize of a ship is the rolling motion. Hence, it is reasonable to investigate the ship motion as a two dimensional problem in *Cyz* plane.

Numerical simulation of the movement of ballast water is very difficult for traditional grid-based numerical models, as it involves not only complex geometries and free surfaces, but also fluid - solid interactions as well as other complex physics in a comparably very large scale. In many circumstances, violent fluid - structure interactions lead to breaking water, air entrapment, and multi-phase flows, where the dynamics of the entrapped air at the impact may play a dominant role during the process and contribute to the high-pressure maxima and pressure oscillations. Though conventional grid-based methods such

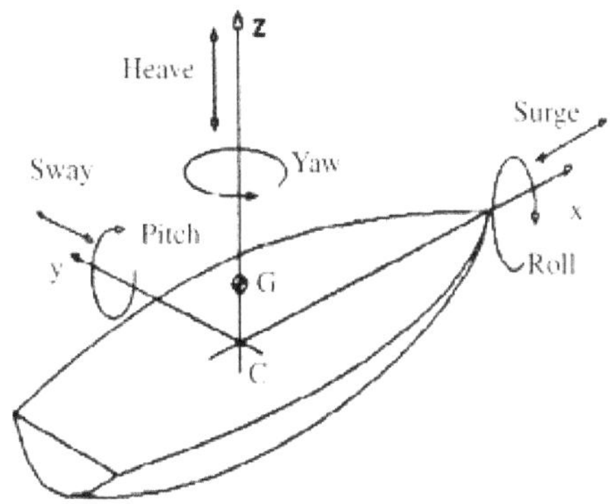

Figure 6.40 Co-ordinate axis and definition of the six degrees of motions (Yang et al., 2012).

as the finite difference method (FDM), the finite volume method (FVM), and the finite element method (FEM) have achieved greatly in simulating fluid flow in ocean hydrodynamics and coastal engineering, there is still a long way to go for practical engineering applications in modeling the movement of ballast water and its interaction with surrounding ship cabins.

Yang et al. investigated the movement of ballast water by using SPH method (Yang et al., 2012). A sketch of a 2D ship carrying ballast water in a wave tank is shown in **Figure 6.41**. A wave maker undergoes simple harmonic

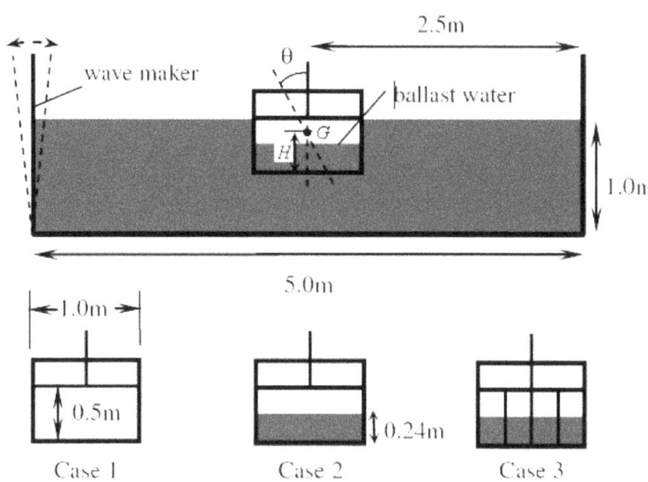

Figure 6.41 Sketch of the computational domain and three cases of ballast water simulations (Case 1: no ballast water; Case 2: carrying ballast water in a single big tank; Case 3: carrying ballast water in four small tanks) (Yang et al., 2012).

motion to produce water waves. The amplitude of the harmonic motion is 5 degree and the period is 1 second. The net weight of the ship is 250 kg, the moment of inertia is 32.55 $kg \cdot m^2$, and the wall thickness is 0.04 m. G is the centre of gravity of the ship without ballast water, H is G's height from the bottom of the ship, and θ is the roll angle. We investigate the stability of the ship by carrying ballast water in three different cases (**Figure 6.41**) and changing the value of H. The initial space is 0.01 m, 47508 fluid particles are used.

Figure 6.42 compares the roll angles between three cases with different H value. For H = 0.25 m, the roll angles approximately satisfy harmonic motion. The angle amplitudes of case 2 and 3 are bigger than that of case 1. This means ballast water increases roll angles, but not too much. For H = 0.375 m, the roll angles of case 2 are much bigger than that of case 1 and 3. For H = 0.5 m, roll angles of case 2 are much bigger than that of case 1 and bigger than 90 degree for $t > 6$ s, that means carrying ballast water in a single big tank makes the ship capsized. Roll angles of case 3 can ignore comparing with case 1, this means carrying ballast water in separated tanks make the ship more stable. For H = 0.625 m, ships in cases 1 and 3 both capsized, but the ship in case 2 is very stable.

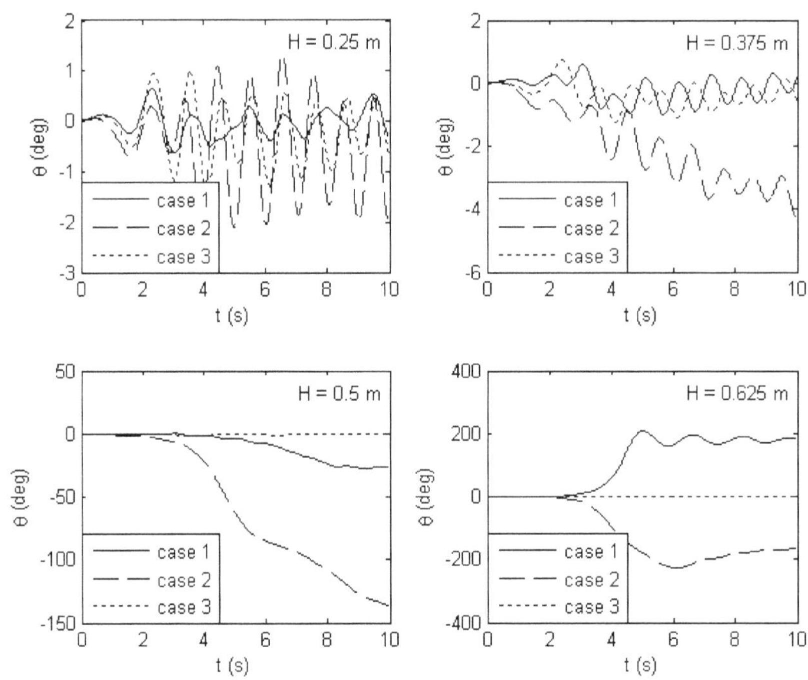

Figure 6.42 Comparisons of roll angle between three cases with different H value (Yang et al., 2012).

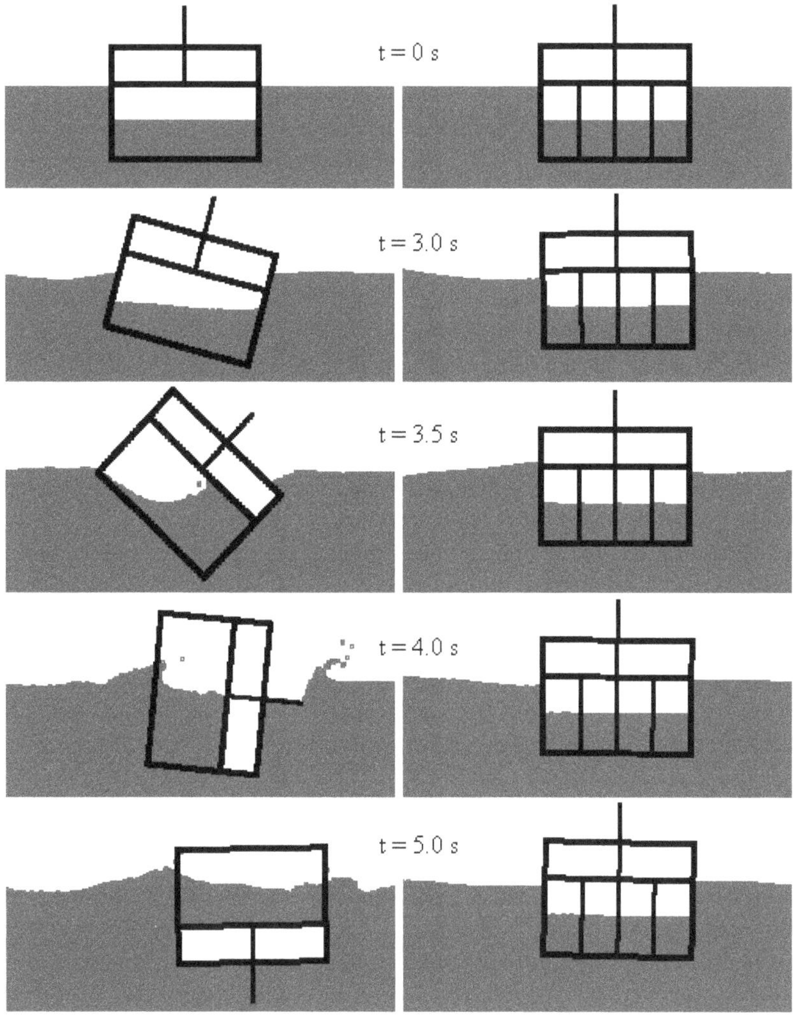

Figure 6.43 Comparison of the configurations of case 2 (left) and case 3 (right) (H = 0.625 m) (Yang et al., 2012).

Figure 6.43 shows that when the ship carrying ballast water in one big tank tilts to the right side, the ballast water tends to shift towards the right side, too. So the center of gravity of the ballast water shifts. Instead of righting the ship, the buoyancy force on the ship turns the ship in the same direction of tilt, and the ship rotates and capsizes. This phenomenon is called free surface effect. The ship carrying ballast water in four tanks doesn't capsize. Although the centre of gravity of individual ballast water shift, but the summation of all the centre of gravities does not shift the centre of the ship that significantly as the ballast water in one big tank.

6.6 Water entry and exit

Water entry and exit can be frequently observed in daily life and engineering and sciences. Typical examples include the entry of re-entry capsule of spacecrafts and submarine, landing of aircraft on water, diving of sportsman and underwater launch of missiles. The phenomenon of water entry is of significant importance both in theory and practice. Numerical simulation of water entry problems is a formidable task and involves rapid movement and breakup of free surfaces, strong turbulence and vortex and violent fluid-solid interaction. Though conventional grid based methods like FDM, FVM and FEM have achieved greatly in computational fluid dynamics, there are still many difficulties in simulating water entry problems.

In this section, the SPH method is applied to simulating water entry and exit. Similarly, the SPH method is also improved with kernel gradient correction (KGC) and density correction to improve the computational accuracy in particle approximation and with Reynolds Averaged Navier-Stokes turbulence model to capture the inherent physics of flow turbulence. Three numerical examples are modeled using this improved SPH method, and the obtained numerical results are compared with experimental observations and results from other sources.

6.6.1 Water exit of a cylinder

In this section, water exit of a cylinder will be numerically simulated using the presented SPH model. Many researchers have conducted similar studies on water exit using different approaches. For example, Greenhow and Moyo provided detailed investigations on water entry and exit of horizontal circular cylinders (Greenhow and Moyo, 1997). Lin also presented a numerical simulation of water exit of a cylinder using a cut-cell technique with a fixed-grid (Lin, 2007). Tyvand and Miloh provided theoretical results for this water exit problem (Tyvand and Miloh, 1995). Their theory works well in modeling the free surface shape before the cylinder exits the bulk water, while it is not valid when the free surface breaks up.

In this work, the problem setup is the same as those in Greenhow and Moyo's, and Lin's work. In the water exit problem, the radius of the horizontal cylinder is chosen to be a = 1.0 m and its center at a distance below the still water surface of d = 1.25 m. The gravitational acceleration is fixed to be g = 1.0 m/s^2 and the impulsive upward velocity is applied to the cylinder at V = 0.39 m/s. Approximately 200,000 particles are used in the SPH simulation with an initial particle spacing of 0.025 m, and a time step $\Delta t = 0.2e\text{-}3$ s. In order to compare the SPH results with existing results (Tyvand and Miloh, 1995; Greenhow and Moyo, 1997; Lin, 2007) at corresponding time instants, the physical time is non-dimensionalized as $T = |Vt/d|$.

Figure 6.44 shows the snapshots of the SPH results (right column) and existing numerical and theoretical results (left column) at T = 0.0, 0.2, 0.4, 0.6, 0.8, 1.0, 2.0 and 3.0. It is clear that SPH results agree well with existing numerical results for the free surface morphology and position of the horizontal cylinder. All numerical results are also close to the theoretical results at earlier stages (before T = 0.4) but difference gradually becomes larger as time marches to 0.6, after which the theoretical results are no longer valid, and the free surface begins to breakup. When time further increases, and the cylinder moves to higher positions, water begins to detach from around the surface of the cylinder, falls down to bulk water and forms water splash with violent free surface deformation (T = 1.0). At around T = 2.0, the cylinder is to exit from the bulk

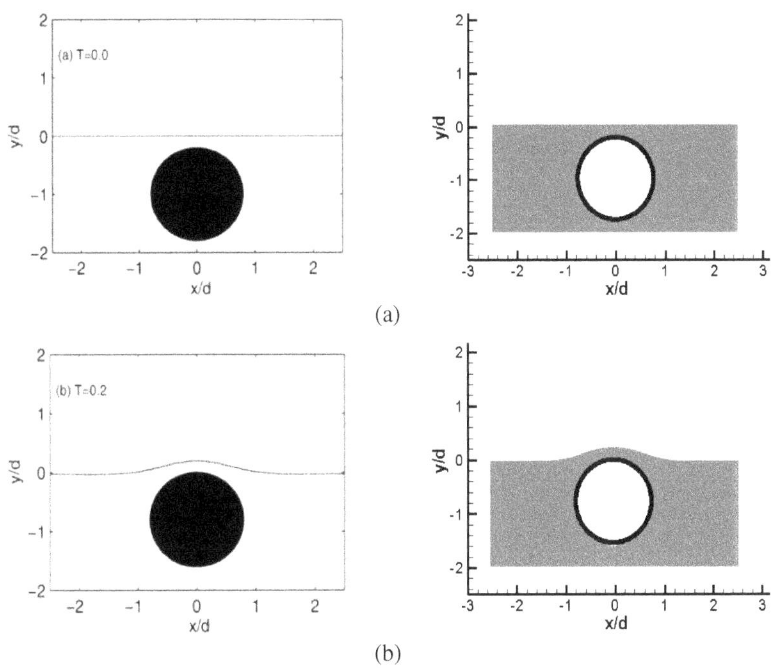

Figure 6.44 Simulation results of the water exit of a horizontal cylinder with comparisons among results from other sources. Left: numerical results by Lin (solid line), numerical results by Greenhow and Moyo (dashed line), and theoretical results by Tyvand and Miloh (dotted line) (Lin, 2007); Right: SPH results.

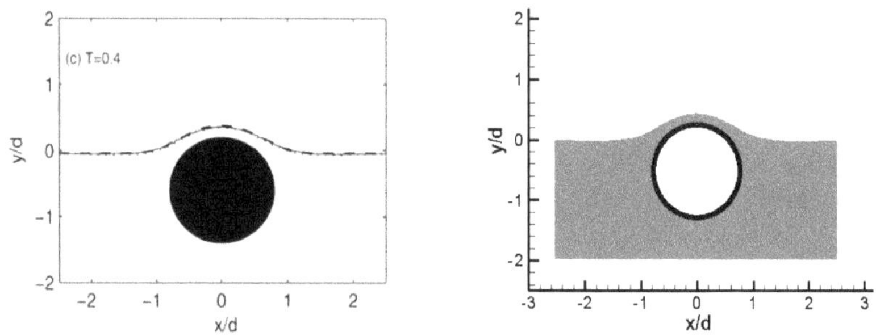

(c)

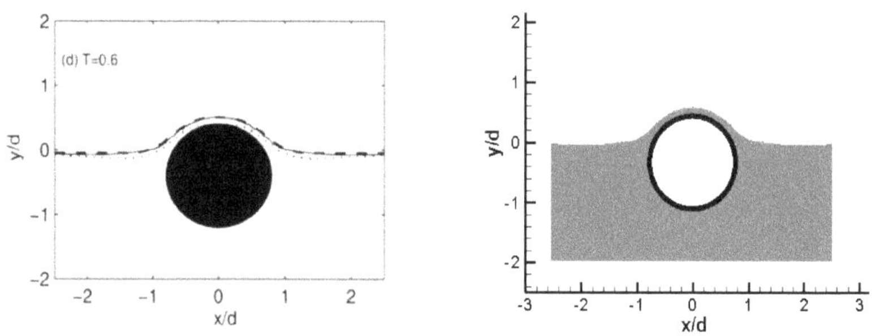

(d)

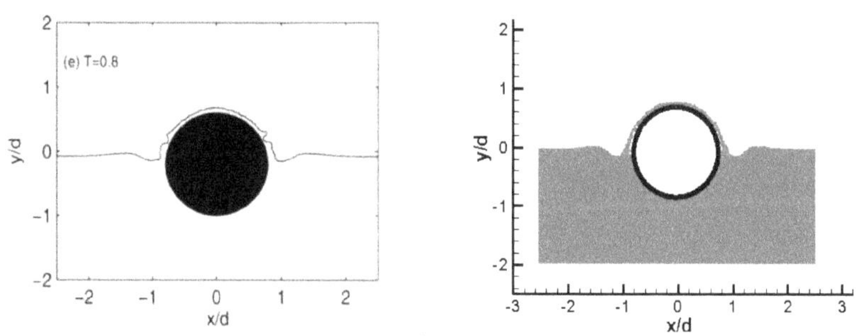

(e)

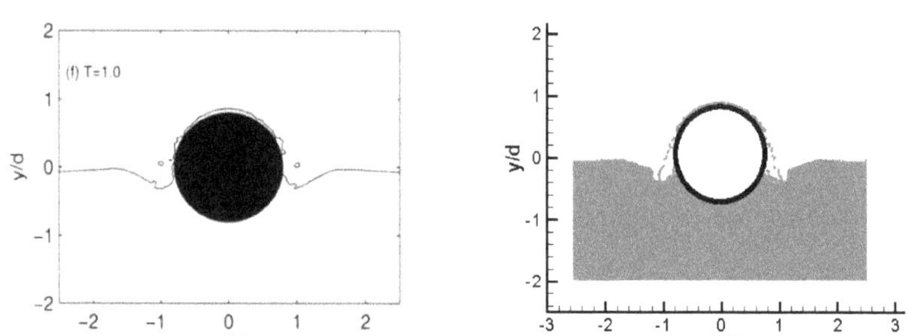

(f)

Figure 6.44 (*Continued*)

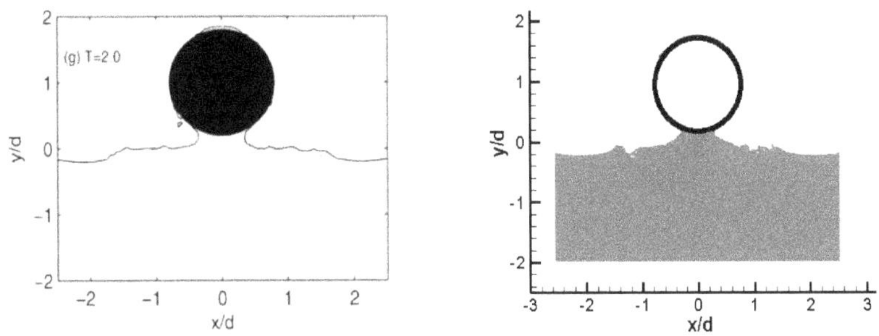

(g)

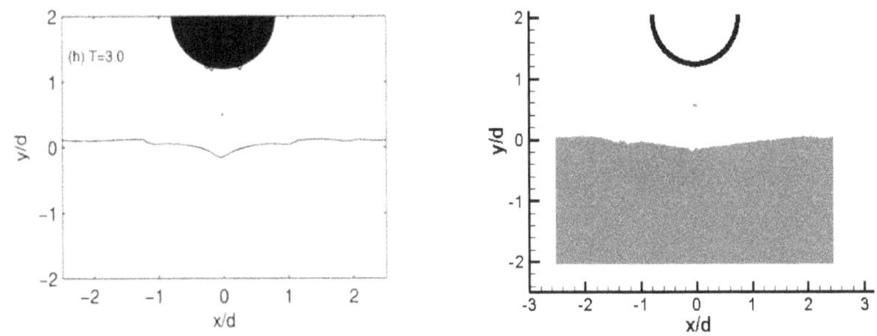

(h)

Figure 6.44 (*Continued*)

water, water moves to the middle space evacuated by the upward moving cylinder, and this leads to the rise of water level in middle region. It is noted that the SPH method can not only well predict the free surface morphology and position of the horizontal cylinder before its exit from water, but also well describe the water detachment, breakup, splash and wave generation during the exit of cylinder from water.

6.6.2 Sinking of a submerged cylinder

The geometry of this example is the same as the above example while the initially submerged cylinder moves in a downward speed of V = 0.39 m/s. **Figure 6.45** shows the snapshots of the SPH results (right column) and existing numerical and theoretical results (left column) at T = 0.0, 0.4, 1.0 and 2.0. Again, the SPH results agree well with existing numerical and theoretical results. With the downward movement of the cylinder, water on the top of the cylinder also falls down and forms an apparent valley (at T = 0.4). As time further increases, water from both sides moves to the center of the valley, forms a region with enhanced water level (at T = 1.0), and generates water waves to both sides (at T = 2.0). It is further demonstrated that the present SPH model can well

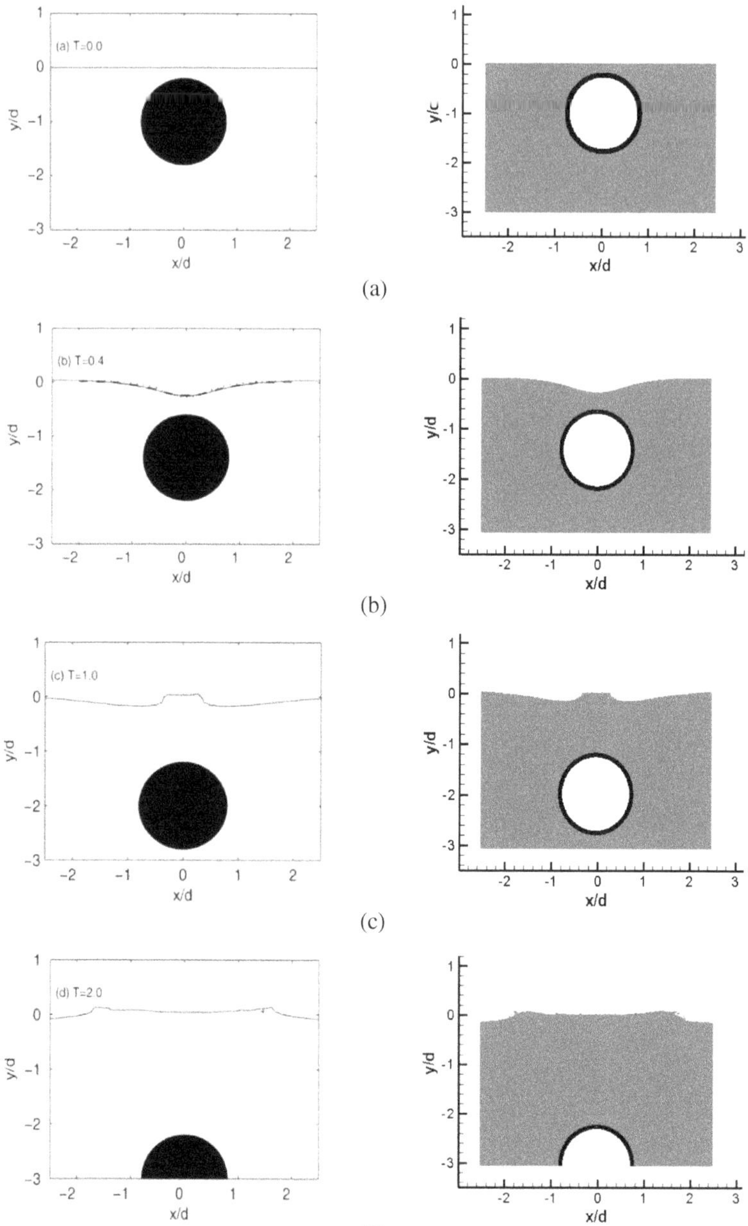

Figure 6.45 Simulation results of the sinking of a submerged cylinder. Left: numerical results by Lin (solid line), numerical results by Greenhow and Moyo (dashed line), and theoretical results by Tyvand and Miloh (dotted line) (Lin, 2007); Right: SPH results.

predict the fluid flow interacting with moving rigid body including free surface deformation with breakup and evolution of the cylinder.

6.6.3 Free falling of a cylinder

This example is similar to the above case except that the cylinder falls freely from a certain height. Free falling cylinder can reach high speed when approaches the water surface, leading to the break up of free surfaces and can even generate cavity. The problem geometry is shown in **Figure 6.46**, the length and height of the water are 200 mm and 50 mm, respectively, and the radius of the cylinder is 5.5 mm, which has the same density as water. The initial downward velocity of the cylinder is 2.955 m/s.

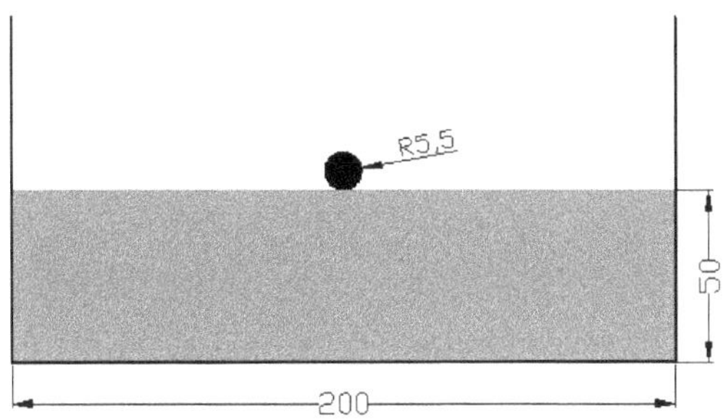

Figure 6.46 Numerical model of the water entry of a cylinder (unit: mm).

Figure 6.47 shows the pressure evolution during the water entry process at 0.006, 0.02 0.03, 0.035, 0.2, 0.26 s. At 0.006 s when the cylinder meets the water surface, a pressure wave produces and then transmits in water. At 0.02 s, after the interaction with the solid wall, the pressure wave changes its direction and forms a reflection wave with a maximum pressure of about 12000 Pa. The whole pressure field is smooth during the pressure wave propagation. The reflection wave meets the falling cylinder at about 0.03 s, and produces a new interaction with the cylinder. With these disturbances, the pressure field is not as smooth as before. However, this effect will gradually disappear as time elapses and the pressure field becomes smooth again.

Figure 6.48 shows the experimental observations (Greenhow and Lin, 1983) and SPH results. It is clear that the obtained SPH results are close to experimental observations, and it shows that the SPH method is valid for problems related to water entry and exit.

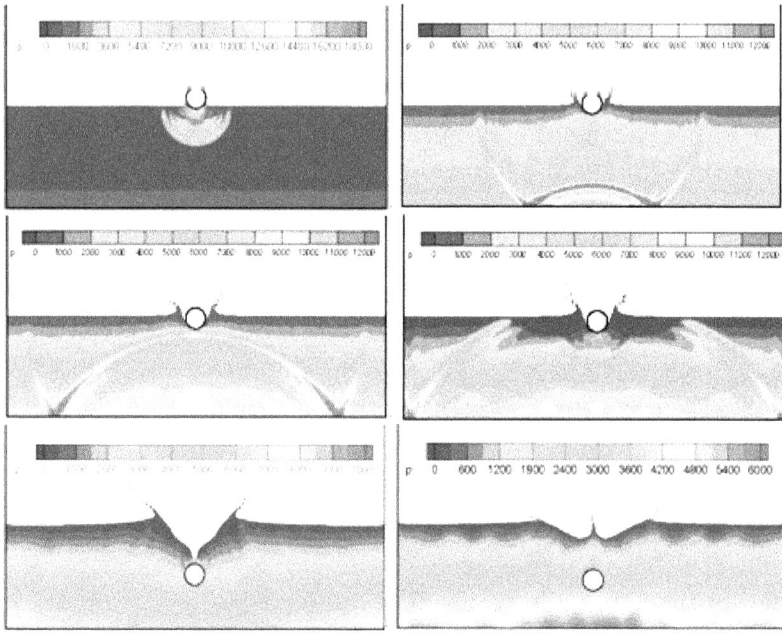

Figure 6.47 Pressure evolution at 0.006, 0.02 0.03, 0.035, 0.2, 0.26 s.

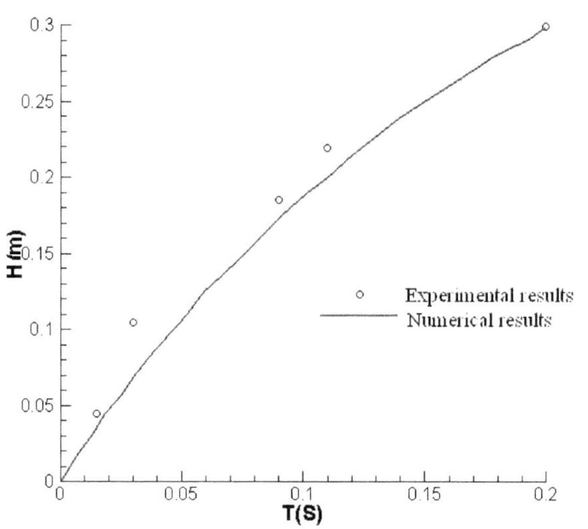

Figure 6.48 Penetration depths obtained from numerical simulation and experimental observation.

6.6.4 Underwater launch

For underwater high speed moving objects, cavitation happens when water moves at an extremely high speed, which causes local water pressure around the moving objects to drop below saturation pressure, creating bubble cavities. Supercavitation occurs if a bubble cavity is large enough to envelop the object and is also strong enough to maintain its integrity. As such, it is possible to greatly reduce the drag of an underwater body, and to enable it moving in a dramatic high speed underwater. This causes the rapidly development of underwater supercavitation weapons and vehicles.

Cavitation induced by high speed moving underwater objects is a complex unsteady and discontinuous or periodic phenomenon with the formation, growth and rapid collapse of bubble cavities, and therefore it is neither reliably assessable nor fully understood yet. Despite the great advances during the last decades, Cavitation and supercavitation are still an ongoing research area, while needs wider and deeper exploration of inherent mechanics.

With the advancement of computer hardware and software, computer modeling with CFD techniques has gradually become a strong tool for understanding cavitations. In this section, the SPH method is used to model the process of underwater launch of a projectile. The object of the research is to examine the shape of the supercavity, and to explore the feasibility of applying SPH to supercavity problems. Different from above examples in which water is incompressible, though a weakly compressible SPH model is used, in this example, as launching speed is quite high, water around the projectile should be treated as compressible. As such, N-S equation with energy equation should be used (Liu et al., 2012b). Also a true equation of state for water such as Gruneisen equation of state should be used as follows

$$p = \frac{\rho_0 C^2 \psi [1 + (1 - \frac{\gamma_0}{2})\psi - \frac{a}{2}\psi^2]}{[1 - (S_1 - 1)\psi - S_2 \frac{\psi^2}{\psi + 1} - S_3 \frac{\psi^3}{(\psi + 1)^2}]^2} + (\gamma_0 + a\psi)e. \tag{6.43}$$

In the case of expansion, the pressure of water is

$$p = \rho_0 C_0^2 \psi + (\gamma_0 + a\psi)e. \tag{6.44}$$

Where ρ_0 is the initial density, η is the ratio of the density after and before disturbance, and $\psi = \eta - 1$. When $\psi > 0$, water is in compressed state, and when $\psi < 0$, water is in expanded state. Some material parameters and coefficients of the Mie-Gruneisen equation of state for water are given in **Table 6.1**.

Table 6.1 Material parameters and coefficients of the Mie-Gruneisen equation of state for water.

Symbol	Meaning	Value
ρ_0	Initial density	1000 Kg/m^3
C_0	Reference sound speed	1480 m/s
γ_0	Gruneisen coefficient	0.5
a	Volume correction coefficient	0
S_1	Fitting coefficient	2.56
S_2	Fitting coefficient	1.986
S_3	Fitting coefficient	1.2268

In the simulation, two cases of water exit of a projectile (or projectile launch) are modeled. One case is associated with projectile launching right from water (**Figure 6.49**a), while the other case is projectile launching from a launch canister (**Figure 6.49**b). The initial launching speed is 100 m/s, which is a subsonic launch.

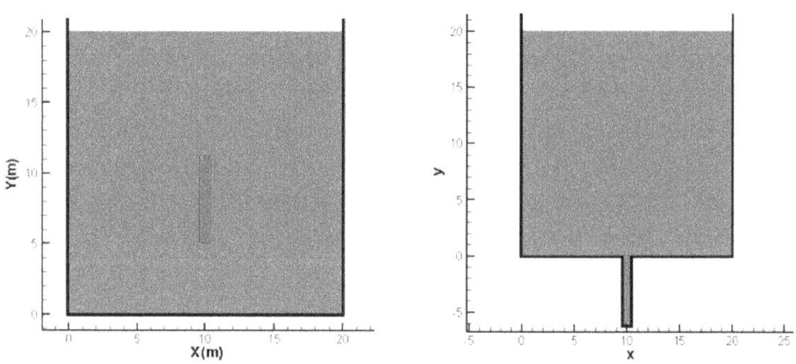

(a) (b)

Figure 6.49 Illustration of problem setup.

Figure 6.50 shows the pressure evolution during projectile launching right from water at 0.0016, 0.004, 0.0064, and 0.009 s. It is clear that right after launching the projectile, a pressure wave is generated, with a cavity formed around the projectile. After then with the cruise of the projectile, pressure wave propagates in water while the bubble cavity gradually grows up. Once the pressure wave reaches the solid wall, it reflects backwards into water region, interplaying with outward propagating pressure wave, and influencing the movement of the projectile.

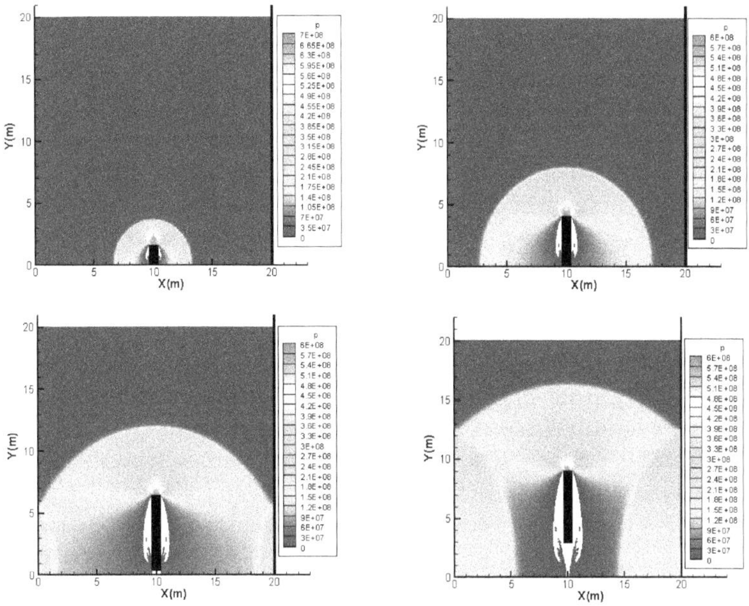

Figure 6.50 Pressure evolution during projectile launching right from water at 0.0016, 0.004, 0.0064, and 0.009 s.

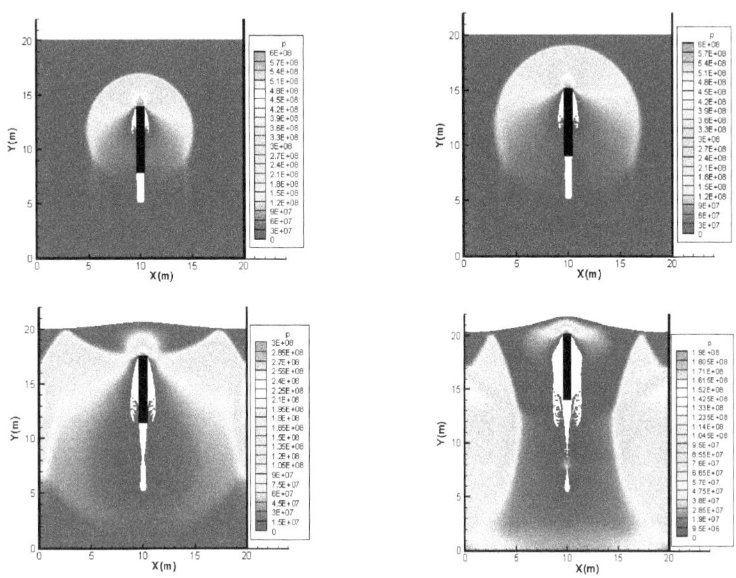

Figure 6.51 Pressure evolution during projectile launching from a launch canister at 0.0028, 0.004, 0.0064 and 0.009 s.

Figure 6.51 shows the pressure evolution during projectile launching from a launch canister at 0.0028, 0.004, 0.0064 and 0.009 s. Similar to projectile launching from water, pressure wave generation, propagation, reflection and interaction can also be observed. It is noted that when the projectile is launched from water, a bubble cavity is generated, and developed bigger to gradually envelop the projectile. In contrast, when the projectile is launched from a canister, there are two bubble cavities, a main bubble cavity enveloping the projectile, and a smaller one around the aft of projection. The circulatory flow is also more obvious.

From the preliminary results, it is clear that the meshfree, particle Lagrangian method is appealing in modeling cavity flows induced from underwater high speed moving objects. For both cases, inherent physics of pressure wave generation, propagation, reflection and interaction as well as bubble cavity formation and evolution can be well described.

It is noted that current setups of projectile are limited to size of the geometry, with which the reflection wave can have significant influence on the dynamics of projectile. For modelling realistic projectile launching, a bigger computational domain or a wave-damping technique is necessary. Also in the presented SPH model, cavitations criterion is not considered. This is acceptable for predicting cavity shape, but is not sufficient in providing interaction of liquid and vapor phases. Future work will also need incorporation of a reliable cavitation model, and need to take account of the collapse of bubble cavity when the projectile exits water.

6.7 Oil spill and boom movement

The world's total oil reserve is around 300 billion tons, among which over 100 billion tons are from offshore oil. During the process of ocean oil extraction and transportation, oil inevitably leaks and spills. According to the statistics of the U.S. National Academy of Sciences, the world annually leaked oil in water is about 170 to 880 million tons (Fingas, 2011). Oil leakage and spill causes pollution of ocean environment, deaths of marine life and other economic losses on marine and coastal areas. Typical examples including oil spill in the Gulf of Mexico in 2010, and in the Bohai Bay in 2011, both leading to severe environment disaster and tremendous economic losses. Therefore, how to confine the spill of oil after its leakage, and prevent oil from spreading to wider areas is a very important task in ocean engineering.

Booms are the most commonly used equipment to concentrate leaked oil and prevent it from spreading. A boom usually has two main basic parts: an upper freeboard to prevent oil form flowing over the top of the boom, and an immerged skirt to prevent oil from being swept underneath the boom (Fingas, 2011). The performance and ability of a boom to contain oil is affected not only by the characteristics of itself, but also by external environmental factors. The boom's characteristic features include the size and design of the freeboard, the height and angle of the skirt, and the momentum of inertia of the boom. External factors include oil type, water currents, water waves, winds, and etc. In some situations, the booms may fail to contain the oil and the oil escapes beneath or over the boom. There are different modes of boom failure, such as entrainment, drainage, critical accumulation, splash-over, submergence, and planing (Goodman et al., 1996; Fingas, 2011).

In order to study the mechanism of oil spill containment and boom failure, experiments were carried out in laboratory water flumes. Brown et al. (Brown et al., 1996) observed oil containment and boom failure in an outdoor flowing water channel and obtained experimental data on boom failure mechanisms. Amini et al. (Amini et al., 2009) experimentally investigated the instability mechanism that can cause the failure of an oil spill barrier. It was reported that the barrier draft and its type are the main factors influencing the velocity in the vicinity of the barrier.

With the rapid development of computer hardware and software as well as numerical methods, numerical simulations of oil spill become gradually popular. However, oil spill in ocean and inland water involves flows with water-oil two-phase flows with free surfaces, and containing oil spill using boom involves strong fluid-structure interaction. Both two-phase and free surface flows, and fluid-solid interaction are important but formidable tasks for numerical simulations as conventional grid-based numerical methods are difficult in simultaneously treating moving and deformable solid objects when tracking free surfaces and fluid interfaces. Currently, existing numerical simulations of oil spill are mainly conducted using commercial CFD software, such as FLUENT and CFX, in which free surfaces are usually treated as a slip walls and the boom does not move (Goodman et al., 1996; Ning and Zhang, 2002; Amini and Schleiss, 2009; Fang et al., 2011). The obtained numerical results are therefore different from practical problems.

Smoothed particle hydrodynamics (SPH), due to its meshfree, Lagrangian particle nature, can be attractive in modeling oil spill and boom movement. Violeau et al. (Violeau et al., 2007) once studied the mechanism of oil leakage by entrainment using conventional SPH and obtained some preliminary results. Yang et al. also applied a multiphase SPH to modeling oil spill and boom movement (Yang and Liu, 2013). In the work, numerical simulations are carried out in a numerical water flume. As shown in **Figure 6.52**, the length of the numerical flume is 18 m, and the water depth is 2.5 m. There is a wave maker (① in **Figure 6.52**) to make wave for investigating wave effects on oil spill containment, and wave-making is implemented by rotating the wave maker at a

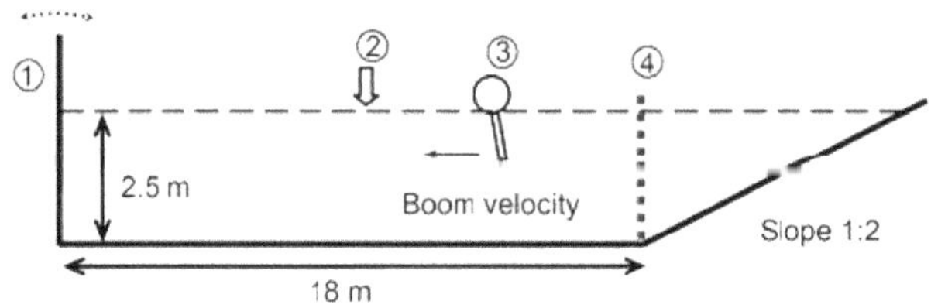

Figure 6.52 Sketch of the numerical flume. ①: wave maker, ②: oil inlet, ③: boom, ④: porous media for absorbing wave energy (Yang and Liu, 2013).

specific angle and frequency. ② in **Figure 6.52** shows an oil inlet while a certain amount of oil leaks at a specific speed for a period of time before the boom arrives at the vicinity of the oil inlet. The boom (③ in **Figure 6.52**) moves leftwards at a constant horizontal velocity U_b controlled by a towing ship, and it can move freely in the vertical direction. The height of skirt is 0.75 m and the boom skirt angle refers to the angle between the boom skirt and the vertical direction. In order to absorb the reflected wave energy from the outlet end of the flume, a layer of porous media was set in front of the right end of the flume (④ in **Figure 6.52**) (Yang and Liu, 2013).

In this section, oil spill containment was numerically simulated under various situations using an in-house SPH code with modified algorithms in improving computational accuracy and enhancing boundary treatment. Oil spill containment is a very complex process involving water-oil two-phase flows and fluid-structure interaction with free surfaces, deformable interfaces and moving structures. It's therefore difficult for numerical simulations and there are very limited reports in simulating oil spill and boom movement. It is even more difficult to give quantitatively agreeable results with experimental observations. As the effectiveness of the in-house SPH code with modified algorithms has been demonstrated in a wide range of applications with free surfaces, moving interfaces and fluid-structure interaction (Shao et al., 2012a; Yang et al., 2012), it is used to investigate the oil spill containment with different influencing factors. In this work, about 20,000 fluid particles are used, and four major factors including oil type, boom velocity, skirt angle, and water waves are considered. Two types of oil, the same as those in (Violeau et al., 2007), are used, and the related properties (density and viscosity) are listed in **Table 6.2**.

Table 6.2 Parameters of oil and water

Type of oil	ρ (kg·m^{-3})	ν (m^{2}s^{-1})
Light oil	850	3.32×10^{-6}
Heavy oil	995	3.00×10^{-2}
Water	1000	1.00×10^{-6}

6.7.1 Effects of oil type

Different types of oil can lead to different flow and spill performances that require different treatments when oil spill is contained with boom. In order to investigate the influence of oil type in oil spill containment, both light oil and heavy oil are studied without considering water wave. In the simulation, the skirt angle is taken as 10° and boom velocity is 0.7 m/s leftwards.

Figure 6.53 shows the flow pattern evolution of oil layer and the velocity field of water flow at different typical instants. When a certain amount of oil leaks from the oil inlet, it gradually accumulates near the leaking area, and spills outwards. When boom moves leftwards, water underneath the boom moves rightwards. Therefore for both light and heavy oil, a vortex is formed around the boom with the vertex center located right behind the boom.

As the density of light oil is smaller than that of water, an oil layer over the water surface is generated. When the boom moves leftwards, at the very beginning, a small portion of oil can escape underneath the boom. With the advancement of the boom, the oil layer becomes longer and thinner, and boom failure (oil escapement) does not happen again (see **Figure 6.53**, left column). In contrast, as the density of heavy oil is close to that of water, it is more likely to form a shorter but thicker oil layer. Hence it is more likely for a boom with the same skirt angle and moving velocity to fail in containing the spilled oil (see **Figure 6.53**, right column). It also can be observed that heavy oil frequently escapes from the bottom of the boom. This boom failure is a mixed form of entrainment and drainage.

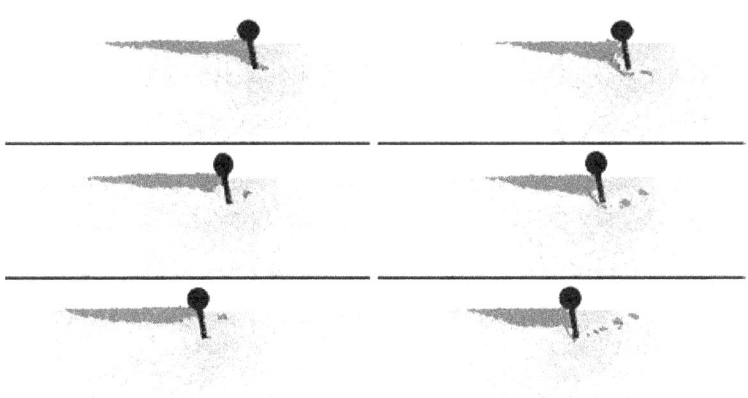

Figure 6.53 Oil spill containment for light oil (left) and heavy oil (right) (skirt angle 10°, boom velocity 0.7 m/s) (Yang and Liu, 2013).

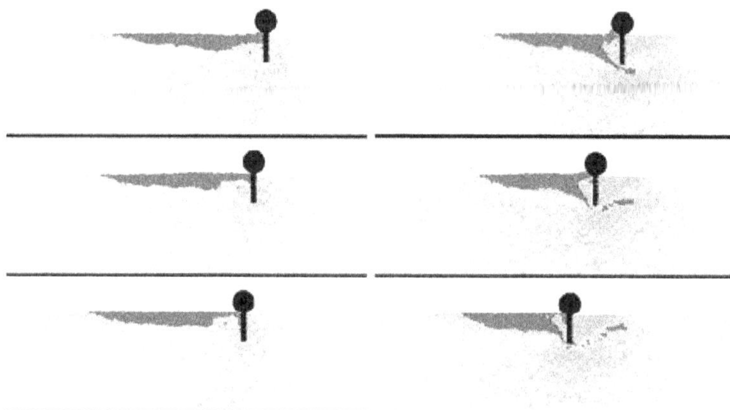

Figure 6.54 Oil spill containment with two boom velocities, 0.3 m/s (left) and 0.7 m/s (right), (heavy oil, skirt angle 0°) (Yang and Liu, 2013).

6.7.2 Effects of boom velocity

In this subsection, the performance of heavy oil spill containment with two different boom velocities, 0.3 m/s and 0.7 m/s, are numerically simulated to investigate the effects of boom velocity while the skirt angle is taken as 0°, and the water wave is not considered. **Figure 6.54** shows the flow pattern evolution of oil layer and the velocity field of water for these two different boom velocities. It is clear that when the boom moves at 0.3 m/s, the spilled oil does not escape and boom failure does not happen (see **Figure 6.54**, left column). Instead, when a boom moves at a higher velocity, it is more likely to form a thicker oil layer in front of the boom and it is therefore more likely for the accumulated oil to escape from the bottom of the boom (see **Figure 6.54**, right column). This is a mixed form of boom failure with drainage and accumulation. It is also observed that a boom with higher velocity can lead to a stronger vortex around the boom.

6.7.3 Effects of skirt angle

For different flow simulations, the skirt angle can be changed to meet the needs of oil spill containment. In this subsection, the effects of skirt angle in containing oil spill are investigated for light oil, with a boom velocity of 0.7 m/s. Water wave effects are not considered. Figure 6 shows the flow pattern evolution of oil layer and the velocity field of water for two skirt angles, –30° and 30°. It is obvious that different deployment of the boom can result in quite different performances in oil spill containment. Different skirt angles lead to different flow fields, especially around the boom. For a negative skirt angle, the flow

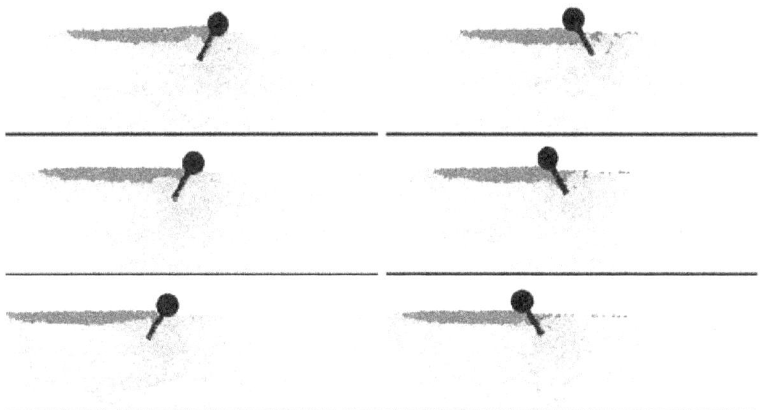

Figure 6.55 Oil spill containment with two skirt angles, –30° (left) and 30° (right), (light oil, boom velocity 0.7 m/s) (Yang and Liu, 2013).

velocity (direction and magnitude) of the water and oil right in front of the boom is close to the velocity of the boom, so it is easier for the oil layer to become longer and thinner. Hence the spilled oil does not escape and boom failure does not happen (see **Figure 6.55**, left column). In contrast, for a positive skirt angle, the flow direction of the water and oil right in front of the boom is different from the direction of the boom movement; it is therefore more likely for oil to escape from the bottom of the boom (see **Figure 6.55**, right column).

6.7.4 Effects of waves

For oil spill in ocean environment, wave effects can be very important. In order to make waves, the rotational angle of the wave maker (① in **Figure 6.52**) in degrees is $\theta = 6\sin(\pi t)$, and the period is 2 seconds. **Figure 6.56** shows the flow pattern evolution of oil layer and the velocity field of water for heavy oil spill containment with and without considering wave effects, while the skirt angle is 0° and the boom velocity is 0.5 m/s. There is no boom failure when water wave is not considered (see **Figure 6.56**, left column). It is revealed that water wave is significant in affecting the flow pattern and velocity field (see **Figure 6.56**, right column). With the advancement of the water wave, the spilled oil layer and boom can move upwards and downwards. If some oil is in a position above the boom, it can escape from the top of the boom (splashover). In contrast, if some oil is in a position below the boom, it can escape from the bottom of the boom (accumulation). More importantly, boom failure periodically happens with the periodical interaction of the water wave and the boom. Therefore, in order to effectively contain oil spill in an environment with water waves, a more suitable design of the boom is necessary.

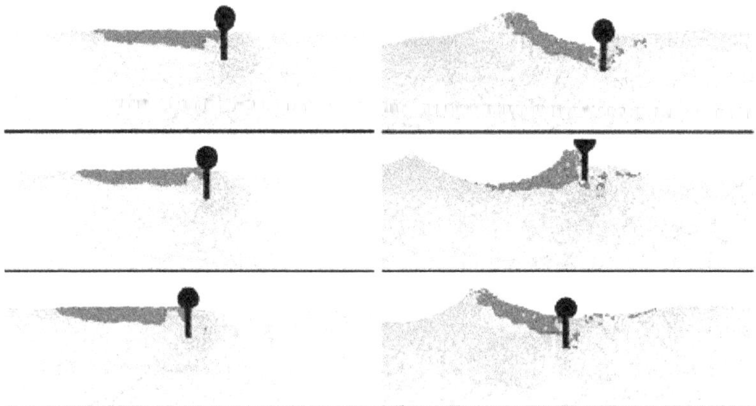

Figure 6.56 Oil spill containment without (left) and with (right) wave effects (heavy oil, skirt angle 0°, boom velocity 0.5 m/s) (Yang and Liu, 2013).

6.8 Hydro-elasticity

Violent free surface flows with strong fluid-structure interactions can be observed widely in hydrodynamics and ocean engineering. They can produce tremendous hydro-pressure load on solid structures and cause the structure to deform elastically or even plastically. These are usually referred to as hydro-elasticity and hydro-plasticity. For example, under extreme weather conditions, the rolling and breaking up of water surface can produce strong slamming effects on hull structures, offshore platforms and nearby buildings, and can further lead to local damages and global instability of structures. The large amplitude liquid sloshing in oil or liquefied natural gas (LNG) ships can result in very high impact pressure on the container, which can damage the hull walls and further lead to the leakage of oil, and even capsizing ships. Therefore, how to effectively model strong fluid-solid interaction with hydro-elasticity is very important for applications in hydrodynamics and ocean engineering.

Recently, the SPH method has been extended to hydro-elastic problems with violent fluid-structure interaction (Liu et al. 2013; Oger et al., 2009). In the SPH simulation, fluid particles are used to model the free surface flows which are governed by Navier-Stokes equations, and solid particles are used to model the movement and deformation of moving solid objects governed by dynamic partial different equations with suitable constitutive equations. The interaction of the neighboring fluid and solid particles renders the fluid-solid interaction, and the non-slip solid boundary conditions. Some typical examples are provided here.

6.8.1 Head-on collision of two rubber rings

In order to validate its effectiveness in dealing with movement and deformation of elastic solid objects, the head-on collision of two rubber rings is modeled firstly. **Figure 6.57** shows the illustration of the head-on collision of two rubber rings. The two rubber rings are from the same material with same geometric dimensions and physical properties. The head-on impacting speed is 50 m/s, and the relative velocity is 100m/s. The inner and outer radius of the rubber rings are 0.03 m and 0.04 m respectively, and the initial distance between the centers is 0.09 m. The density of the rubber is 1010 Kg/m^3, the shear modulus is 1.6×10^8 Pa, and the initial sound speed is 852 m/s. About 18000 particles are used with an initial particle spacing of 0.0005 m. The time step is taken as 0.5e-7 s, and the coefficient of the artificial stress is taken as $e = 0.3$, $q = 4$.

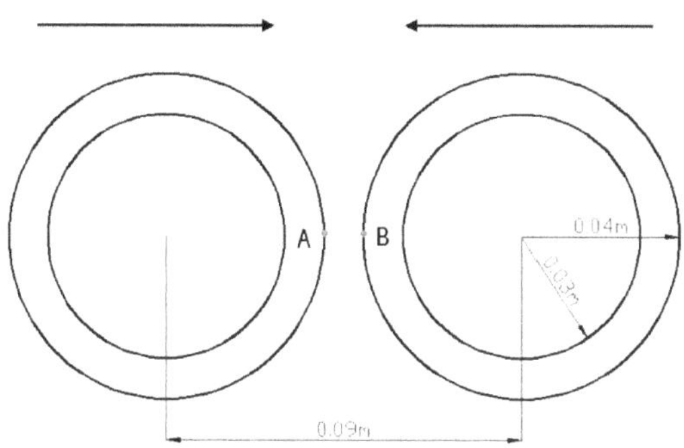

Figure 6.57 Illustration of the head-on collision of two rubber rings (Liu et al. 2013).

Figure 6.58 shows the simulation snapshots of the head-on collision of two rubber rings using the improved SPH method. As shown in **Figure 6.58**, as two rubber rings approaching and impacting onto each other, large deformation occurs, and the initial circular ring in the interface area is quickly flattened (**Figure 6.58**b). As the stress wave propagates in the two rubber rings, the initial circular rings are elongated vertically and turn to be elliptic (**Figure 6.58**c). Due to the elastic nature of the solid objects, the two rubber rings tend to bounce back, gradually restore their original shape (**Figure 6.58**d and e) and are further elongated horizontally (**Figure 6.58**f). The elongation in vertical and horizontal direction lasts several periods with gradually decayed amplitude, while the two rubber rings finally restore their original circular shape.

To further investigate the head-on collision process, the positions of particle *A* and *B* (see **Figure 6.57**) are tracked. **Figure 6.59** shows the horizontal and vertical displacement of particle *A* and *B*. It is clear that during the head-on collision process, particle *A* and *B* stay on the horizontal line (**Figure 6.59**b).

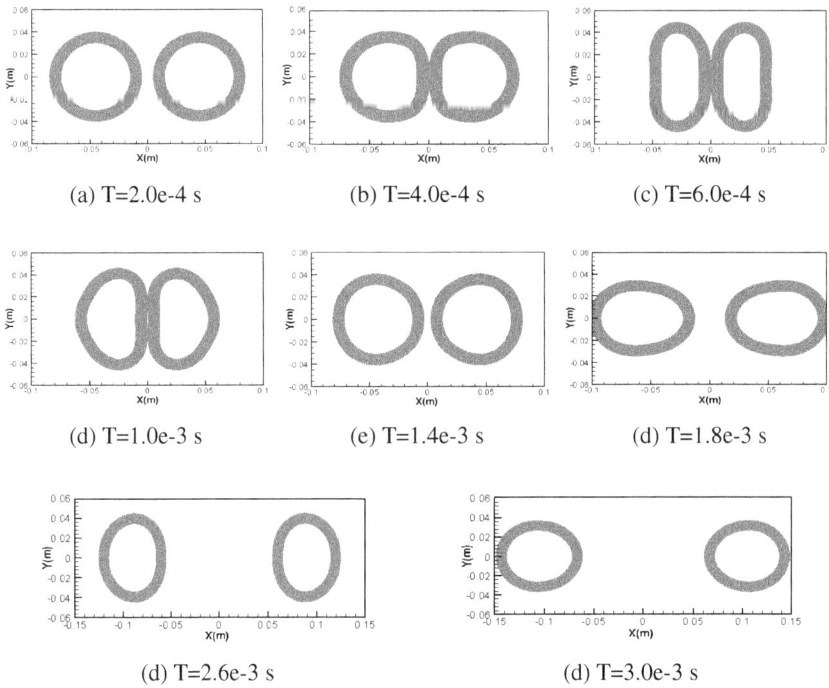

(a) T=2.0e-4 s (b) T=4.0e-4 s (c) T=6.0e-4 s

(d) T=1.0e-3 s (e) T=1.4e-3 s (d) T=1.8e-3 s

(d) T=2.6e-3 s (d) T=3.0e-3 s

Figure 6.58 Simulation snapshots of the head-on collision of two rubber rings (Liu et al. 2013).

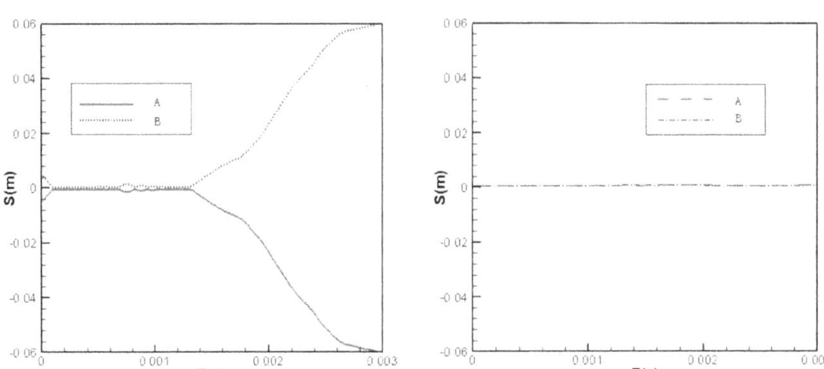

Figure 6.59 Horizontal (left) and vertical (right) displacements of particle A and B (Liu et al. 2013).

Right after the sudden collision, particle *A* and *B* stay on the same position (with small oscillations) for a long period of time to accumulate deformation. Later as the two rubber rings tend to bounce back and separate from each other, particle *A* moves along the negative direction and particle *B* moves along the positive direction. The obtained horizontal displacements of particle *A* and *B* are therefore anti-symmetric.

6.8.2 Dam break with an elastic gate

In this case, a dam break is modeled, in which an elastic gate is placed at the exit (see **Figure 6.60**). The water pressure from dam break can cause movement and deformation of the elastic plate. The improved SPH method is used to simulate this typical hydro-elastic problem. As shown in **Figure 6.60**, the top of the gate is clamped and the bottom is free. The height and width of the water are 0.14m and 0.1 m separately. The elastic gate is 0.079 m high and 0.005 m thick. The density and compressibility modulus of water are 1000 Kg/m^3 and 2×10^6 N/m^2 respectively. The density, bulk and shear modulus of the elastic gate are 1100 Kg/m^3, 2×10^7 and 4.27×10^6 N/m^2 respectively. In the simulation, the time step is 2.5×10^{-6} s, the coefficient of the artificial stress is taken as $e = 0.3$, $q = 4$, and around 22000 particles are used.

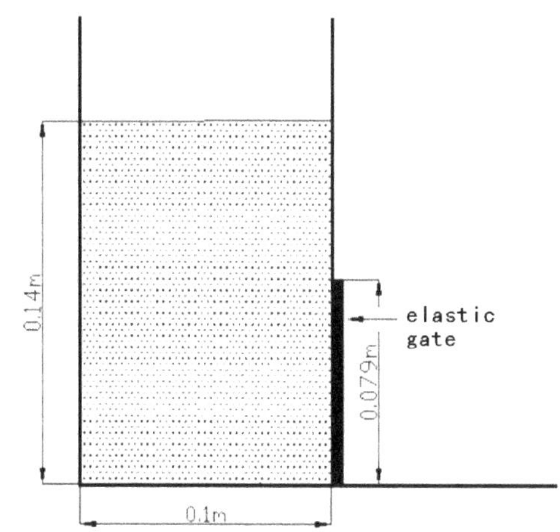

Figure 6.60 Illustration of dam-break with an elastic gate at the exit (Liu et al. 2013).

Figure 6.61 shows the SPH simulation snapshots and the corresponding experimental observations (Antoci et al., 2007). It is shown that before dam break both water and the elastic gate are still. After the sudden removal of the clamp, under static water pressure, the elastic gate will deform and gradually open. The contained water is discharged from the gate and this leads to bigger displacement of the elastic gate with bigger water currents out of the exit. **Figure 6.62** shows the horizontal and vertical displacements of the free end of the elastic gate with experimental observations. It is noted that as the movement of elastic gate, more water flow out of the exit with bigger pressure load on the elastic gate, which further increase the movement and deformation of the elastic gate until maximal displacements are reached. Later as the water height reduces, the pressure load on the elastic gate also reduces, leading to smaller displacements. As such, in **Figure 6.62**, there are peak values both in the horizontal and vertical displacements. **Figure 6.63** shows the comparisons of water level from SPH simulation and experimental observation (Antoci et al., 2007).

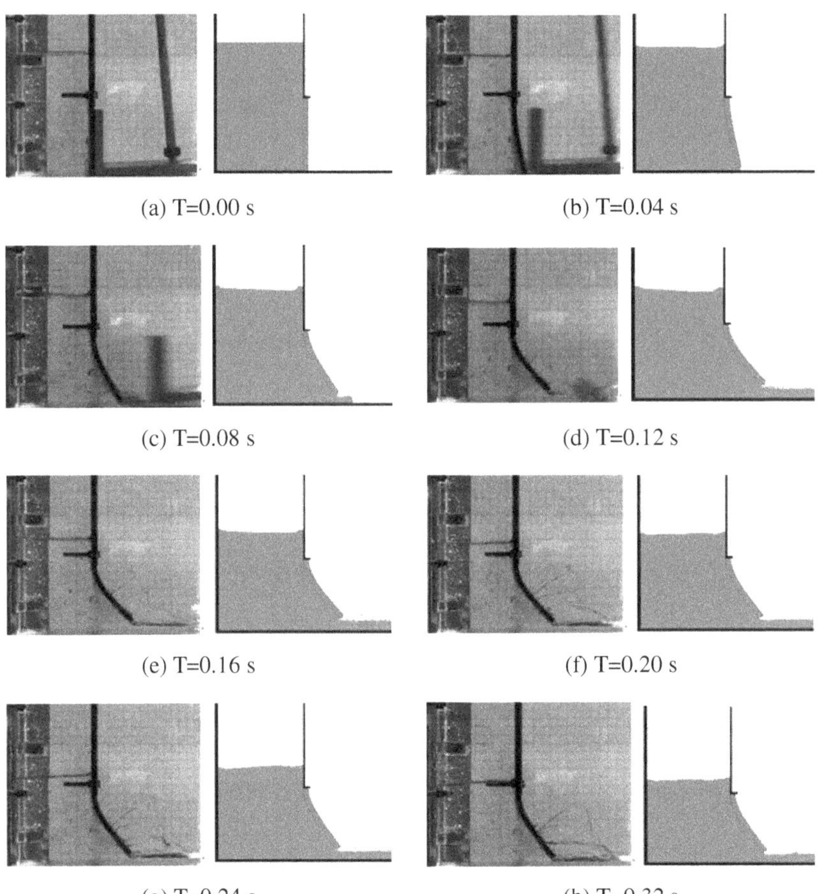

Figure 6.61 SPH simulation snapshots and experimental observations (Liu et al. 2013).

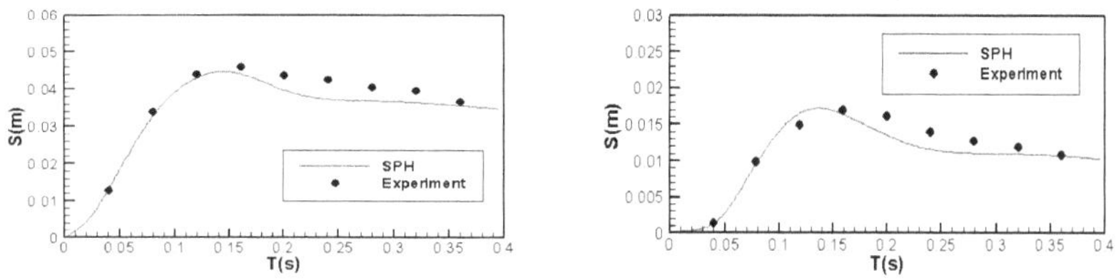

Figure 6.62 Horizontal (left) and vertical (right) displacements of the free end of the elastic gate (Liu et al. 2013).

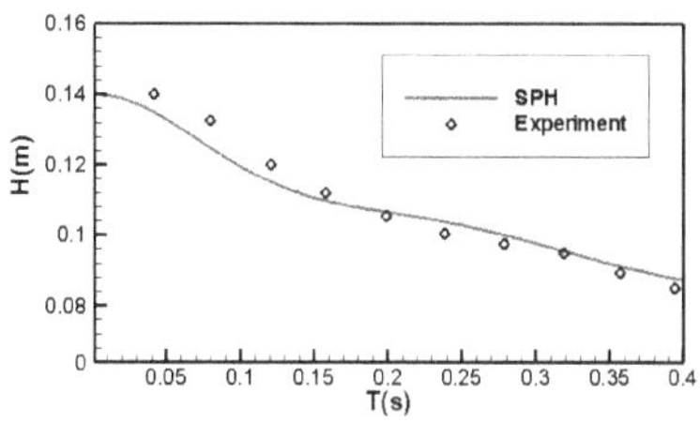

Figure 6.63 Time history of water level (Liu et al. 2013).

6.8.3 Water impact onto a forefront elastic plate

This example involves water flow from a dam break impacting onto a forefront elastic plate. Idelsohn et al. had modeled the same problem using the Particle Finite Element method (PFEM) (Idelsohn et al., 2008). **Figure 6.64** is an illustration of the problem setup. The height and width of the water are 0.292 m, 0.146 m separately, and the elastic gate is 0.08 m high and 0.012 m thick. The density of the elastic gate is 2500 Kg/m^3 with a bulk modulus of 0.33×10^6 N/m^2 and a shear modulus of 0.5×10^6 N/m^2. In the SPH simulation, the time step is 5.0×10^{-6} s, and about 15000 particles are used.

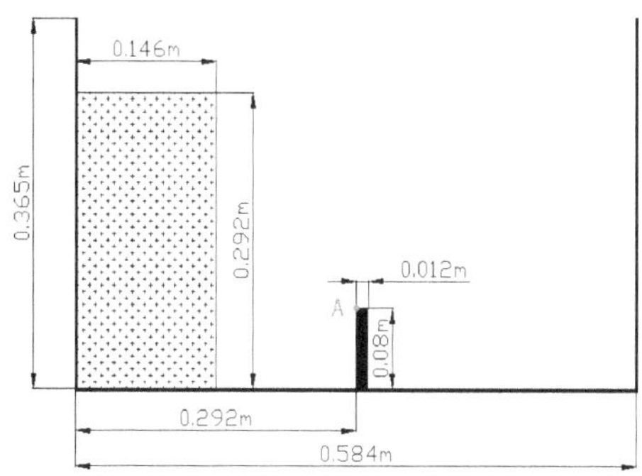

Figure 6.64 Illustration of the water impact onto a forefront elastic plate (Liu et al. 2013).

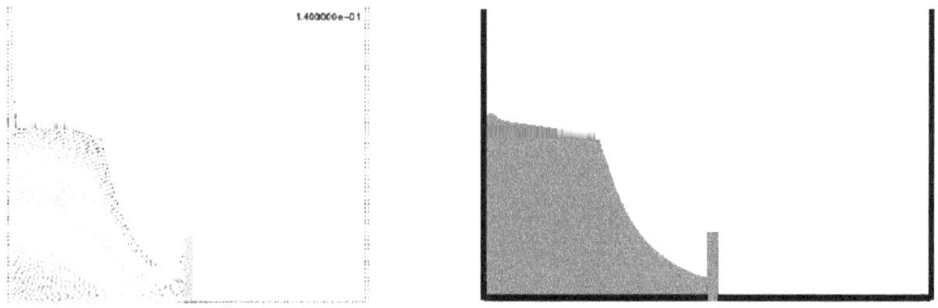

(a) T=0.14 s

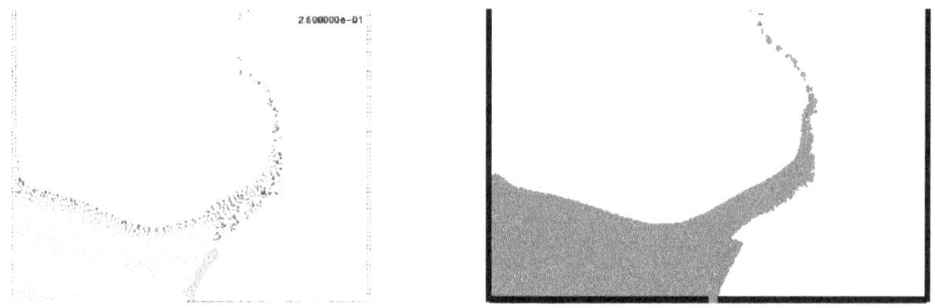

(b) T=0.26 s

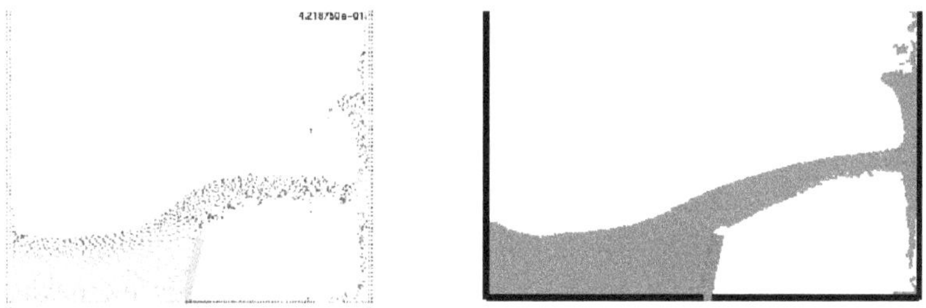

(c) T=0.42 s

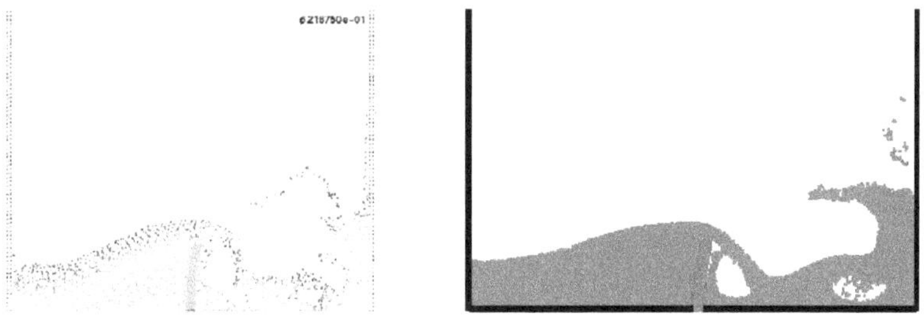

(d) T=0.62 s

Figure 6.65 Simulation snapshots from PFEM (left) and SPH (right) at different instants (Liu et al. 2013).

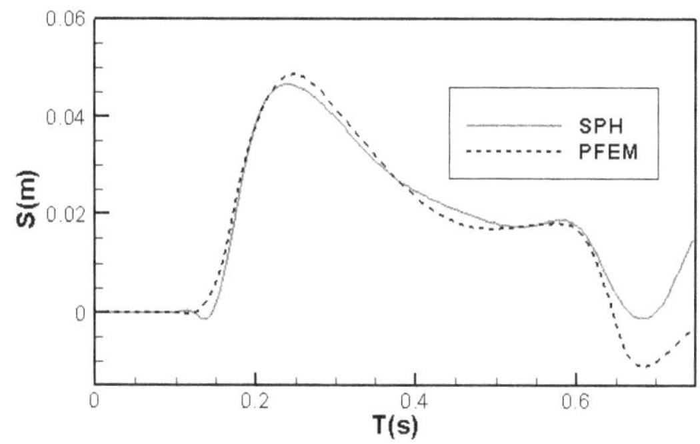

Figure 6.66 Horizontal displacement of the free end of the elastic plate (Liu et al. 2013).

Figure 6.65 shows the simulation snapshots from PFEM (left) and SPH (right) at different instants. **Figure 6.66** shows the time history of horizontal displacement of the free end of the elastic plate. From **Figure 6.65** and **Figure 6.66**, at about 0.14 s, the dam break flow impacts the elastic plate, causing its movement and deformation. At about 0.26 s, the displacement of the elastic plate reaches a maximum value. Later as the water level gradually reduces, the pressure impact on the elastic plate also reduces, leading to smaller displacements. At even later stages, as water particles bounce back from the right solid wall, and move leftwards in the container, the elastic plate can demonstrate negative displacement (move leftwards). It is clear that for both the flow pattern and the displacement of the elastic plate, the obtained numerical results from the present SPH method in general agree well with those from PFEM (Idelsohn et al., 2008). At very later stages ($T > 0.6$ s), there are some discrepancies, basically due to the complex turbulence and cavity effects.

6.9 Concluding remarks

This chapter addresses the application of SPH method to modeling incompressible flows that usually involve violent free surface deformation, strong turbulence and fluid-structure interaction. These fluid phenomena can be frequently observable and important in hydrodynamics and ocean engineering.

There are basically two approaches, weakly compressible SPH (WCSPH) and incompressible SPH (ISPH), to model incompressible fluid flows. In these two approaches, the governing equations are the same and the only difference is the calculation of pressure. In WCSPH, the pressure is obtained from an artificial equation of state. In contrast, in ISPH, the pressure is obtained from solving a Poisson pressure equation using some kind of prediction correction algorithm.

Early WCSPH methods generally produce pressure oscillations and can be inferior to ISPH which generally lead to smooth pressure field. However, with some advanced treatments such as the use of higher order approximation schemes (e.g., density and kernel correction) and artificial viscosity, it is possible to get accurate results with smooth pressure field.

This chapter also provides a comparative study of the ISPH method and an improved weakly compressible SPH method in modeling incompressible free surface flows. Both methods solve the Navier-Stokes equations in Lagrangian form and no artificial viscosity is used. The ISPH algorithm presented is based on the classical SPH projection method with some common treatments on solid boundaries and free surfaces. The WCSPH model only includes density correction. For three typical numerical examples, including a benchmark problem with hydrostatic pressure, a dam breaking problem with large impact pressure, and a liquid sloshing problem with periodically changing hydrodynamic pressure, it is found that WCSPH with density correction is more accurate than ISPH in spatial and temporal pressure distributions, while ISPH is more sensitive to numerical oscillations and is easier to cause instability. This implies that WCSPH with improved treatments may be more suitable for problems with violent water impact and fluid-structure interactions.

The WCSPH model with advanced algorithms (e.g., density and kernel correction, and RANS turbulence model) is then applied to several typical classes of incompressible fluid flows including free surface flows such as the dam breaking problem against a vertical wall and against a sharp edged obstacle. The SPH method is also applied to model free surface flows interacting with rigid bodies. Typical examples include liquid sloshing, water entry and exit, oil spill and boom movement. In the simulation, fluid particles are used to model the free surface flows which are governed by Navier-Stokes equations, and solid particles are used to model the movement (translation and rotation) of moving rigid objects. The interaction of the neighboring fluid and solid particles renders the fluid-solid interaction, and the non-slip solid boundary conditions. The force and momentum on the rigid body can be accumulated from the force and momentum on each solid particles interacting with fluid particles, and then drive the motion of the rigid body.

Finally the SPH method is applied to hydro-elastic problems in which structures may experience tremendous hydro-pressure load causing elastic deformation. In the simulation, fluid particles are used to model the fluid flows which are governed by Navier-Stokes equations, and solid particles are used to model the movement and deformation of moving solid objects governed by dynamic partial different equations with suitable constitutive equations. Numerical examples of free surface flows, fluid-structure interaction and hydro-elastic problems demonstrate the effectiveness of SPH in modeling incompressible fluid flows in hydrodynamics and ocean engineering.

References

Akyildiz, H. Unal N. E. (2006) Sloshing in a three-dimensional rectangular tank: numerical simulation and experiemntal validation. *Ocean Engineering*, **33**(16): 2135–2149.

Ala, G., Francomano, E., Tortofici, A., Toscano, E. and Viola, F. (2007a) Corrective meshless particle formulations for time domain maxwell's equations, *Journal of Computational And Applied Mathematics* **210**(1–2): 34–46.

Ala, G., Francomano, E., Tortorici, A., Toscano, E. and Viola, F. (2006) Smoothed particle electromagnetics: A mesh-free solver for transients, *Journal Of Computational And Applied Mathematics* **191**(2): 194–205.

Ala, G., Francomano, E., Tortorici, A., Toscano, E., Viola, F., di Ingegneria Elettrica, D. and e delle Telecomunicazioni, E. (2007b) A mesh-free particle method for transient full-wave simulation, *IEEE Transactions on Magnetics* **43**(4): 1333–1336.

Alia, A. and Souli, M. (2006) High explosive simulation using multi-material formulations, *Applied Thermal Engineering* **26**(10): 1032–1042.

Amini, A., De Cesare, G. and Schleiss, A. J. (2009) Velocity profiles and interface instability in a two-phase fluid: Investigations using ultrasonic velocity profiler, *Experiments in Fluids* **46**(4): 683–692.

Amini, A. and Schleiss, A. J. (2009) Numerical modeling of oil-water multiphase flow contained by an oil spill barrier, *Engineering Applications of Computational Fluid Mechanics* **3**(2): 207–219.

Anghileri, M., Castelletti, L. M. L. and Tirelli, M. (2005) Fluid structure interaction of water filled tanks during the impact with the ground, *International Journal of Impact Engineering* **31**(3): 235–254.

Antoci, C., Gallati, M. and Sibilla, S. (2007) Numerical simulation of fluid-structure interaction by SPH, *Computers & Structures* **85**(11–14): 879–890.

Apfel, R. E., Tian, Y., Jankovsky, J., Shi, T., Chen, X., Holt, R. G., Trinh, E., Croonquist, A., Thornton, K. C. and Sacco, J. A. (1997) Free oscillations and surfactant studies of superdeformed drops in microgravity, *Physical Review Letters* **78**(10): 1912–1915.

Bøckmann, A., Shipliova, O. and Skeie, G. (2012) Incompressible SPH for free surface flows, *Computers & Fluids* **67**: 138–151.

Benz, W. (1988) Applications of smooth particle hydrodynamics (SPH) to astrophysical problems, *Computer Physics Communications* **48**(1): 97–105.

Benz, W. and Asphaug, E. (1995) Simulations of brittle solids using smooth particle hydrodynamics, *Computer Physics Communications* **87**(1): 253–265.

Berczik, P. (2000) Modeling the star formation in galaxies using the chemo-dynamical SPH code, *Astrophysics and Space Science* **271**(2): 103–126.

Bonet, J. and Kulasegaram, S. (2000) Correction and stabilization of smooth particle hydrodynamics methods with applications in metal forming simulations, *International Journal for Numerical Methods in Engineering* **47**(6): 1189–1214.

Borve, S., Omang, M. and Trulsen, J. (2001) Regularized smoothed particle hydrodynamics: A new approach to simulating magnetohydrodynamic shocks, *Astrophysical Journal* **561**(1): 82–93.

Bromm, V., Yoshida, N. and Hernquist, L. (2003) The first supernova explosions in the universe, *Astrophysical Journal* **596**(2): L135–L138.

Brown, H. M., Goodman, R. H., An, C. F. and Bittner, J. (1996) Boom failure mechanisms: Comparison of channel experiments with computer modelling results, *Spill Science & Technology Bulletin* **3**(4): 217–220.

Bui, H. H., Fukagawa, R. and Sako, K. (2006). Smoothed particle hydrodynamics for soil mechanics, 275, Taylor & Francis.

Bui, H. H., Fukagawa, R., Sako, K. and Ohno, S. (2008) Lagrangian meshfree particles method (SPH) for large deformation and failure flows of geomaterial using elastic-plastic soil constitutive model, *International Journal for Numerical And Analytical Methods in Geomechanics* **32**(12): 1537–1570.

Bui, H. H., Sako, K. and Fukagawa, R. (2007) Numerical simulation of soil-water interaction using smoothed particle hydrodynamics (SPH) method, *Journal of Terramechanics* **44**(5): 339–346.

Bulgarelli, U. P. (2005) The application of numerical methods for the solution of some problems in free-surface hydrodynamics, *Journal of Ship Research* **49**(4): 288–301.

Bursik, M., Martinez-Hackert, B., Delgado, H. and Gonzalez-Huesca, A. (2003) A smoothed-particle hydrodynamic automaton of landform degradation by overland flow, *Geomorphology* **53**(1–2): 25–44.

Busegnies, Y., Francois, J. and Paulus, G. (2007) Unidimensional SPH simulations of reactive shock tubes in an astrophysical perspective, *Shock Waves* **16**(4–5): 359–389.

Cariou, A. and Casella, G. (1999) Liquid sloshing in ship tanks: A comparative study of numerical simulation, *Marine structures* **12**(3): 183–198.

Chen, B. F. (2005) Viscous fluid in tank under coupled surge, heave, and pitch motions, *Journal of waterway, port, coastal, and ocean engineering* **131**: 239.

Chen, B. F. and Nokes, R. (2005) Time-independent finite difference analysis of fully non-linear and viscous fluid sloshing in a rectangular tank, *Journal of Computational Physics* **209**(1): 47–81.

Chen, J. K. and Beraun, J. E. (2000a) A generalized smoothed particle hydrodynamics method for nonlinear dynamic problems, *Computer Methods in Applied Mechanics and Engineering* **190**(1–2): 225–239.

Chen, J. K. and Beraun, J. E. (2000b) A generalized smoothed particle hydrodynamics method for nonlinear dynamic problems, *Computer Methods in Applied Mechanics and Engineering* **190**: 225–239.

Chen, J. K., Beraun, J. E. and Carney, T. C. (1999) A corrective smoothed particle method for boundary value problems in heat conduction, *International Journal for Numerical methods in Engineering* **46**: 231–252.

Chen, J. S., Pan, C., Roque, C. and Wang, H. P. (1998) A lagrangian reproducing kernel particle method for metal forming analysis, *Computational Mechanics* **22**(3): 289–307.

Chen, Z., Zong, Z., Li, H. T. and Li, J. (2013a) An investigation into the pressure on solid walls in 2d sloshing using SPH method, *Ocean Engineering* **59**: 129–141.

Chen, Z., Zong, Z., Liu, M. and Li, H. (2013b) A comparative study of truly incompressible and weakly compressible SPH methods for free surface incompressible flows, *International Journal for Numerical Methods in Fluids* **73**(9): 813–829.

Chikazawa, Y., Koshizuka, S. and Oka, Y. (2001) A particle method for elastic and visco-plastic structures and fluid-structure interactions, *Computational Mechanics* **27**(2): 97–106.

Cleary, P., Ha, J., Alguine, V. and Nguyen, T. (2002) Flow modelling in casting processes, *Applied Mathematical Modelling* **26**(2): 171–190.

Cleary, P. W. (1998) Modelling confined multi-material heat and mass flows using SPH, *Applied Mathematical Modelling* **22**(12): 981–993.

Cleary, P. W. and Ha, J. (2000) Three dimensional modelling of high pressure die casting, *International Journal of Cast Metals Research* **12**(6): 357–365.

Cleary, P. W., Ha, J. and Ahuja, V. (2000) High pressure die casting simulation using smoothed particle hydrodynamics, *International Journal of Cast Metals Research* **12**(6): 335–355.

Cleary, P. W. and Prakash, M. (2004) Discrete-element modelling and smoothed particle hydrodynamics: Potential in the environmental sciences, *Philosophical Transactions: Mathematical, Physical and Engineering Sciences* **362**(1822): 2003–2030.

Cleary, P. W., Prakash, M. and Ha, J. (2006) Novel applications of smoothed particle hydrodynamics (SPH) in metal forming, *Journal of Materials Processing Technology* **177**(1–3): 41–48.

Cleary, P. W., Prakash, M., Ha, J., Stokes, N. and Scott, C. (2007) Smooth particle hydrodynamics: Status and future potential, *Progress in Computational Fluid Dynamics* **7**(2–4): 70–90.

Colagrossi, A. and Landrini, M. (2003) Numerical simulation of interfacial flows by smoothed particle hydrodynamics, *Journal of Computational Physics* **191**(2): 448–475.

Crespo, A. J. C., Gomez-Gesteira, M., Carracedo, P. and Dalrymple, R. A. (2008a) Hybridation of generation propagation models and SPH model to study severe sea states in galician coast, *Journal of Marine Systems* **72**(1–4): 135–144.

Crespo, A. J. C., Gomez-Gesteira, M. and Dalrymple, R. A. (2007) 3d SPH simulation of large waves mitigation with a dike, *Journal of Hydraulic Research* **45**(5): 631–642.

Crespo, A. J. C., Gomez-Gesteira, M. and Dalrymple, R. A. (2008b) Modeling dam break behavior over a wet bed by a SPH technique, *Journal of Waterway Port Coastal and Ocean Engineering-ASCE* **134**(6): 313–320.

Cummins, S. J. and Rudman, M. (1999) An SPH projection method, *Journal of Computational Physics* **152**: 584–607.

Dalrymple, R. A. and Rogers, B. D. (2006) Numerical modeling of water waves with the SPH method, *Coastal Engineering* **53**(2–3): 141–147.

Delorme, L., Iglesias, A. S. and Perez, S. A. (2005). Sloshing loads simulation in lng tankers with SPH, *International Conference on Computational Methods in Marine Engineering*, Barcelona.

Dilts, G. A. (1999) Moving-least-squares-particle hydrodynamics—i. Consistency and stability, *International Journal for Numerical Methods in Engineering* **44**(8): 1115–1155.

Dilts, G. A. (2000) Moving-least-squares-particle hydrodynamics—ii. Conservation and boundaries, *International Journal for Numerical Methods in Engineering* **48**(10): 1503–1524.

Dolag, K., Bartelmann, M. and Lesch, H. (1999) SPH simulations of magnetic fields in galaxy clusters, *Astronomy and Astrophysics* **348**(2): 351–363.

Faltinsen, O. (1978) A numerical nonlinear method of sloshing in tanks with two-dimensional flow, *Journal of Ship Research* **22**(3).

Faltinsen, O., Landrini, M. and Greco, M. (2004) Slamming in marine applications, *Journal of Engineering Mathematics* **48**(3): 187–217.

Faltinsen, O. and Timokha, A. N. A multimodal method for liquid sloshing in a two-dimensional circular tank, *Journal of Fluid Mechanics* **665**: 457–479.

Faltinsen, O. M., Rognebakke, O. F., Lukovsky, I. A. and Timokha, A. N. (2000) Multidimensional modal analysis of nonlinear sloshing in a rectangular tank with finite water depth, *Journal of Fluid Mechanics* **407**: 201–234.

Faltinsen, O. M. and Timokha, A. N. (2001) An adaptive multimodal approach to nonlinear sloshing in a rectangular tank, *Journal of Fluid Mechanics* **432**: 167–200.

Faltinsen, O. M. and Timokha, A. N. (2009) Sloshing. Cambridge University Press, Cambridge.

Fang, H. S., Bao, K., Wei, J. A., Zhang, H., Wu, E. H. and Zheng, L. L. (2009) Simulations of droplet spreading and solidification using an improved SPH model, *Numerical Heat Transfer, Part A: Applications* **55**(2): 124–143.

Fang, X., Wu, W. and Wu, W. (2011). Numerical simulation technology of oil containment by boom. 2nd International Conference on Environmental Science and Technology, Singapore.

Fang, Z., Duan, M. and Zhu, R. (2007) Numerical simulation of liquid sloshing in a liquid tank based on level-set method, *Journal of Ship Mechanics* **11**(1): 62.

Fingas, M. (2011) Oil spill science and technology: Prevention, response, and cleanup. Gulf Professional Publishing.

Francomano, E., Tortorici, A., Toscano, E., Ala, G. and Viola, F. (2009) On the use of a meshless solver for PDEs governing electromagnetic transients, *Applied Mathematics and Computation* **209**(1): 42–51.

Frederic, A. R. and James, C. L. (1999) Smoothed particle hydrodynamics calculations of stellar interactions, *Journal of Computational And Applied Mathematics* **109**: 213–230.

Gallati, M., Braschi, G. and Falappi, S. (2005) Sph simulations of the waves produced by a falling mass into a reservoir, *Nuovo Cimento Della Societa Italiana Di Fisica C-Geophysics and Space Physics* **28**(2): 129–140.

Garg, R., Narayanan, C., Lakehal, D. and Subramaniam, S. (2007) Accurate numerical estimation of interphase momentum transfer in lagrangian-eulerian simulations of dispersed two-phase flows, *International Journal of Multiphase Flow* **33**(12): 1337–1364.

Ge, W., Chen, F. G., Gao, J., Gao, S. Q., Huang, J., Liu, X. X., Ren, Y., Sun, Q. C., Wang, L. M., Wang, W., Yang, N., Zhang, J. Y., Zhao, H., Zhou, G. Z. and Li, J. H. (2007) Analytical multi-scale method for multi-phase complex systems in process engineering — bridging reductionism and holism, *Chemical Engineering Science* **62**(13): 3346–3377.

Ghazali, J. N. and Kamsin, A. (2008). A real time simulation and modeling of flood hazard, *12th WSEAS International Conference on Systems*, 438–443, Heraklion, Greece.

Gomez-Gesteira, M. and Dalrymple, R. A. (2004) Using a three-dimensional smoothed particle hydrodynamics method for wave impact on a tall structure, *Journal of Waterway Port Coastal And Ocean Engineering-ASCE* **130**(2): 63–69.

Gong, K. and Liu, H. (2009) Water entry of a wedge based on SPH model with an improved boundary treatment, *Journal of Hydrodynamics, Ser. B* **21**(6): 750–757.

Goodman, R. H., Brown, H. M., An, C. and Rowe, R. D. (1996) Dynamic modelling of oil boom failure using computational fluid dynamics, *Spill Science & Technology Bulletin* **3**(4): 213–216.

Gotoh, H. and Sakai, T. (2006) Key issues in the particle method for computation of wave breaking, *Coastal Engineering* **53**(2–3): 171–179.

Gotoh, H., Shao, S. D. and Memita, T. (2004) SPH-les model for numerical investigation of wave interaction with partially immersed breakwater, *Coastal Engineering* **46**(1): 39–63.

Greenhow, M. and Moyo, S. (1997) Water entry and exit of horizontal circular cylinders, *Philosophical Transactions of the Royal Society of London. Series A: Mathematical, Physical and Engineering Sciences* **355**(1724): 551–563.

Guilcher, P. M., Ducrozet, G., Doring, M., Alessandrini, B. and Ferrant, P. (2006). Numerical simulation of wave-body interactions using a modified

SPH solver, *Proceedings of the Sixteenth International Offshore and Polar Engineering Conference*, 348–355, San Francisco, California.

Gutfraind, R. and Savage, S. B. (1998) Flow of fractured ice through wedge-shaped channels: Smoothed particle hydrodynamics and discrete element simulations, *Mechanics Of Materials* **29**(1): 1–17.

Ha, J. and Cleary, P. W. (2000) Comparison of SPH simulations of high pressure die casting with the experiments and vof simulations of schmid and klein, *International Journal Of Cast Metals Research* **12**(6): 409–418.

Ha, J. and Cleary, P. W. (2005) Simulation of high pressure die filling of a moderately complex industrial object using smoothed particle hydrodynamics, *International Journal of Cast Metals Research* **18**(2): 81–92.

Herrera, P. A., Massabo, M. and Beckie, R. D. (2009) A meshless method to simulate solute transport in heterogeneous porous media, *Advances in Water Resources* **32**(3): 413–429.

Hieber, S. E. (2004) Remeshed smoothed particle hydrodynamics simulation of the mechanical behavior of human organs, *Technology And Health Care* **12**(4): 305–314.

Hirt, C. and Nichols, B. (1981) Volume of fluid (VOF) method for the dynamics of free boundaries* 1, *Journal of computational physics* **39**(1): 201–225.

Hong, J. M., Lee, H. Y., Yoon, J. C. and Kim, C. H. (2008) Bubbles alive, *Acm Transactions on Graphics* **27**(3).

Hosseini, S. M. and Amanifard, N. (2007) Presenting a modified SPH algorithm for numerical studies of fluid-structure interaction problems, *IJE Trans B: Applications* **20**: 167–178.

Hosseini, S. M. and Feng, J. J. (2009) A particle-based model for the transport of erythrocytes in capillaries, *Chemical Engineering Science* **64**(22): 4488–4497.

Hu, C. and Kashiwagi, M. (2004) A CIP-based method for numerical simulations of violent free-surface flows, *Journal of Marine Science and Technology* **9**(4): 143–157.

Hu, D., Long, T., Xiao, Y., Han, X. and Gu, Y. (2014) Fluid-structure interaction analysis by coupled FE-SPH model based on a novel searching algorithm, *Computer Methods in Applied Mechanics and Engineering* **276**: 266–286.

Hu, W., Yao, L. G. and Hua, Z. Z. (2007) Parallel point interpolation method for three-dimensional metal forming simulations, *Engineering Analysis with Boundary Elements* **31**(4): 326–342.

Hu, X. Y. and Adams, N. A. (2006) A multi-phase SPH method for macroscopic and mesoscopic flows, *Journal of Computational Physics* **213**(2): 844–861.

Hu, X. Y. and Adams, N. A. (2007) An incompressible multi-phase SPH method, *Journal of Computational Physics* **227**(1): 264–278.

Hu, X. Y. and Adams, N. A. (2009) A constant-density approach for incompressible multi-phase SPH, *Journal of Computational Physics* **228**(6): 2082–2091.

Hui, H. H., Fukagawa, R. and Sako, K. (2006) Smoothed particle hydrodynamics for soil mechanics, *Terramechanics* **26**: 49–53.

Hultman, J. and Pharayn, A. (1999) Hierarchical, dissipative formation of elliptical galaxies: Is thermal instability the key mechanism? Hydrodynamic simulations including supernova feedback multi-phase gas and metal enrichment in cdm: Structure and dynamics of elliptical galaxies, *Astronomy and Astrophysics* **347**: 769–798.

Ibrahim, R. A. (2005) Liquid sloshing dynamics: Theory and applications. Cambridge University Press, Cambridge.

Ibrahim, R. A., Pilipchuk, V. N. and Ikeda, T. (2001) Recent advances in liquid sloshing dynamics, *Applied Mechanics Reviews* **54**: 133–199.

Idelsohn, S., Marti, J., Limache, A. and Onate, E. (2008) Unified lagrangian formulation for elastic solids and incompressible fluids: Application to fluid–structure interaction problems via the PFEM, *Computer Methods in Applied Mechanics and Engineering* **197**(19): 1762–1776.

Iglesias, A. S., Rojas, L. P. and Rodriguez, R. Z. (2004) Simulation of anti-roll tanks and sloshing type problems with smoothed particle hydrodynamics, *Ocean Engineering* **31**(8–9): 1169–1192.

Issa, R. and Violeau, D. (2008) Modelling a plunging breaking solitary wave with eddy-viscosity turbulent SPH models, *Computers Materials & Continua* **8**(3): 151–164.

Jeong, J. H., Jhon, M. S., Halow, J. S. and Van Osdol, J. (2003) Smoothed particle hydrodynamics: Applications to heat conduction, *Computer Physics Communications* **153**(1): 71–84.

Ji, S. Y., Li, H., Shen, H. T., Wang, R. X. and Yue, Q. J. (2007) A hybrid lagrangian-eulerian numerical model for sea-ice dynamics, *Acta Oceanologica Sinica* **26**: 12–24.

Ji, S. Y., Shen, H. T., Wang, Z. L., Shen, H. H. and Yue, Q. J. (2005) A viscoelastic-plastic constitutive model with mohr-coulomb yielding criterion for sea ice dynamics, *Acta Oceanologica Sinica* **24**(4): 54–65.

Jiang, F. and Sousa, A. C. M. (2006a) SPH numerical modeling for ballistic-diffusive heat conduction, *Numerical Heat Transfer, Part B: Fundamentals* **50**(6): 499–515.

Jiang, F. M., Oliveira, M. S. A. and Sousa, A. C. M. (2006) Sph simulation of transition to turbulence for planar shear flow subjected to a streamwise magnetic field, *Journal of Computational Physics* **217**(2): 485–501.

Jiang, F. M. and Sousa, A. C. M. (2006b) Sph numerical modeling for ballistic-diffusive heat conduction, *Numerical Heat Transfer Part B-Fundamentals* **50**(6): 499–515.

Khayyer, A., Gotoh, H. and Shao, S. D. (2008) Corrected incompressible SPH method for accurate water-surface tracking in breaking waves, *Coastal Engineering* **55**(3): 236–250.

Kim, Y. (2007) Experimental and numerical analyses of sloshing flows, *Journal of Engineering Mathematics* **58**(1–4): 191–210.

Kipfer, P. and Westermann, R. (2006). Realistic and interactive simulation of rivers, *Graphics Interface 2006*, 41–48, Quebec, Canada

Koshizuka, S., Nobe, A. and Oka, Y. (1998) Numerical analysis of breaking waves using the moving particle semi implicit method, *International Journal for Numerical Methods in Fluids* **26**(7): 751–769.

Koshizuka, S. and Oka, Y. (1996) Moving-particle semi-implicit method for fragmentation of incompressible fluid, *Nuclear science and engineering* **123**(3): 421–434.

Krill, C. E. and Chen, L. Q. (2002) Computer simulation of 3-D grain growth using a phase-field model, *Acta Materialia* **50**(12): 3057–3073.

Laigle, D., Lachamp, P. and Naaim, M. (2007) SPH-based numerical investigation of mudflow and other complex fluid flow interactions with structures, *Computational Geosciences* **11**(4): 297–306.

Lee, E.-S., Moulinec, C., Xu, R., Violeau, D., Laurence, D. and Stansby, P. (2008) Comparisons of weakly compressible and truly incompressible algorithms for the SPH mesh free particle method, *Journal of Computational Physics* **227**: 8417–8436.

Lee, T. H., Zhou, Z. Q. and Cao, Y. S. (2002) Numerical simulations of hydraulic jumps in water sloshing and water impacting, *Journal of Fluids Engineering* **124**: 215–226.

Lee, W. H. (2000) Newtonian hydrodynamics of the coalescence of black holes with neutron stars — iii. Irrotational binaries with a stiff equation of state, *Monthly Notices of The Royal Astronomical Society* **318**(2): 606–624.

Lee, W. H. and Kluzniak, W. (1999) Newtonian hydrodynamics of the coalescence of black holes with neutron stars ii. Tidally locked binaries with a soft equation of state, *Monthly Notices of The Royal Astronomical Society* **308**(3): 780–794.

Libersky, L. D. and Petschek, A. G. (1991). Smooth particle hydrodynamics with strength of materials, *Advances in the Free-Lagrange Method Including Contributions on Adaptive Gridding and the Smooth Particle Hydrodynamics Method: Proceedings of the Next Free-Lagrange Conference*, 248–257, Jackson Lake Lodge, Moran, WY, USA.

Libersky, L. D., Petschek, A. G., Carney, T. C., Hipp, J. R. and Allahdadi, F. A. (1993) High strain lagrangian hydrodynamics: A three-dimensional SPH code for dynamic material response, *Journal of Computational Physics* **109**(1): 67–75.

Lin, P. (2007) A fixed-grid model for simulation of a moving body in free surface flows, *Computers & Fluids* **36**(3): 549–561.

Liu, D. and Lin, P. (2008) A numerical study of three-dimensional liquid sloshing in tanks, *Journal of Computational Physics* **227**(8): 3921–3939.

Liu, G. R. and Liu, M. B. (2003) Smoothed particle hydrodynamics: A meshfree particle method. World Scientific, Singapore.

Liu, J., Koshizuka, S. and Oka, Y. (2005a) A hybrid particle-mesh method for viscous, incompressible, multiphase flows, *Journal of Computational Physics* **202**(1): 65–93.

Liu, M. B. and Liu, G. R. (2004) Smoothed particle hydrodynamics: Some recent developments in theory and applications, *J. Beijing Polytech. Univ.* **30**: 61–71.

Liu, M. B. and Liu, G. R. (2005) Meshfree particle simulation of micro channel flows with surface tension, *Computational Mechanics* **35**(5): 332–341.

Liu, M. B. and Liu, G. R. (2006) Restoring particle consistency in smoothed particle hydrodynamics, *Applied Numerical Mathematics* **56**(1): 19–36.

Liu, M. B. and Liu, G. R. (2010) Smoothed particle hydrodynamics (SPH): An overview and recent developments, *Archives of Computational Methods in Engineering* **17**(1): 25–76.

Liu, M. B., Liu, G. R. and Lam, K. Y. (2002) Investigations into water mitigation using a meshless particle method, *Shock Waves* **12**(3): 181–195.

Liu, M. B., Liu, G. R. and Lam, K. Y. (2003a) Comparative study of the real and artificial detonation models in underwater explosions, *Electronic Modeling* **25**(2): 113–124.

Liu, M. B., Liu, G. R. and Lam, K. Y. (2003b) A one-dimensional meshfree particle formulation for simulating shock waves, *Shock Waves* **13**(3): 201–211.

Liu, M. B., Liu, G. R., Lam, K. Y. and Zong, Z. (2003c) Computer simulation of shaped charge detonation using meshless particle method, *Fragblast* **7**(3): 181–202.

Liu, M. B., Liu, G. R., Lam, K. Y. and Zong, Z. (2003d) Meshfree particle simulation of the detonation process for high explosives in shaped charge unlined cavity configurations, *Shock Waves* **12**(6): 509–520.

Liu, M. B., Liu, G. R., Lam, K. Y. and Zong, Z. (2003e) Smoothed particle hydrodynamics for numerical simulation of underwater explosion, *Computational Mechanics* **30**(2): 106–118.

Liu, M. B., Liu, G. R., Zong, Z. and Lam, K. Y. (2003f) Computer simulation of high explosive explosion using smoothed particle hydrodynamics methodology, *Computers & Fluids* **32**(3): 305–322.

Liu, M. B., Shao, J. L. and Chang, J. Z. (2012a) On the treatment of solid boundary in smoothed particle hydro-dynamics, *Science China Technological Sciences* **55**(1): 244–254.

Liu, M. B., Shao, J. R. and Li, H. Q. (2013) Numerical simulation of hydro-elastic problems with smoothed particle hydrodynamics method, *Journal of Hydrodynamics, Ser. B* **25**(5): 673–682.

Liu, M. B., Shao, J. R. and Shang, Z. (2012b). SPH modeling of supercavity induced by underwater high speed objects, *the 8th International Symposium on Cavitation (CAV2012)*, 939–942, Singapore.

Liu, M. B., Xie, W. P. and Liu, G. R. (2005b) Modeling incompressible flows using a finite particle method, *Applied Mathematical Modelling* **29**(12): 1252–1270.

Liu, X., Xu, H. H., Shao, S. D. and Lin, P. Z. (2013) An improved incompressible SPH model for simulation of wave-structure interaction, *Computers & Fluids* **71**: 113–123.

Lohner, R., Yang, C. and Onate, E. (2006) On the simulation of flows with violent free surface motion, *Computer Methods in Applied Mechanics and Engineering* **195**(41–43): 5597–5620.

Lopez, H., Sigalotti, L. and Di, G. (2006) Oscillation of viscous drops with smoothed particle hydrodynamics, *Physical Review E* **73**(5): 51201.

Maeda, K. and Sakai, H. (2007). Seepage failure analysis with evolution of air bubbles by SPH, New Frontiers in Chinese and Japanese Geotechniques (Proceedings of the 3rd Sino-Japan Geotechnical Symposium). Chongqing, China: 402–409.

Mair, H. U. (1999) Review: Hydrocodes for structural response to underwater explosions, *Shock and Vibration* **6**(2): 81–96.

McDougall, S. and Hungr, O. (2004) A model for the analysis of rapid landslide motion across three-dimensional terrain, *Canadian Geotechnical Journal* **41**(6): 1084–1097.

McDougall, S. and Hungr, O. (2005) Dynamic modelling of entrainment in rapid landslides, *Canadian Geotechnical Journal* **42**(5): 1437–1448.

Meglicki, Z. (1994) Verification and accuracy of smoothed particle magnetohydrodynamics, *Computer Physics Communications* **81**(1–2): 91–104.

Mei, C. (1989) The applied dynamics of ocean surface waves. World Scientific, Singapore.

Melean, Y. and Sigalotti, L. D. (2005) Coalescence of colliding van der waals liquid drops, *International Journal of Heat & Mass Transfer* **48**: 4041–4061.

Ming, F. R., Sun, P. N. and Zhang, A. M. (2014) Investigation on charge parameters of underwater contact explosion based on axisymmetric SPH method, *Applied Mathematics and Mechanics-English Edition* **35**(4): 453–468.

Mitra, S., Upadhyay, P. P. and Sinhamahapatra, K. P. (2008) Slosh dynamics of inviscid fluids in two-dimensional tanks of various geometry using finite element method, *International Journal for Numerical Methods in Fluids* **56**(9): 1625–1651.

Monaghan, J. (2012) Smoothed particle hydrodynamics and its diverse applications, *Annual Review of Fluid Mechanics* **44**: 323–346.

Monaghan, J. J. (1990) Modelling the universe, *Astronomical Society of Australia Proceedings* **8**(3): 233–237.

Monaghan, J. J. (1992) Smooth particle hydrodynamics, *Annual Review of Astronmy and Astrophysics* **30**: 543–574.

Monaghan, J. J. (1994) Simulating free surface flows with SPH, *Journal of Computational Physics* **110**(2): 399–406.

Monaghan, J. J. and Kocharyan, A. (1995) SPH simulation of multi-phase flow, *Computer Physics Communications* **87**: 225–235.

Monaghan, J. J., Kos, A. and Issa, N. (2003) Fluid motion generated by impact, *Journal of Waterway Port Coastal and Ocean Engineering-ASCE* **129**(6): 250–259.

Monaghan, J. J. and Lattanzio, J. C. (1991) A simulation of the collapse and fragmentation of cooling molecular clouds, *Astrophysical Journal* **375**(1): 177–189.

Moresi, L., Muhlhous, H. and Dufour, F. (2001). An overview of numerical methods for earth simulations, 113-119, http://www.earthbyte.org/Events/ChapmanConference/AbstractsReceived-/TalkFiles./Moresi-et-al.pdf.

Morris, J., Fox, P. and Zhu, Y. (1997) Modeling low reynolds number incompressible flows using SPH, *Journal of computational physics* **136**(1): 214–226.

Morris, J. P., Zhu, Y. and Fox, P. J. (1999) Parallel simulations of pore-scale flow through porous media, *Computers and Geotechnics* **25**(4): 227–246.

Muller, M. (2004) Interactive blood simulation for virtual surgery based on smoothed particle hydrodynamics, *Technology and Health Care* **12**(1): 25–31.

Ning, C. and Zhang, Z. (2002) Numerical simualtion of oil-boom failure with cfx4.3, *Journal of Beijing University of Chemical Technology* **29**(5): 25–29.

Nugent, S. and Posch, H. A. (2000) Liquid drops and surface tension with smoothed particle applied mechanics, *Physical Review E* **62**(4): 4968–4975.

Oger, G., Doring, M., Alessandrini, B. and Ferrant, P. (2007) An improved SPH method: Towards higher order convergence, *Journal of Computational Physics* **225**: 1472–1492.

Oger, G., Guilcher, P. M., Jacquin, E., Brosset, L., Deuff, J. B. and Le Touz, D. (2009) Simulations of hydro-elastic impacts using a parallel SPH model, *ISOPE* 2009-FD-08.

Oger, L. and Savage, S. B. (1999) Smoothed particle hydrodynamics for cohesive grains, *Computer Methods in Applied Mechanics and Engineering* **180**(1–2): 169–183.

Pan, X., Zhang, H. and Lu, Y. (2008) Numerical simulation of viscous liquid sloshing by moving-particle semi-implicit method, *Journal of Marine Science and Application* **7**(3): 184–189.

Panizzo, A. (2005). Sph modelling of underwater landslide generated waves, *Coastal Engineering 2004 (Proceedings of the 29th International Conference)*, 1147–1159, Lisbon, Portugal.

Password, F. (2003) Cosmological smoothed particle hydrodynamics simulations: A hybrid multiphase model for star formation, *Monthly Notices of The Royal Astronomical Society* **339**(2): 289–311.

Pastor, M., Haddad, B., Sorbino, G., Cuomo, S. and Drempetic, V. (2009) A depth-integrated, coupled SPH model for flow-like landslides and related phenomena, *International Journal for Numerical and Analytical Methods in Geomechanics* **33**(2).

Peregrine, H. (2003) Water-wave impact on walls, *Annual Review in Fluid Mechanics* **35**: 23–43.

Pimenta, L. C. A., Mendes, M. L., Mesquita, R. C. and Pereira, G. A. S. (2007) Fluids in electrostatic fields: An analogy for multirobot control, *IEEE Transactions on Magnetics* **43**(4): 1765–1768.

Prakash, M., Cleary, P. W., Grandfield, J., Rohan, P. and Nguyen, V. (2007a) Optimisation of ingot casting wheel design using SPH simulations, *Progress in Computational Fluid Dynamics* **7**(2–4): 101–110.

Prakash, M., Cleary, P. W., Ha, J., Noui-Mehidi, M. N., Blackburn, H. and Brooks, G. (2007b) Simulation of suspension of solids in a liquid in a mixing tank using SPH and comparison with physical modelling experiments, *Progress in Computational Fluid Dynamics* **7**(2): 91–100.

Price, D. J. and Monaghan, J. J. (2004a) Smoothed particle magnetohydrodynamics — i. Algorithm and tests in one dimension, *Monthly Notices Of The Royal Astronomical Society* **348**(1): 123–138.

Price, D. J. and Monaghan, J. J. (2004b) Smoothed particle magnetohydrodynamics — ii. Variational principles and variable smoothing-length terms, *Monthly Notices Of The Royal Astronomical Society* **348**(1): 139–152.

Price, D. J. and Monaghan, J. J. (2004c) Smoothed particle magnetohydrodynamics: Some shocking results, *Astrophysics and Space Science* **292**(1): 279–283.

Qiu, L. C. (2008) Two-dimensional SPH simulations of landslide-generated water waves, *Journal of Hydraulic Engineering-ASCE* **134**(5): 668–671.

Rafiee, A. and Thiagarajan, K. P. (2009) An SPH projection method for simulating fluid-hypoelastic structure interaction, *Computer Methods in Applied Mechanics and Engineering* **198**: 2785–2795.

Randles, P. W. and Libersky, L. D. (1996) Smoothed particle hydrodynamics: Some recent improvements and applications, *Computer Methods in Applied Mechanics and Engineering* **139**(1): 375–408.

Rhee, S. H. and Engineer, L. (2005) Unstructured grid based reynolds-averaged navier-stokes method for liquid tank sloshing, *Journal of Fluids Engineering* **127**: 572.

Ritchie, B. W. and Thomas, P. A. (2001) Multiphase smoothed-particle hydrodynamics, *Monthly Notices of The Royal Astronomical Society* **323**(3): 743–756.

Rogers, B. D. and Dalrymple, R. A. (2005). Sph modeling of breaking waves, *Coastal Engineering 2004 (Proceedings of the 29th International Conference)*, 415–427, Lisbon, Portugal.

Rook, R., Yildiz, M. and Dost, S. (2007) Modeling transient heat transfer using SPH and implicit time integration, *Numerical Heat Transfer Part B-Fundamentals* **51**(1): 1–23.

Rosswog, S. and Wagner, P. (2002) Towards a macroscopic modeling of the complexity in traffic flow, *Physical Review E* **65**(3): 36106.

Sakai, H. and Maeda, K. (2006). Seepage failure of granular ground accounting for soil-water-gas interaction, Geomechanics and geotechnics of particulate media, *Proceedings of the International Symposium on Geomechanics and Geotechnics of Particulate Media*. Ube, Yamaguchi, Japan: 273–279.

Schafer, C., Speith, R. and Kley, W. (2007) Collisions between equal-sized ice grain agglomerates, *Astronomy & Astrophysics* **470**(2): 733–739.

Senz, D. G., Bravo, E. and Woosley, S. E. (1999) Single and multiple detonations in white dwarfs, *Astronomy And Astrophysics* **349** 177–188.

Sethian, J. (2003) Level set methods and fast marching methods, *Journal of Computing and Information Technology* **11**(1): 1–2.

Shadloo, M. S., Zainali, A., Yildiz, M. and Suleman, A. (2012) A robust weakly compressible SPH method and its comparison with an incompressible SPH, *International Journal for Numerical Methods in Engineering* **89**: 939–956.

Shao, J. R., Li, H. Q., Liu, G. R. and Liu, M. B. (2012a) An improved SPH method for modeling liquid sloshing dynamics, *Computers & Structures* **100–101**(18–26).

Shao, J. R., Liu, M. B., Yang, X. F. and Cheng, L. (2012b) Improved smoothed particle hydrodynamics with rans for free surface flow problems, *International Journal of Computational Methods* **9**(1): 1240001.

Shao, S. and Gotoh, H. (2005) Turbulence particle models for tracking free surfaces, *Journal of Hydraulic Research* **43**(3): 276.

Shao, S. D. (2006) Incompressible SPH simulation of wave breaking and overtopping with turbulence modelling, *International Journal For Numerical Methods In Fluids* **50**(5): 597–621.

Shao, S. D. (2009) Incompressible SPH simulation of water entry of a free-falling object, *International Journal For Numerical Methods In Fluids* **59**(1): 91–115.

Shao, S. D. and Gotoh, H. (2004) Simulating coupled motion of progressive wave and floating curtain wall by SPH-LES model, *Coastal Engineering Journal* **46**(2): 171–202.

Shao, S. D., Ji, C. M., Graham, D. I., Reeve, D. E., James, P. W. and Chadwick, A. J. (2006) Simulation of wave overtopping by an incompressible SPH model, *Coastal Engineering* **53**(9): 723–735.

Shao, S. D. and Lo, E. Y. M. (2003) Incompressible SPH method for simulating newtonian and non-newtonian flows with a free surface, *Advances in Water Resources* **26**(7): 787–800.

Shen, H. T., Su, J. S. and Liu, L. W. (2000) SPH simulation of river ice dynamics, *Journal of Computational Physics* **165**(2): 752–770.

Sigalotti, L. D. G. and Lopez, H. (2008) Adaptive kernel estimation and SPH tensile instability, *Computers & Mathematics with Applications* **55**(1): 23–50.

Smagorinsky J. (1963) General circulation experiments with the primitive equations, *Monthly Weather Review* **91**(3): 99–164.

Sousa, A. C. M. and Jiang, F. M. (2007) SPH as an inverse numerical tool for the prediction of diffusive properties in porous media, *Diffusion in Solids and Liquids: Heat Transfer - Microstructure & Properties* **553**: 171–189.

Souto Iglesias, A., Delorme, L., Perez-Rojas, L. and Abril-Perez, S. (2006) Liquid moment amplitude assessment in sloshing type problems with smooth particle hydrodynamics, *Ocean Engineering* **33**(11–12): 1462–1484.

Swegle, J. W. (1992). Report at sandia national laboratories.

Swegle, J. W. and Attaway, S. W. (1995) On the feasibility of using smoothed particle hydrodynamics for underwater explosion calculations, *Computational Mechanics* **17**(3): 151–168.

Tanaka, N. and Takano, T. (2005) Microscopic-scale simulation of blood flow using SPH method, *International Journal of Computational Methods* **2**(4): 555–568.

Tartakovsky, A. and Meakin, P. (2005a) Modeling of surface tension and contact angles with smoothed particle hydrodynamics, *Physical Review E* **72**(2): 26301.

Tartakovsky, A. M. and Meakin, P. (2005b) Simulation of unsaturated flow in complex fractures using smoothed particle hydrodynamics, *Vadose Zone Journal* **4**(3): 848–855.

Tartakovsky, A. M. and Meakin, P. (2006) Pore scale modeling of immiscible and miscible fluid flows using smoothed particle hydrodynamics, *Advances in Water Resources* **29**(10): 1464–1478.

Thacker, R. J. and Couchman, H. M. P. (2001) Star formation, supernova feedback, and the angular momentum problem in numerical cold dark matter cosmogony: Halfway there, *Astrophysical Journal* **555**(1): L17–L20.

Tsubota, K., Wada, S. and Yamaguchi, T. (2006) Simulation study on effects of hematocrit on blood flow properties using particle method, *Journal of Biomechanical Science and Engineering* **1**(1): 159–170.

Tyvand, P. A. and Miloh, T. (1995) Free-surface flow due to impulsive motion of a submerged circular cylinder, *Journal of Fluid Mechanics* **286**(1): 67–101.

Veldman, A. E. P., Gerrits, J., Luppes, R., Helder, J. A. and Vreeburg, J. P. B. (2007) The numerical simulation of liquid sloshing on board spacecraft, *Journal of Computational Physics* **224**(1): 82–99.

Violeau, D., Buvat, C., Abed-Meraim, K. and de Nanteuil, E. (2007) Numerical modelling of boom and oil spill with SPH, *Coastal Engineering* **54**(12): 895–913.

Wang, R. X., Ji, S. Y., Shen, H. T. and Yue, Q. J. (2005) Modified pic method for sea ice dynamics, *China Ocean Engineering* **19**(3): 457–468.

Wang, W., Huang, Y., Grujicic, M. and Chrisey, D. B. (2008) Study of impact-induced mechanical effects in cell direct writing using smooth particle hydrodynamic method, *Journal of Manufacturing Science and Engineering-Transactions of the ASME* **130**(2).

Wu, G. X., Ma, Q. W. and Eatock Taylor, R. (1998) Numerical simulation of sloshing waves in a 3D tank based on a finite element method, *Applied Ocean Research* **20**(6): 337–355.

Yang, G., Fu, Y., Hu, D. and Han, X. (2013) Feasibility analysis of SPH method in the simulation of condensed explosives detonation with ignition and growth model, *Computers & Fluids* **88**: 51–59.

Yang, X. F. and Liu, M. B. (2013) Numerical modeling of oil spill containment by boom using SPH, *SCIENCE CHINA Physics, Mechanics & Astronomy* **56**(2): 315–321.

Yang, X. F., Peng, S. L., Liu, M. B. and Shao, J. R. (2012) Numerical simulation of ballast water by SPH method, *International Journal of Computational Methods* **9**(1): 1240002.

Yim, S. C., Yuk, D., Panizzo, A., Di Risio, M. and Liu, P. L. F. (2008) Numerical simulations of wave generation by a vertical plunger using rans and SPH models, *Journal of Waterway Port Coastal And Ocean Engineering-ASCE* **134**(3): 143–159.

Zhang, A. M., Cao, X. Y., Ming, F. R. and Zhang, Z. F. (2013a) Investigation on a damaged ship model sinking into water based on three-dimensional SPH method, *Applied Ocean Research* **42**: 24–31.

Zhang, A. M., Ming, F. R. and Wang, S. P. (2013b) Coupled SPH-BEM method for transient fluid-structure interaction and applications in underwater impacts, *Applied Ocean Research* **43**: 223–233.

Zhang, A. M., Yang, W. S., Huang, C. and Ming, F. R. (2013c) Numerical simulation of column charge underwater explosion based on SPH and bem combination, *Computers & Fluids* **71**: 169–178.

Zhang, A. M., Yang, W. S. and Yao, X. L. (2012) Numerical simulation of underwater contact explosion, *Applied Ocean Research* **34**: 10–20.

Zhang, M. Y., Zhang, H. and Zheng, L. L. (2007) Application of smoothed particle hydrodynamics method to free surface and solidification problems, *Numerical Heat Transfer Part A-Applications* **52**(4): 299–314.

Zhang, M. Y., Zhang, H. and Zheng, L. L. (2008) Simulation of droplet spreading, splashing and solidification using smoothed particle hydrodynamics method, *International Journal of Heat & Mass Transfer* **51**(13–14): 3410–3419.

Zhang, M. Y., Zhang, H. and Zheng, L. L. (2009) Numerical investigation of substrate melting and deformation during thermal spray coating by SPH method, *Plasma Chemistry and Plasma Processing* **29**(1): 55–68.

Zheng, X., Duan, W. Y. and Ma, Q. W. (2012) A new scheme for identifying free surface particles in improved SPH, *Science China Physics, Mechanics and Astronomy* **55**(8): 1454–1463.

Zhou, C. E., Liu, G. R. and Lou, K. Y. (2007) Three-dimensional penetration simulation using smoothed particle hydrodynamics, *International Journal of Computational Methods* **4**(4): 671–691.

Zhou, G., Chen, Z., Ge, W. and Li, J. (2010) SPH simulation of oil displacement in cavity-fracture structures, *Chemical Engineering Science* **65**(11): 3363–3371.

Zhou, G. Z., Ge, W. and Li, J. H. (2008) A revised surface tension model for macro-scale particle methods, *Powder Technology* **183**(1): 21–26.

Zhu, Y. and Fox, P. J. (2002) Simulation of pore-scale dispersion in periodic porous media using smoothed particle hydrodynamics, *Journal of Computational Physics* **182**(2): 622–645.

Zhu, Y., Fox, P. J. and Morris, J. P. (1999) A pore-scale numerical model for flow through porous media, *International Journal For Numerical And Analytical Methods In Geomechanics* **23**(9): 881–904.

Zou, S. and Dalrymple, R. A. (2005). Sediment suspension modeling by smoothed particle hydrodynamics, *Coastal Engineering 2004 (Proceedings of the 29th International Conference)*, 1948–1958, Lisbon, Portugal.

Chapter 7

Three Typical Particle Methods

In Chapter 1, the literature review has shown that various meshfree and particle methods have been quickly developed as the next generation computational techniques. Among these meshfree particle methods, there is a class of particle methods in which particles are used not only to represent the state of a system as material particles, but also serve as interpolation or approximation points. For different scales, there are different particle methods: the classic molecular dynamics (MD) for micro-scale, the dissipative particle dynamics (DPD) method for meso-scale, and the smoothed particle hydrodynamics (SPH) method for macro-scale. Detailed formulation for MD, DPD and SPH have been presented in the previous chapters.

This chapter addresses three typical particle methods for macro-scale applications, i.e., the particle-in-cell (PIC) method, the material point method (MPM), and the moving-particle semi-implicit (MPS) method. PIC, MPM and MPS are currently widely used particle methods. They share some similarities but differ in a number of aspects, and will be introduced concisely.

This chapter is outlined as follows.

- In Section 7.1, the history and development, basic concept and implementation procedure of PIC are introduced. Features of PIC are compared with those from SPH.
- In Section 7.2, the history and development, basic concept and implementation procedure of MPM are introduced. Governing equations and the corresponding discretization as well as the mapping between background mesh and movable material particles are provided.
- In Section 7.3, the history and development, basic concept and implementation procedure of MPS are introduced. Governing equations and the corresponding discretization are discussed.
- In Section 7.4, some remarks and conclusion are given.

7.1 Particle-in-cell method

7.1.1 History and development

Particle-in-Cell method is a widely applied particle method, and it is a combined Eulerian-Lagrangian computing method suitable for solving multi-material problems involving large fluid distortions. The PIC method was initiated in the Fluid Dynamics Group at the Las Alamos National Laboratory (LANL) in the late 50s and 60s to solve complex computational fluid dynamic problems including reactive flows, multi-material flows, multi-phase flows and flows with spatial discontinuities (Johnson, 1996). The motivation is to invent a new approach, which can effectively avoid the disadvantages of mesh entanglement in the Lagrangian simulation of multi-materials under high pressure, in which solids behave like fluids. Harlow and his co-workers in LANL proposed and developed the PIC method (Evans and Harlow, 1957; Harlow, 1957), where Lagrangian particles are used to carry material mass, position and other information on a 2D uniform Eulerian mesh. The PIC method is therefore a dual description method with both Lagrangian and Eulerian features: Lagrangian description to move the mass particles, while Eulerian descriptions to interpolate information between mass particles and Eulerian nodes. Due to its special advantages and success, the PIC method was comprehensively investigated and widely applied to various areas due to its special advantages. Different variants were developed including: (1) Fluid-in-Cell (FLIC) (Gentry et al., 1966) to address the particle fluctuations and large memory requirements of PIC, (2) Vorticity and Stream Function Method (Fromm and Harlow, 1963) for incompressible flows, (3) Marker-and-Cell (MAC) (Harlow and Welch, 1965) for free surface flow, etc. Some recent developments of PIC include the work of Brackbill and his colleagues (Brackbill and Ruppel, 1986; Brackbill et al., 1988), and the material point method (MPM) by Sulsky, Chen and their colleagues (Sulsky et al., 1994; Sulsky et al., 1995). Recently MPM has been developed quickly and has been applied to many challenging problems in solid dynamics, especially for problems with large deformation and intense loadings (Liang et al., 2013). In this section, the basic concept of the PIC method is provided. More details on PIC can be found in (Franz, 2001) and many other related publications.

7.1.2 Basic concept

Consider a general fluid dynamic problem in 2D space, discretize the time as $t_{n+1} = t_n + \Delta t$ and discretize the problem domain with an Eulerian mesh with a number of particles enclosed in each mesh cell, as shown in **Figure 7.1**. Each particle represents a fluid element and carries properties such as position and

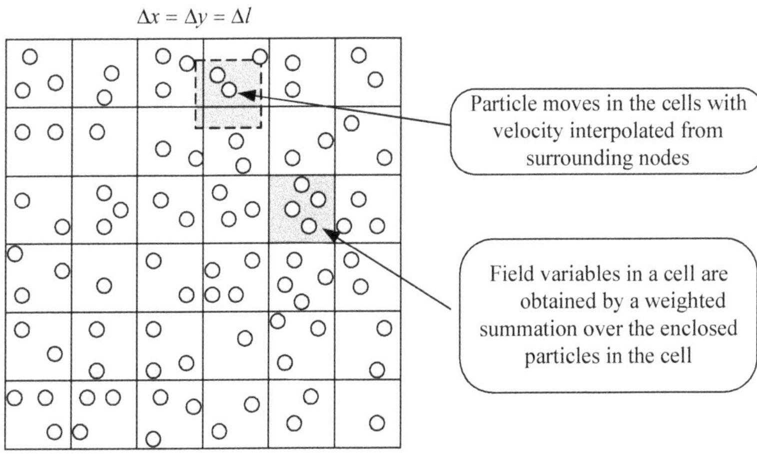

Figure 7.1 Domain discretization in the particle-in-cell (PIC) method.

velocity. For mesh cell (i, j), the density and velocity can be obtained by a summation over the particles within the cell.

$$\rho_{i,j}^{n} = \frac{1}{\Delta l^2}\sum_{k}^{N} m_k \delta[\mathbf{x}_k^n(i, j)], \tag{7.1}$$

$$\mathbf{v}_{i,j}^{n} = \frac{1}{\sum_{k}^{N} m_k}\sum_{k}^{N} m_k \mathbf{v}_k^n \delta[\mathbf{x}_k^n(i, j)], \tag{7.2}$$

where $\rho_{i,j}^{n}$ and $\mathbf{v}_{i,j}^{n}$ are the density and velocity of cell (i, j) at time step n, m_k is the mass of particle k, and N is the total number particles in the (i, j).

$$\delta[\mathbf{x}(i, j)] = \delta[\mathrm{int}(\frac{x}{\Delta l} - i)]\delta[\mathrm{int}(\frac{y}{\Delta l} - j)], \tag{7.3}$$

where x and y are the components of the position vector $\mathbf{x}$ in a two dimensional space.

For a general momentum equation, we have

$$\frac{\partial(\rho\mathbf{v})}{\partial t} = -\nabla p - \nabla\cdot(\rho\mathbf{v}\mathbf{v}). \tag{7.4}$$

The contribution from pressure term in RHS of above equation can be approximated as

$$\begin{cases} vx_{i,j}^{n+1} = vx_{i,j}^{n} - \dfrac{\Delta t}{2\Delta l \rho_{i,j}^{n}}(p_{i+1,j}^{n} - p_{i-1,j}^{n}) \\ vy_{i,j}^{n+1} = vy_{i,j}^{n} - \dfrac{\Delta t}{2\Delta l \rho_{i,j}^{n}}(p_{i,j+1}^{n} - p_{i,j-1}^{n}) \end{cases}, \tag{7.5}$$

where vx and vy are the velocity component in x and y directions respectively. Pressure in the RHS of equation (7.4) can be obtained from an equation of state.

It is the second Lagrangian part in equation (7.4) that involves particle movement. Its treatment is the essence of the PIC method, in which the velocity for particle k is accumulated by the following weighted summation

$$\mathbf{v}_k^{n+\frac{1}{2}} = \frac{1}{2}\sum_m \frac{s_m}{\Delta l^2}[\mathbf{v}_m^{n+1} + \mathbf{v}_m^{n}]. \tag{7.6}$$

The summation is based on the mth mesh cells overlapped with a square of width Δl centered at particle k considered. s is the overlapped area of the neighboring cells with the measure square.

After determining the particle velocity, the particle position in the next time step is

$$\mathbf{x}_k^{n+1} = \mathbf{x}_k^{n+1} + \Delta t \mathbf{v}_k^{n+\frac{1}{2}}. \tag{7.7}$$

7.1.3 Implementation procedure

Therefore a general implementation procedure for the PIC method is

1. Introduce an Eulerian grid with a number of fluid particles within each grid cell;
2. Solve the momentum equation by a standard FDM scheme to obtain the velocity at each grid cell (e.g. equation (7.5));
3. Accumulate the velocity by a weighted summation to obtain the particle velocity (equation (7.6));
4. Move the particle to a new position using the obtained particle velocity (equation (7.7));
5. Calculate the cell properties over summation of the enclosed particles (equations (7.1) and (7.2)).
6. Repeat stage 2-5 until the stopping criteria are satisfied.

Table 7.1 Comparison of SPH and PIC

	SPH	PIC
Method description	Lagrangian	Dual (Eulerian and Lagrangian)
Mesh/grid	No	Yes
Function of the Particles	Material particles and field function approximation	Material particles
Particle information	Directly obtained by solving the conservation equation using the SPH approximations.	Interpolated from or to the Eulerian mesh cells.
Interaction between particles	Yes	No

7.1.4 Comparison of SPH and PIC

It is clear that the PIC method is a dual description method with Lagrangian particles to move in the Eulerian grid cells. The inherent background Eulerian mesh determines that PIC method is not a truly meshfree particle method.

As summed up in **Table 7.1**, except that the particles in SPH and PIC both represent material blocks, the SPH method and the PIC method are quite different. The SPH method is a Lagrangian description, while the PIC method is an Eulerian-Lagrangian description. The SPH method is a truly meshfree method, while the PIC method is inherent with an Eulerian mesh for calculating the pressure gradient. The particles in SPH method not only represent material particles, but also act as computational frame for approximating both the field functions and their derivatives. In the SPH method, the particle information is obtained by solving the conservation equation using the SPH approximations. While in the PIC method, the particle information is interpolated from the background Eulerian mesh. The particles within the influence domain in the SPH method interact with each other in the form of particle approximations. The particles in the PIC method do not interact with each other, but exchange information with a background mesh.

7.2 Material point method

7.2.1 History and development

The material point method can be regarded as an extension of FLIP (Brackbill and Ruppel, 1986; Brackbill et al., 1988) from applications in fluid dynamics to solid mechanics problems. MPM was originally developed by Sulsky, Chen and their colleagues (Sulsky et al., 1994; Sulsky et al., 1995), and it was later applied to different areas such as explosion and impact (Zhang et al., 2006; Lian et al., 2011), ice dynamics (Sulsky et al., 2007), multiphase flow (Zhang et al., 2008), and fluid-structure interactions (York and Sulsky, 2000).

Similar to SPH, in MPM, the material domain can be discretized and represented with a set of particles, each associated with local field variables such as mass, velocity and other properties. The particles are then refereed to as material points and can move in a Lagrangian frame. The movement of the particles describes the material deformation and moving features. But different to SPH, in MPM, a background mesh is used to provide an Eulerian description of the material domain and it is predefined to calculate the gradient and integrate the momentum equation. The particles and the background mesh are connected at each time step. Field variables are firstly mapped from particles to grid nodes to establish and then solve the momentum equation in an Eulerian frame. After solving the momentum equation, the solutions are then mapped from the grid nodes to particles for updating their positions and velocities in a Lagrangian frame. At the end of each step, the connectivity of the particles with the background mesh needs to re-establish. For a fixed background mesh, there are particles flowing in or departing from certain mesh cells. For movable background mesh, the original mesh is discarded and a new background mesh needs to be set up for the next time step. As such there is no mesh distortion or mesh entanglement as in FEM.

7.2.2 Basic concept

Figure 7.2 illustrates the domain discretization in the material point method in a two dimensional space. Each particle is associated with field variables such

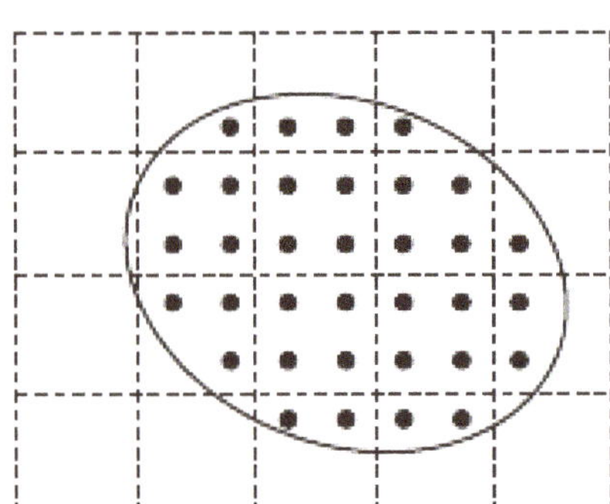

Figure 7.2 Domain discretization in the material point method in a two dimensional space. Dashed lines represent the background mesh, .filled particles denote the material points and the solid line describes the material domain.

as position, velocity, mass, density, stress, and others. Particles move in the background mesh depending on the internal (inter-particle interactions) and external force (body forces or external loadings). Since mass on each particle keeps constant, mass of the whole system is rigorously conserved. Also as the particles are basically lumped material points, density on a grid node can be obtained using summation.

The background mesh can be fixed just as the background mesh in PIC or freely movable just as the background mesh in the particle finite element method (PFEM) (Idelsoh et al., 2004; Onate et al., 2004). For the sake of convenience, the background mesh is usually taken as regular quadrilateral mesh in two dimensional spaces, and hexahedral mesh in three dimensional spaces. The background mesh is used to compute gradients in the momentum equation.

For solid dynamics problems, the momentum equation in a Lagrangian frame can be written as

$$\rho \frac{d\mathbf{v}}{\mathrm{dt}} = \nabla \cdot \boldsymbol{\sigma} + \rho \mathbf{F}, \tag{7.8}$$

where $\boldsymbol{\sigma}$ and $\mathbf{F}$ are stress tensor and body force respectively.

The boundary conditions and initial conditions can be written as

$$\boldsymbol{\sigma}(\mathbf{x},t) \cdot n \big| \Gamma_t = \bar{\mathbf{t}}(t), \quad \mathbf{u}(\mathbf{x},t) \big| \Gamma_t = \bar{\mathbf{u}}(t), \tag{7.9}$$

$$\mathbf{u}(\mathbf{x},0) = \mathbf{u}_0(\mathbf{x}), \quad \mathbf{v}(\mathbf{x},0) = \mathbf{v}_0(\mathbf{x}), \tag{7.10}$$

where Γ_t denotes the boundary with a prescribed traction vector $\bar{\mathbf{t}}$, Γ_u denotes the boundary with a prescribed traction vector $\bar{\mathbf{u}}$, n is the outward normal to the boundary surface Γ_t, $\mathbf{u}_0(\mathbf{x})$ and $\mathbf{v}_0(\mathbf{x})$ are the initial displacement and velocity respectively.

One notable difference between conventional SPH and MPM is that SPH is based on strong form formulations and directly approximates the governing partial differential equations. In contrast, MPM approximation is based on weak form formulations. For example, the Galerkin weak form of equation (7.8) can be written as

$$\int_\Omega \rho \mathbf{a} \cdot \mathbf{w} d\Omega + \int_\Omega \boldsymbol{\sigma} : \nabla \mathbf{w} d\Omega - \int_\Omega \rho \mathbf{w} \cdot \mathbf{F} d\Omega - \int_\Omega \mathbf{w} \cdot \bar{\mathbf{t}} d\Omega = 0, \tag{7.11}$$

where $\mathbf{a}$ and $\mathbf{w}$ are the acceleration vector and test function respectively, Ω is the configuration of the continuum.

If assuming the total number of particles to be N_p, density can be approximated as

$$\rho(\mathbf{x}, t) = \sum_{p=1}^{N_p} m_p \delta(\mathbf{x} - \mathbf{x}_p), \tag{7.12}$$

where m_p is the mass of particle p, δ is the Dirac delta function. Substituting equation (7.12) to equation (7.11), and converting the integrals over an infinitesimal domain $d\Omega$ to summation over the particles (with corresponding particle volume of $\frac{m_p}{\rho_p}$), we can obtain

$$\sum_{p=1}^{N_p} m_p \mathbf{a}_\mathrm{p} \cdot \mathbf{w}_p + \sum_{p=1}^{N_p} \frac{m_p}{\rho_p} \boldsymbol{\sigma}_\mathrm{p} : \nabla \mathbf{w}_p - \sum_{p=1}^{N_p} m_p \mathbf{w}_\mathrm{p} \cdot \mathbf{F}_p - \sum_{p=1}^{N_p} \frac{m_p}{\rho_p h} \vec{\mathbf{t}}_\mathrm{p} \cdot \mathbf{w}_p = 0, \tag{7.13}$$

where h is the boundary layer thickness.

As in MPM, particles and the background mesh are rigidly connected, it is feasible to establish the mapping of the information between particles and grid nodes by using a shape function, $N_I(\mathbf{x}_i)$. The shape function is similar to that in FEM and is established on grid nodes. The coordinates, displacements and test functions of a particle p in a cell can be approximated using the grid nodal displacements as

$$\mathbf{x}_p = \sum_{I=1}^{N_g} N_{Ip} \mathbf{x}_I, \tag{7.14}$$

$$\mathbf{u}_p = \sum_{I=1}^{N_g} N_{Ip} \mathbf{u}_I, \tag{7.15}$$

and

$$\mathbf{w}_p = \sum_{I=1}^{N_g} N_{Ip} \mathbf{w}_I, \tag{7.16}$$

where N_{Ip} $(= N_I(\mathrm{x}_\mathrm{p}))$ is the shape function associated with grid node I evaluated at particle p, and N_g is the nodal number.

Substituting equations (7.14)-(7.16) to equation (7.13), we can obtain the following equation

$$\dot{\mathbf{p}}_I = \mathbf{f}_I^{\mathrm{int}} + \mathbf{f}_I^{\mathrm{ext}}, \tag{7.17}$$

where $\mathbf{p}_I$ is the momentum on grid node I, $\mathbf{f}_I^{\text{int}}$ and $\mathbf{f}_I^{\text{ext}}$ are the internal and external forces respectively. $\mathbf{p}_I$ is written as

$$\mathbf{p}_I = \sum_{J=1}^{N_g} m_{IJ}\mathbf{v}_J = m_I\mathbf{v}_I , \tag{7.18}$$

$$m_{IJ} = \sum_{p=1}^{N_p} m_p N_{IP} N_{JP} , \tag{7.19}$$

$$m_I = \sum_{p=1}^{N_p} m_p N_{IP} , \tag{7.20}$$

where m_{IJ} and m_I are the mass matrix and lumped mass matrix. $\mathbf{f}_I^{\text{ext}}$ is written as

$$\mathbf{f}_I^{\text{ext}} = \sum_{p=1}^{N_p} m_p N_{IP}\mathbf{F}_P + \sum_{p=1}^{N_p} \frac{m_p}{\rho_p h} N_{IP}\vec{\mathbf{t}}_{\text{p}} , \tag{7.21}$$

and $\mathbf{f}_I^{\text{int}}$ is written as

$$\mathbf{f}_I^{\text{int}} = -\sum_{p=1}^{N_p} \frac{m_p}{\rho_p} \boldsymbol{\sigma}_{\text{p}} \cdot \nabla N_I |_{\mathbf{x}_p} . \tag{7.22}$$

Integrating equation (7.17) with an explicit time integration algorithm (e.g., the simple Euler scheme, Leap-frog scheme or others) can produce the momentum for the next step. The field variables on material points can be obtained through mapping the solutions on grid nodes. It is noted that though m_p keeps constant, m_I needs to be computed at each time step.

7.2.3 Implementation procedure

If field variables at time instant t^k are known, the field variables at time instant t^{k+1} can be obtained according to the following procedure (for a simple Euler integration scheme).

1. update variables (e.g., mass, momentum and nodal forces) at grid nodes through mapping from particles,

$$m_I^k = \sum_{p=1}^{N_p} m_p N_{IP}^k , \tag{7.23}$$

$$\mathbf{p}_I^k = \sum_{p=1}^{N_p} m_p \mathbf{v}_p^k N_{IP}^k , \tag{7.24}$$

$$\mathbf{f}_I^k = \mathbf{f}_I^{\text{int},k} + \mathbf{f}_I^{\text{ext},k} . \tag{7.25}$$

2. integrate momentum equation on background mesh and enforce boundary condition,

$$\mathbf{p}_I^{k+1} = \mathbf{p}_I^k + \mathbf{f}_I^k \Delta t . \tag{7.26}$$

3. update the field variables (velocity and position) on particles,

$$\mathbf{v}_p^{k+1} = \mathbf{v}_p^k + \mathbf{a}_p^k \Delta t = \mathbf{x}_p^k + \sum_{I=1}^{N_I} \frac{\mathbf{f}_I^k}{m_I^k} N_{IP}^k \Delta t , \tag{7.27}$$

$$\mathbf{x}_p^{k+1} = \mathbf{x}_p^k + \mathbf{v}_p^k \Delta t = \mathbf{x}_p^k + \sum_{I=1}^{N_I} \frac{\mathbf{p}_I^{k+1}}{m_I^k} N_{IP}^k \Delta t . \tag{7.28}$$

4. map particle velocity back to grid nodes,

$$\mathbf{v}_p^{k+1} = \frac{\sum_p m_p \mathbf{v}_I^{k+1} N_{IP}^k}{m_I^k} . \tag{7.29}$$

5. calculate strain and vorticity increment,
6. and update field variables (e.g., density and stress) on particles, and then repeat step 1 for the next step.

It is noted that the calculation of stress and strain (on particles) can be conducted before or after the calculating of nodal forces and the integration of momentum equation. These lead to two different updating schemes, update stress first (USF), update stress last (USL) and modified update stress last (MUSL) (Bardenhagen, 2002; Nairn, 2003; Liang et al., 2013).

7.2.4 Comparison of SPH and MPM

Similar to PIC, MPM is also a dual description method with Lagrangian particles to move in the background mesh. As a background mesh is required, MPM is not a truly meshfree particle method.

As summed up in **Table 7.2**, except that the particles in SPH and MPM both represent material blocks, the SPH method and the MPM method are different. The SPH method is a Lagrangian description, while MPM is an dual description (Eulerian-Lagrangian if the background mesh is fixed or Lagrangian-Lagrangian if the background mesh is movable). In SPH, particles are used not only to represent the state of a system, but also to approximate the governing particle differential equations. In MPM, the movement of the material particles describes the material deformation, while momentum equation and related gradients are computed on the background mesh. This is usually more accurate and the treatment of solid boundaries is much easier. In SPH, particles interact with each other within a pre-defined cut-off distance and an efficient algorithm to search the nearest neighbor particle is necessary. In MPM, in general, material particles do not interact with each, and particle-pair searching is avoided. Therefore, MPM can be more cost-effective than SPH.

Table 7.2 Comparison of SPH and MPM

	SPH	**MPM**
Method description	Lagrangian description	Dual description with background mesh and movable particles
Governing equations	Strong form	Weak form
Background mesh	No	Yes (for computing momentum equation and related gradients)
Function of the Particles	Material particles and field function approximation	Material particles
Particle information	Directly obtained by solving the conservation equation using the SPH approximations.	Interpolated from or to the background grid nodes.
Interaction between particles	Yes	No

7.3 Moving-particle semi-implicit method

7.3.1 History and development

The moving-particle semi-implicit method is a computational method originally developed for modeling incompressible, free surface flows by Koshizuka and his co-workers (Koshizuka et al., 1995; Koshizuka and Oka,

1996; Koshizuka et al., 1998; Heo et al., 2001). As a macroscopic and deterministic method, MPS uses a set of particles to represent the state of a modeling system, and therefore MPS conserves mass exactly. The particles possess material properties, interact with each other within a range controlled by a weight function (usually referred to as kernel function), and move according to inter-particle interactions and external forces. In MPS, there is no explicit interface tracking for multiphase flows – the motion of the fluid is represented by the motion of the particles, and fluid surfaces or fluid-fluid interfaces move with particles representing their phase defined at the initial stage.

7.3.2 Basic concept

For a viscous, incompressible fluid, the governing equations are expressed by the conservation of continuity (or mass) and momentum in Lagrangian form as

$$\frac{d\rho}{dt} = 0, \tag{7.30}$$

and

$$\frac{d\mathbf{v}}{dt} = -\frac{1}{\rho}\nabla p + \upsilon\nabla^2\mathbf{v} + \mathbf{F}, \tag{7.31}$$

where ρ and p are density and pressure, υ is laminar kinetic viscosity, t is time, $\mathbf{v}$ and $\mathbf{F}$ are velocity and external force vectors.

It is noted that the left hand side of equation (7.31) denotes the Lagrangian differentiation involving the advection term, calculated through the tracking of particle motion in MPS (and also in SPH). The above continuity and momentum equations are discretized by use of differential operator models, namely the gradient and Laplacian operators, through using a scalar kernel function. The kernel function serves as a weight function, which describe the interaction of a particle with its neighboring particles within a compact support domain. A typical MPS kernel function is

$$w(r) = \begin{cases} \frac{r_c}{r} - 1 & 0 \le r < r_c \\ 0 & r \ge r_c \end{cases}, \tag{7.32}$$

where r is the distance between two neighboring particles, and r_c is the cut-off distance for the compact support domain. As shown in **Figure 7.3**, the kernel function showed in equation (7.32) approaches infinite when two interacting particles approach each other infinitely. This may cause numerical difficulties. Actually, the kernel functions frequently used in SPH (Liu and Liu, 2003) may also be applicable in MPS. Ataie-Ashtiani and Farhadi comparatively analyzed the performance of 6 kernel functions including the standard kernel (equation (7.32)) by simulating a dam-break flow, and it was reported that the B-spline kernel performed better than other choices (Ataie-Ashtiani and Farhadi, 2006).

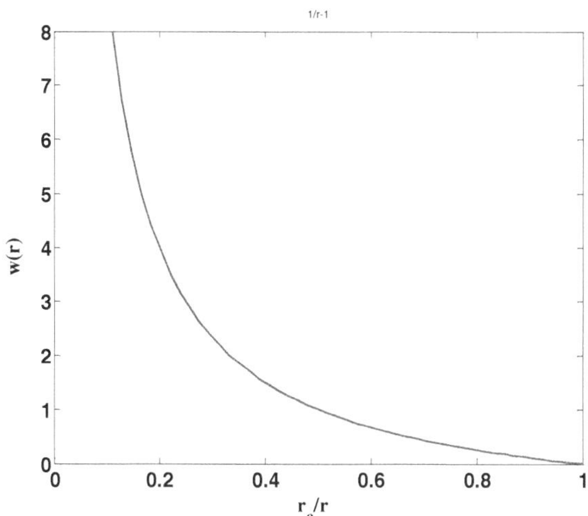

Figure 7.3 Kernel function in MPS.

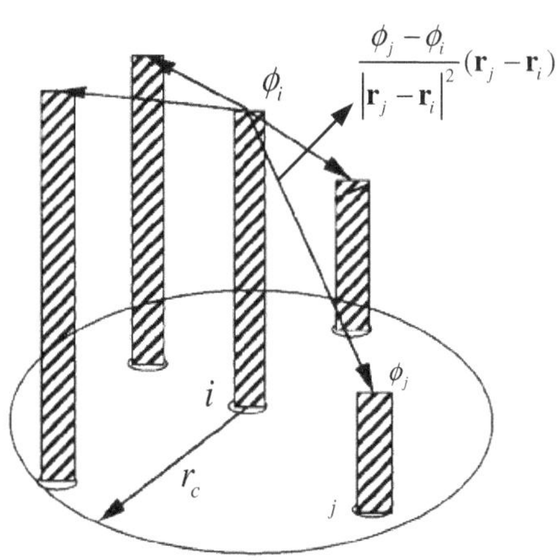

Figure 7.4 Concept of gradient in MPS.

Equation (7.32) shows that in MPS, the inter-particle interactions are restricted to the cut-off distance r_c, and this leads to a finite number of neighboring particles. In order to compute the inter-particle interactions, a fast algorithm is required to search the nearest neighboring particles, just as in SPH method. Therefore, the nearest neighbor searching algorithms in SPH (Liu and Liu, 2003), such as the link list and tree search algorithm are also applicable to MPS.

Following (Koshizuka et al., 1998), the gradient operator for an arbitrary scalar function ϕ, is a local weighted average of the gradient vectors between particle i and its neighboring particles j, as shown in **Figure 7.4**

$$< \nabla\phi >_i = \frac{d}{n^0}\sum_{j\neq i}\frac{\phi_j - \phi_i}{\left|\mathbf{r}_j - \mathbf{r}_i\right|^2}(\mathbf{r}_j - \mathbf{r}_i)w(\mathbf{r}_j - \mathbf{r}_i)\,, \tag{7.33}$$

where d is the number of space, and n^0 is the constant particle number density (for incompressible fluids). It is found that if replacing ϕ_i with the minimum value of ϕ_j among the neighboring particles can avoid possible unphysical particle acceleration. For example, the pressure gradient is obtained by replacing ϕ with the minimal value of $\overline{p}_i$ among neighboring particles as

$$< \nabla p >_i = \frac{d}{n^0}\sum_{j\neq i}\frac{p_j - \overline{p}_i}{\left|\mathbf{r}_j - \mathbf{r}_i\right|^2}(\mathbf{r}_j - \mathbf{r}_i)w(\mathbf{r}_j - \mathbf{r}_i)\,, \tag{7.34}$$

$$\overline{p}_i = \min_{J\subset j}(p_j) \quad for \quad J == \left\{ j : w(\mathbf{r}_j - \mathbf{r}_i) \neq 0 \right\}. \tag{7.35}$$

Similarly following (Koshizuka et al., 1998), the Laplacian operator is modeled as follows

$$< \nabla^2\phi >_i = \frac{2d}{n^0\lambda}\sum_{j\neq i}\left(\phi_j - \phi_i\right)w(\mathbf{r}_j - \mathbf{r}_i)\,, \tag{7.36}$$

where

$$\lambda = \frac{\sum_{j\neq i}\left(\left|\mathbf{r}_j - \mathbf{r}_i\right|^2\right)w(\mathbf{r}_j - \mathbf{r}_i)}{\sum_{j\neq i}w(\mathbf{r}_j - \mathbf{r}_i)}\,. \tag{7.37}$$

For a specific particle i and its neighbors j located at positions $\mathbf{r}_i$ and $\mathbf{r}_j$, the particle number density n at particle i can be defined as

$$< n >_i = \sum_{i \neq j} w(\left|\mathbf{r}_j - \mathbf{r}_i\right|). \tag{7.38}$$

The number of particles in a unit volume (N) can thus be approximated as the ratio of the particle number density to the integration of the kernel function over the whole area (Ω)

$$< N >_i = \frac{< n >_i}{\int_{\Omega} w(r) d\Omega}. \tag{7.39}$$

It is noted that due to the compact supportness of kernel function, the integration of the kernel function over the whole area is actually restricted to the area within the cut-off distance r_c.

7.3.3 Implementation procedure

The implementation procedure in MPS is basically an iterative process with two main steps, a first prediction step and a second correction step. In the prediction step, under given viscosity and external forces, an explicit simulation is conducted to obtain intermediate velocities. These intermediate velocities may not satisfy mass conservation and the resultant intermediate number densities, n^*, may be different from the initial particle number density, n^0. Therefore, in the second correction step, the intermediate velocities are updated implicitly through solving the following Pressure Poisson Equation (PPE) (Koshizuka et al., 1998)

$$< \nabla^2 p^{k+1} >_i = -\frac{\rho}{\Delta t^2} \frac{< n^* >_i - n^0}{n^0}, \tag{7.40}$$

where Δt and k denote time step and the number of time step in calculation. The right hand side (RHS) of (7.40) represents the deviation of the intermediate particle number density from the initial number density. The left hand side (LHS) of (7.40) can be discretized using equation (7.36) to get a set of linear equations, which can be solved with some kind of linear equation solver such as the incomplete Cholesky decomposition conjugate gradient (ICCG) method. Once the pressure field is obtained, the velocity field can be obtained by substituting the pressure field into the velocity correction equation at the second correction step.

As can see from the above discussions that the meshfree nature of MPS method remove the difficulties due to large deformations since MPS uses movable particles rather than mesh as a computational frame to approximate related governing equations. Therefore, during the last decade, the MPS has been applied to different areas including nuclear engineering (Koshizuka and Oka, 1996; Koshizuk et al., 1999; Xie et al., 2005), ocean and coastal engineering (Koshizuka et al., 1998; Khayyer and Gotoh, 2009) and some others (Sun et al., 2009b, a).

7.3.4 Comparison of SPH and MPS

MPS and SPH share many similarities. Firstly, they are both meshfree, Lagrangian and particle methods since movable material particles are used to represent the modeling system and to approximate governing equation. Secondly, both MPS and SPH provide approximations to the strong form of partial differential equations on the basis of integral interpolants. Thirdly, numerical approximations in MPS and SPH are based on local averaged summations. Some numerical techniques such as the free surface and solid boundary treatment algorithms are also similar.

There are basically two main differences between MPS and SPH. Firstly, MPS applies simplified differential operator models solely based on a local weighted averaging process without taking the gradient of a kernel function. While in SPH, the differential operator models are approximated using averaged summation based on the kernel gradient (refer to Chapter 5 for more details). Secondly, the solution process of the original MPS method differs to that of the original SPH method as the solutions to the PDEs are obtained through a semi-implicit prediction-correction process rather than the fully explicit one in the original SPH method. The semi-implicit approach is effective in obtaining smoother pressure field. However, with the development of MPS and SPH, they both borrow ideas from each other. As such, some SPH methods are associated with semi-implicit algorithms to resolve pressure. In contrast, MPS is also extendable to be a fully explicit method, which uses an artificial equation of state to model incompressible flows as slightly compressible and thus computational cost in the modified MPS can be much smaller than that in the original MPS.

As summed up in **Table 7.3**, SPH and MPS are very much similar. Both methods use particles to represent the state of a modeling system and to approximate governing equations in strong form. Both methods do not use any background mesh, and hence they are purely meshfree particles. In both methods particles interact with each other within a pre-defined cut-off distance and an efficient algorithm to search the nearest neighbor particle is necessary.

Table 7.3 Comparison of SPH and MPS

	SPH	**MPS**
Method description	Lagrangian description	Lagrangian description
Governing equations	Strong form	Strong form
Background mesh	No	No
Function of the Particles	Material particles and field function approximation	Material particles and field function approximation
Particle information	Directly obtained by solving the conservation equation using the SPH approximations.	Directly obtained by solving the conservation equation using the MPM approximations.
Inter-particle interaction	Yes	Yes
Kernel and kernel gradients	Both kernel and kernel gradients are used for numerical approximation	Kernel gradients are not used for numerical approximation
Time integration	Explicit	Semi-explicit

7.4 Concluding remarks

In this chapter, three widely used particle methods, PIC, MPM and MPS are briefly introduce. Their histories, basic concepts and implementation procedures are described. In summarizing the contents in this chapter, the following remarks can be made

1. As MPM is an extension of FLIP, an improved version of PIC, MPM and PIC are more alike. Both are of dual description with material particles moving in a Lagrangian frame and background mesh for solving momentum equation and computing gradients. In both methods, Information mapping between material particles and background mesh nodes is required. The difference between PIC and MPM is that PIC solves strong form governing equations, while MPM solves weak form governing equations. Therefore PIC is more frequently used for modeling fluid mechanics problems, while MPM is more frequently used for modeling solid mechanics problems.
2. The MPS method is more like SPH. They are both purely meshfree, Lagrangian particle methods for solving strong form PDEs. The

differences between MPS and SPH are the gradient computation and time integration algorithm, other numerical aspects such as free surface, solid boundary treatment, and nearest neighboring particle search algorithm are similar.

3. With the development of MPS and SPH, they often borrow ideas from each other. For example, the originally explicit SPH method is later modified to be semi-implicit, which is just the same as the original MPS. In contrast, the original semi-implicit MPS is modified to be explicit, just as the conventional SPH.
4. In MPS and SPH, nearest neighboring particle search algorithms are required to compute the inter-particle interactions, which are generally time-consuming and are usually not as accurate and stabe as the grid-based methods. Solid boundary treatment in MPS and SPH is usually a tough issue. As such, it would be helpful to integrate some numerical aspects from grid-based methods to MPS and SPH. For example, coupling MPS and SPH with FEM/FDM/FVM, or developing MPS and SPH with background mesh (either FEM style or FDM/FVM style) can be possible directions for both MPS and SPH.

References

Ataie-Ashtiani B. and Farhadi, L. (2006) A stable moving particle semi-implicit method for free surface flows, *Fluid Dynamics Research* **38**: 241–256.

Bardenhagen S. G. (2002) Energy conservation error in the material point method for solid mechanics, *Journal of Computational Physics* **180**: 383–403.

Brackbill J. U., Kothe, D. B. and Ruppel, H. M. (1988) FLIP: a low-dissipation, particle-in-cell method for fluid flow, *Computer Physics Communications* **48**: 25–38.

Brackbill J. U. and Ruppel, H. M. (1986) FLIP:A method for adaptively zoned, particle-in-cell calculations in two dimensions, Journal of Computational Physics, *Journal of Computational Physics* **65**(2): 314–343.

Evans M. W. and Harlow, F. H. (1957). The particle-in-cell method for hydrodynamic calculations, Los Alamos National Laboratory Report LA-2139.

Franz J. V. (2001) Computational physics, an introduction. Kluwer Academic/Plenum Publishers, New York.

Fromm J. and Harlow, F. H. (1963) Numerical solution of the problem of Vortex Sheet development, *Physics of Fluids* **6**: 975.

Gentry R. A., Martin, R. E. and Daly, B. J. (1966) An Eulerian differencing method for unsteady compressible flow problems, *Journal of Computational Physics* **1**: 87–118.

Harlow F. H. (1957) Hydrodynamic problems involving large fluid distortion, *Journal of the Association for Computing Machinery* **4**: 137.

Harlow F. H. and Welch, J. E. (1965) Numerical calculation of time-dependent viscous incompressible flow of fluids with free surface, *Phyics of Fluids* **8**: 22182.

Heo S., Koshizuka, S., Oka, Y. and Obata, H. (2001) Numerical analysis of boiling on high heat-flux and high subcolloing condition using MPS-MAFL, *International Journal of Heat and Mass Transfer* **45**: 2633–2642.

Idelsoh S. R., Oñate, E. and Pin, F. D. (2004) The particle finite element method: a powerful tool to solve incompressible flows with free-surfaces and breaking waves, *International Journal for Numerical Methods in Engineering* **61**: 964–989.

Johnson N. L. (1996). The legacy and future of CFD at Los Alamos, *Proceedings of the 1996 Canadian CFD Conference*, Ottawa, Canada.

Khayyer A. and Gotoh, H. (2009) Modified moving particle semi-implicit methods for the prediction of 2D wave impact pressure, *Coastal Engineering* **56**: 419–440.

Koshizuk S., Ikeda, H. and Oka, Y. (1999) Numerical analysis of fragmentation mechanisms in vapor explosion, *Nuclear Engineering and Design* **189**: 423–433.

Koshizuka S., Nobe, A. and Oka, Y. (1998) Numerical analysis of breaking waves using the moving particle semi-implicit method, *International Journal for Numerical Methods in Fluids* **26**(7): 751–769.

Koshizuka S. and Oka, Y. (1996) Moving-particle semi-implicit method for fragmentation of incompressible fluid, *Nuclear Science and Engineering* **123**(3): 421–434.

Koshizuka S., Tamako, H. and Y., O. (1995) A particle method for incompressible viscous flow with fluid fragmentation, *Journal of Computational Fluid Dynamics* **4**: 29–46.

Lian Y. P., Zhang, X., Zhou, X. and Z. T. Ma (2011) A FEMP method and its application in modeling dynamic response of reinforced concrete subjected to impact loading, *Computer Methods in Applied Mechanics and Engineering* **200**(17): 1659–1670.

Liang Y. P., Zhan, F., Liu, Y. and Zhang, X. (2013) Material point method and its applications, *Advances in Mechanics* **43**: 237–264.

Liu G. R. and Liu, M. B. (2003) Smoothed particle hydrodynamics: a meshfree particle method. World Scientific, Singapore.

Nairn J. A. (2003) Material point method calculations with explicit cracks, *Computer Modeling in Engineering and Sciences* **4**: 649–663.

Onate E., Idelsohn, S. R., Del Pin, F. and Aubry, R. (2004) The particle finite element method. An overview, *International Journal of Computational Methods* **1**(2): 267–307.

Sulsky D., Chen, Z. and Schreyer, H. L. (1994) A particle method for history-dependent materials, *Computer Methods in Applied Mechanics and Engineering* **118**: 179–196.

Sulsky D., J., Z. S. and Schreyer, H. L. (1995) Application of the particle-in-cell method to solid mechanics, *Computer Physics Communications* **87**: 236–252.

Sulsky D., Schreyer, H., Peterson, K., Kwok, R. and M. Coon (2007) Using the material-point method to model sea ice dynamics, *Journal of Geophysical Research: Oceans (1978–2012)* **112**((C2)): 1–18.

Sun Z., Xi, G. and Chen, X. (2009a) Mechanism study of deformation and mass transfer for binary droplet collisions with particle method, *Phyics of Fluids* **21**: 032106.

Sun Z., Xi, G. and Chen, X. (2009b) A numerical study of stir mixing of liquids with particle method, *Chemical Engineering Science* **64**: 341–350.

Xie H., Koshizuk, S. and Oka, Y. (2005) Simulation of drop deposition process in annular mist flow using three-dimensional particle method, *Nuclear Engineering and Design* **235**: 1687–1697.

York A. R. and Sulsky, D. (2000) Fluid-membrane interaction based on the material point method, *International Journal for Numerical Methods in Engineering* **48**(6): 901–924.

Zhang D. Z., Zou, Q. S., VanderHeyden, W. B. and X. Ma (2008) Material point method applied to multiphase flows, *Journal of Computational Physics* **227**(6): 3159–3173.

Zhang X., Sze, K. Y. and S. Ma (2006) An explicit material point finite element method for hyper-velocity impact *International Journal for Numerical Methods in Engineering* **66**(4): 689–706.

Index

K

L

M

N

www.ingramcontent.com/pod-product-compliance
Ingram Content Group UK Ltd.
Pitfield, Milton Keynes, MK11 3LW, UK
UKHW020401150926
8624IPUK00004B/280

9 789814 571692